Studies in the Methodology of Natural and Social Sciences

Igor Hanzel

Studies in the Methodology of Natural and Social Sciences

PETER LANG
Frankfurt am Main · Berlin · Bern · Bruxelles · New York · Oxford · Wien

Bibliographic Information published by the Deutsche Nationalbibliothek
The Deutsche Nationalbibliothek lists this publication in the Deutsche Nationalbibliografie; detailed bibliographic data is available in the internet at http://dnb.d-nb.de.

Cover Design:
Olaf Gloeckler, Atelier Platen, Friedberg

ISBN 978-3-631-60865-4
© Peter Lang GmbH
Internationaler Verlag der Wissenschaften
Frankfurt am Main 2010
All rights reserved.

All parts of this publication are protected by copyright. Any utilisation outside the strict limits of the copyright law, without the permission of the publisher, is forbidden and liable to prosecution. This applies in particular to reproductions, translations, microfilming, and storage and processing in electronic retrieval systems.

www.peterlang.de

To Maria, Jakub and Martin

Content

Introduction 11

Chapter 1: Carnap, Hempel, Popper and Transparent Intensional Logic 17
    1.1 The D-N Model 18
    1.2 Rudolf Carnap 23
    1.3 The "Standard Conception" of scientific theories 30
    1.4 Carnap and language: From Vienna to Santa Fe 39
    1.5 Transparent Intensional Logic 63

Chapter 2: Experimentation, Empirical Knowledge and Measurement 69
    2.1 Carnap and Hempel on measurement statements 69
    2.2 Quine's attempt at reduction 73
    2.3 Kyburg Jr. on basic magnitudes 76
    2.4 Experimentation, cycles of knowledge and the differentiation between the already observed and not as yet observed 77

Chapter 3: Cyclical Method of Theory Construction I:
The Retreat into and Coming out from the Formal Ground 93
    3.1 The cyclical method of theory construction: from appearances to the formal ground and "back" 94
    3.2 Scientific laws and scientific explanations based on formal ground 139
    3.3 The nature of nomic measurement 171
    3.4 Three attempts at reduction 180

Chapter 4: Cyclical Method of Theory Construction II:
The Retreat into and Coming out from the Real Ground 191
    4.1 The three contexts of Marx's theory of value 191
    4.2 Marx, Ricardo, and Bailey on value: A comparison 235
    4.3 Measurement 248

Chapter 5: What is Grounded Theory about? Methodological Reflections and Explications 269
    5.1 Introduction 269
    5.2 Grounded Theory – An Overview 271
    5.3 "Categories" and categories, "properties" and properties – A semantical clarification 280
    5.4 Grounded Theory's black eyes: Deduction, enumerative incomplete Induction, abduction and the enumerative thought-universal 282
    5.5 Grounded Theory's feathers in the cap: Cyclical method, unit busting and the nonenumerative thought-universal 289

Chapter 6: Beyond the Qualitative Quantitative Divide in the Social Sciences 305
    6.1 The qualitative-quantitative divide 305
    6.2 The qualitative-quantitative divide: Back to the roots 310
    6.3 To new shores 316
    6.4 Explanation based on Ohm's law 324

Chapter 7: Historical Sociology vs. Rational Choice Theory: The Adventures of Nominalism 327
    7.1 Introduction 327
    7.2 Historical Sociology 328
    7.3 Rational Choice Theory 333
    7.4 Mechanism, path dependence, conjuncture and all that 336
    7.5 Attempt at synthesis: Beyond the nominalism of RCT and Skocpol's *States and Social Revolutions* 340

Chapter 8: Reflections about Metareflections: Roy Bhaskar and Wesley C. Salmon 347
    8.1 Can ontology (of the social world) be a substitute for epistemology and methodology of (social) science? 347
    8.2 Salmon versus Hegel on causation, principle of common cause, and theoretical explanation 363

Conclusion 381

References 387

Acknowledgements

Several persons institutions and were helpful in writing this book. Chapter 3 was written at the University of Potsdam in Germany, where I spent the academic year 2000/2001 as a DAAD visiting scholar at the invitation of Professor Hans-Peter Krüger. Subchapter 1. 4 was written at Loras College, Dubuque (IA), where I spent the Fall and Summer terms 2005/2006 as a Fulbright Scholar in Residence at the invitation of Professor Roman T. Ciapalo. All other parts of the book were written at my home institution, Comenius University. Professor Jozef Viceník and Professor Václav Černík from the Philosophical Institute of the Slovak Academy of Science read the drafts of several chapters and suggested many improvements.

I am especially grateful to Professor Roman T. Ciapalo, who over the years read the drafts of the book and made numerous suggestions for improving them; without his support it would have never been written.

I thank the editor of the *Journal for the General Philosophy of Science* for the permission to use in Chapter 3 parts of my paper published in this journal in 2008, Vol. 39, pp. 273–301, under the title "Idealizations and Concretizations in Laws and Explanations in Physics."

Introduction

B. Fay and J. D. Moon in their article (Fay and Moon 1977) formulated the following three questions which, they claim, must be answered by any compelling account of social science: "First, what is the *relationship between interpretation and explanation in social science?*; second, what is the *nature of social scientific theory?*; and third, what is the *role of critique?*" (Fay and Moon 1977, 209).

In this book I will try to show that by answering the second question one finds the answer at the same time to the third question, and, simultaneously, one finds the conditions that are necessary to find the answer to the first question. I will answer the second question in such a way that I first answer a more general question, namely, how should *any* scientific, that is, either a natural or social science, theory be constructed in order that it could *fulfill a critical function*, namely, explain the origin of distorted, false, and one-sided views about the external (natural or social) reality. By investigating the methods of construction of scientific theories I will leave the level of particular sciences and will move at the *metascientific* level. In order to accomplish my investigation at this level, I draw on several conceptual resources. First, I bring in the views developed in the framework of the philosophy of science; here I mean, primarily, the views of Rudolf Carnap, Karl R. Popper, Carl G. Hempel, Leszek Nowak, Jan Such, Roy Bhaskar, and Wesley C. Salmon. Second, I bring in the instrument of analysis and explication of *epistemic categories* as given in the works of G. W. F. Hegel. What I understand with respect to epistemic categories I will explain at the very end of this Introduction. I realize that it is highly unusual, to say the least, to appeal to Hegel in the field of philosophy of science, but it is my contention that by means of categories it becomes possible to enlarge the concept framework of philosophy of science. Third, I bring in the apparatus of modern semantics, primary that of G. Frege, R. Carnap, and P. Tichý.

These three conceptual resources I will apply at the level of scientific disciplines to the following scientific theories: Newton's mechanics, classical mechanics, J. Perrin's computation of Avogadro's constant N, Marx's economic theory, Ricardo's theory of value, Grounded Theory, Historical Sociology, and Rational Choice Theory. In addition, I reconstruct the methods measurement associated with some of these theories.

Such an application I do not accomplish in this book as a unidirectional movement from pregiven schemes to a given subject matter but as a bidirectional movement that enables us both to criticize and modify the conceptual resources and to acquire at the same time a new view about the subject matter under investigation. Exempted from such a reflexive attitude to the applied conceptual resources will be the semanticists of Frege, Carnap, and Tichý which I apply here purely as an instrument without reflecting about their possible limitations with respect to the particular special sciences.[1]

The bidirectional, and, in fact, multidirectional movement, as performed in this book can be represented in Figure 0.1:

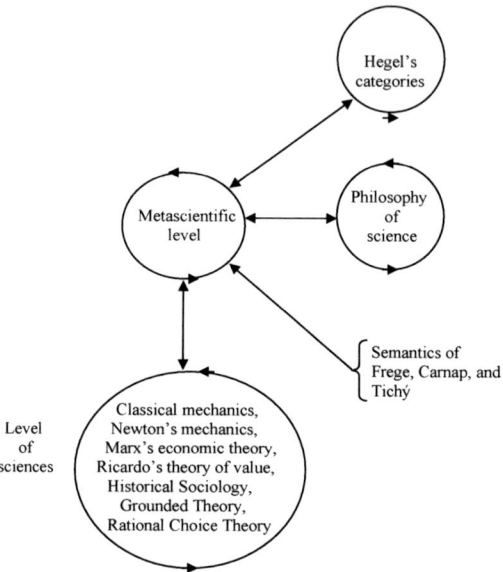

Figure 0.1 The architecture of this book

Chapter 1 has a threefold aim. The first aim is to show the incompleteness of the exclusively logico-syntactical and logico-semantical approaches to the core issues of the philosophy of science, namely, universal scientific laws, scientific explanation based on the latter, and theory construction as given in the works of Carnap, Hempel, and Popper. This result I will use as a point of departure for my investigations into the nature of scientific laws, scientific explanation, and theory construction as given in the chapters to follow. The second aim is to reconstruct

---

[1] Here I mean, especially, the approach of Tichý's semantics, explained in chapter 1, to magnitudes. As shown in tables 1.9 and 1.10 in section 1.5, magnitudes like numerals should have the same extension, namely, numbers.

the development of Carnap's views on language in the years 1931–1947. The results of this reconstruction I will use in my analysis of Carnap's and Hempel's views on the structure of scientific theories. My third aim here is to introduce the central concepts of Transparent Intensional Logic, which will enable me to deal in the chapters to follow with the issue of meaning change of terms given in scientific theories.

In chapter 2, I scrutinize the structure of empirical knowledge, experimentation and measurement. I start by analyzing Carnap's, Hempel's, and Quine's approach to measurement-statements. Finally, I deal with the cyclical nature of empirical knowledge and provide a detailed typology of measurement at the level of empirical knowledge together with the central categories enabling us to deal with the differentiation between the already observed and the as yet not observed.

The first type of thought-movement going beyond the realm of empirical knowledge, experimentation, and direct measurement I present in chapter 3. I start with a detailed reconstruction of the method of theory-construction as given in Book I and Book III of Newton's *Principia*, and I then draw certain conclusions pertaining to the internal structure of the *Principia* by bringing in Hegel's categories of theory construction from his *Science of Logic*. These categories enable me to show that the method of *Principia* is a specific type of cyclical method of theory construction. I then reconstruct the specific type of scientific law, as well as specific types of scientific explanations corresponding to the structure of scientific laws and scientific explanations given in Newton's *Principia* and in classical mechanics. I compare this reconstruction with the approaches of B. Ellis and J. Bigelow; J. Woodward and C. Hitchcock; and C. Liu to scientific laws, scientific explanations and idealizations as well as with the views of M. Thalos on analytic mechanics. Then I apply transparent intensional logic to the harmonic law of planetary motion in order to deal with the problem of meaning change of the terms of the language of science. Then I reconstruct the structure of the nomic, sometimes also called "derived" form of measurement. Finally, I deal with the futile attempts of C. G. Hempel, B. Ellis and E. Mach to dispose of "metaphysically biased" concepts of science.

In chapter 4, I reconstruct yet another type of the cyclical method of theory construction, namely, that given in Marx's *Capital*. I start with a reconstruction of the three contexts of Marx's theory of value as given in chapter 1 of *Capital*, Volume I, and reconstruct at the same time epistemic categories relevant to this chapter together with the method employed in its construction. Then I compare those categories with categories employed by Ricardo in his theory of value. Finally, I reconstruct the structure of measurement based on Marx's approach to the concepts of political economy like value, surplus-value, production price and average rate of profit given in *Capital*, Volume I, and the manuscript of Volume III.

In chapters 5 through 7, I deal with the methodological reflections related to social science, namely, to the sociological Grounded Theory, to the qualitative-quantitative divide in the social sciences, and to the dispute between Historical Sociology and Rational Choice Theory. Here I reconstruct the methods of construction applied in them and show their highly specific features like realism, opposed to nominalism, with respect to structures underlying the phenomena to be explained, as well the consecutive shift in meaning of terms they apply in that construction.

In chapter 8, I perform, based on the results of the previous chapters, an analysis of Roy Bhaskar's and Wesley C. Salmon's approaches in the framework of philosophy of science. I try to show that their approaches display certain deficits regarding the application of categories as a tool of a philosophico-methodological analysis of scientific theories as well as with respect to the methods of explanation and measurement given in scientific theories.

In the conclusion of the book, I perform reflections on the place of the method interpretation in social science and try to clarify how it is possible to link the methods of interpretation and explanation in social science.

Let me now clarify what I understand in regard to *categories*. I view terms like "causality," "causal process," phenomena," and so on, following Hegel,[2] as terms the meanings of which stand for the objective[3] epistemic/cognitive content of human knowledge and where these meanings are the result of the cooperated thought-activity of human beings. At the same time these meanings serve as the basis for successive thought-activities generating a new objective, intersubjectively valid epistemic/cognitive content. In fact, the possession of these meanings enables the very process of human thinking to yield new objective content. And because these meanings stand for an *objective* content, they can be viewed as characterizing human knowledge/cognition of real objects. They can, therefore, reproduce – by means of thought-operations – a really existing object as an object in thought, that is, as a thought-object. What has to be emphasized here is that the meanings of terms like "causality," "causal process," phenomena," and so on, are different from the meanings of terms like "gene," "molecule of oxygen," or "wave function" because their reference is not immediately an object in the external world, but a general type of thought-object that is an invariant for several different thought-objects given in natural and social sciences.

At the same time, the meanings of those terms can be viewed, at least in human society in which science is given, as the determinants and coordinators of

---

2   Here I draw on (Walsh 1953/1954) and (Hanzel and Černík and Viceník 1994).
3   By claiming that categories have not only an epistemological but also an ontological dimension, i.e., objective content, I follow Hegel's and not Kant's approach to categories. For differences between Kant's and Hegel's approaches to categories see (Horstmann 1995).

scientific research, the framework determining the creation of scientific theories. The meanings of those terms express what is common to several natural and social sciences of a certain historic period, what unifies them and what is general and necessary for them. Without these meanings it is impossible to think a single object; human beings need them for the creation of thought-objects given in natural and social sciences and expressed in their own specific terms. According to Kant, by means of categories we can not only "think an object" (§ 10, A80/B106; Kant 1965, 114) but also "[w]e cannot think an object save through categories" (§ 27, B165; Kant 1965, 173).

I label those meanings *categories of thinking* or *categories*, for short, and regard them as enabling both the epistemic/cognitive relation of human beings to objects of their cognition, and the development of this relation. They should thus be understood as thought-expressions or abstractions of the real determinations. Even if they are higher-level thought-expressions or abstractions of meanings of the terms of natural and social sciences, still they stand for knowledge about the external reality. They are however *not* thought-expressions of the real determinations of external reality by itself (these are given by the meaning of the terms of natural and social sciences), nor of the thinking by itself, but are the abstractions of the determinations of real relations between thinking, nature, and society. *The givenness of these relations is the essential condition for the existence of categories of thinking; they are given only and only where the relation between being and cognition, subject and object exists.*

And since I am dealing in this book primarily with scientific *theories*, that is, with what Marx labeled in the introduction to the manuscript *Grundrisse* as "theoretical" or "speculative" conduct of the human consciousness (Marx 1976b, 37; 1973, 102), I deal, in this book, only with those categories that function as *thought-expressions of the relations really given between the cognizing thinking of human beings and the cognized reality, as the modes of the reproduction and reconstruction of the objective in the subjective consciousness of human beings, as modes of their thought-appropriation of the reality that is external to thinking.* If I would have dealt in this book also with what Marx labeled as the "practico-spiritual" (*praktisch-geistige*) appropriation of the external world by human consciousness (Marx 1976b, 37; Marx 1973, 101), then I would have to bring in also those categories that are involved in the creation of *thought*-projects of future *practical* activities transforming the real world.

Chapter 1: Carnap, Hempel Popper and Transparent Intensional Logic

The aim of this chapter is threefold. *First*, to show by a detailed analysis the incompleteness of the exclusively logico-syntactical and logico-semantical approaches to the core issues of philosophy of science, namely, universal scientific laws, scientific explanation based on the latter, and theory construction as given in the works of Rudolf Carnap, Carl G. Hempel, and Karl R. Popper. The results of this analysis I will use as a point of departure for my investigations into the nature of scientific laws, scientific explanation and theory construction as given in the chapters to follow.

The *second* aim is to reconstruct the development of Carnap's views on language in the years 1931–1947. I will use the results of this reconstruction in my analysis of Carnap's and Hempel's views on the structure of scientific theories. The *third* aim is to introduce the central concepts of Transparent Intensional Logic, which will enable me to deal in the chapters to follow with the issue of meaning change of terms that occurred during the process of constructions of scientific theories.

I start with a brief exposition of the main characteristics of Hempel's approach (labeled "the D-N model") to deductive explanations based on universal scientific laws and analyze the problems and paradoxes inherent in this approach. Next, I will trace these characteristics back to Hempel's and Carnap's attempts to ground the concepts of scientific law and explanation exclusively on logic (i.e., logical syntax and/or logical semantics), which led to a highly normative approach alienated from the practice of real science. Then I reconstruct the so-called Standard Conception, labeled also as the "Received View" of scientific theories as it was developed in the framework of logical empiricism. Next, I reconstruct as completely as possible, by drawing upon the unpublished works and correspondence of R. Carnap,[4] the development of his semantics, starting from his semantical turn accomplished under the influence of A. Tarski's remarks during the Paris conference in September 1935, up to his (yet another) turn to the intension/extension distinction in 1943–1947 under the influence of Alonzo Church.[5] Finally, I bring in the central concepts of Transparent Intensional Logic.

---

[4] I would like to thank the Archives of Scientific Philosophy at the University of Pittsburgh for permission to quote from the unpublished letters and materials of R. Carnap. All rights reserved.

[5] I would like to thank the archives at Princeton University for the permission to quote from the unpublished letters and materials of A. Church. All rights reserved.

1.1 The D-N model

Restricting my interest to explanations based on universal scientific laws, Hempel's contribution to the issue of scientific explanation can be described by stating the main features of his D-N model. He claims that "explanation may be construed as an argument in which the explanandum is deduced from the explanans" (Hempel 1959, 299), and states the corresponding condition of adequacy he imposes on proposed scientific explanations (Hempel and Oppenheim 1948, 247–248):

(R1) The explanandum must be a logical consequence of the explanans.
      He adds also the following two conditions:
(R2) The explanans must contain scientific laws.
(R3) The explanans must have empirical content.

Conditions (R1), (R2), and (R3) are classified by Hempel as *logical* conditions[6] of adequacy. This means that in Hempel's approach to scientific explanation based on universal laws, logic should be one of the central instruments of its philosophical reconstruction. The only extra-logical condition of adequacy imposed by Hempel on proposed scientific explanations is the following:

(R4) The sentences constituting the explanans must be true.

It is a well-known fact that Hempel's D-N model with its three logical conditions of adequacy faces serious problems. Let me briefly mention six of them.

First, Hempel, in his reconstruction in the article *Studies in the Logic of Explanation* (Hempel and Oppenheim 1948) gives only one very superficial example of a real scientific explanation of a particular event (Hempel and Oppenheim 1948, 246) and then, bypassing any detailed analysis of real scientific explanations, proceeds immediately to the explanation of a scientific law, claiming that "the explanation of a general regularity consists in subsuming it under another, more comprehensive regularity, under a more general law" (Hempel and Oppenheim 1948, 247).

Second, the seemingly unproblematic shift from the reconstruction of explanation of a particular event to that of a scientific law accomplished in Part I of (Hempel and Oppenheim 1948) encounters serious trouble when Hempel reconstructs scientific explanation by means of a lower functional calculus in Part III

---

6 For attempts to improve the logical conditions of adequacy see, e.g., (Kim 1963), (Ackermann 1965), (Ackermann and Stenner 1966), and (Tuomela 1972).

of the article. Here he is forced, as stated in footnote 33 (Hempel and Oppenheim 1948, 273), to restrict his reconstruction only to the case of explanation of a particular event; the D-N model fails to reconstruct the case of explanation of a scientific law L_{r+1} from laws $L_1, \ldots, L_r$.

Third, even if Hempel were to put a special emphasis on the concept of scientific law and even state a special logical condition (R2) for it, his approach to real scientific laws would nevertheless avoid any attempt at their detailed analysis. Instead of using the operational capacities of the lower functional calculus he himself applies, for example, in Part III of (Hempel and Oppenheim 1948), he *oversimplifies* – as we will see below – the structure of scientific laws. So, for example, he states the law (Hempel 1965a, 338)

"All gases expand when heated under constant pressure,"

and gives its structure as that of the universal conditional form:[7]

$(x)(Fx \rightarrow Gx)$ /1/.

And he also offers the following example (Hempel 1966, 54):

"Whenever a body falls freely from rest in a vacuum near the surface of the earth, the distance it covers in t seconds is $16t^2$ feet",

which he reconstructs again in the same way as /1/, namely as (Hempel 1966, 54):

"Whenever and wherever conditions of a specific kind F occur, then so will always and without any exception certain conditions of another kind, G".

Fourth, Hempel was forced after nearly a quarter of a century, even with the logically transparent reconstruction /1/ of scientific laws, to acknowledge that his approach to scientific laws had failed. This failed approach views scientific laws as displaying the characteristics of being *true*, of being *universal* and as expressing a *general regularity* (Hempel 1942, 231; Hempel and Oppenheim 1948, 247). Nevertheless, Hempel states in the *Postscript (1964)* to his *Studies in the Logic of Explanation* that "it remains an important desideratum to find a satisfactory version of the scope condition which requires more of lawlike sentence than that it must be essentially universal" (Hempel 1964, 293), and later claims as well that "a scientific law cannot be adequately defined as a true statement of universal

---

[7] The same approach was given in his paper "Le problème de la verité" (Hempel 1937, 219) and in his "The Function of General Laws in History" (Hempel 1942, 231–232).

form: this characterization expresses a necessary, but not sufficient, condition for laws" (Hempel 1966, 55). Once this fact was acknowledged in the community of philosophers of science, post-regularity[8] approaches to the concept of scientific law were proposed.

With respect to these post-regularity approaches to the concept of scientific law, it is worth pointing to Van Fraassen's critique of them in his book (Van Fraassen 1989) in which he proves, convincingly, at least in my view, that they *fail as philosophical accounts of scientific laws*.[9] So, what W. Stegmüller stated in the late sixties still holds, namely, "that as yet no precise sufficient condition for an adequate concept of lawlikeness is known" (1969, 460).

The root of the failure of the regularity and post-regularity approaches to the concept of scientific law is, in my view, the fact that they are based on a *one-sided normative* approach to this concept *alienated* from the practice of real science. This point is made by S. D. Mitchell, who distinguishes between a *normative*, a *paradigmatic*, and a *pragmatic* approach to the concept of scientific law as follows (Mitchell 1997, S469):

> [In the normative approach] one begins with a norm or definition of lawfulness and then each candidate generalization ... is reviewed to see if the specified conditions are met. ... The paradigmatic approach begins with a set of *exemplars* of laws (characteristically in physics) and compare these to the generalizations ... The pragmatic approach focuses on the *role* of laws in science, and queries ... generalizations to see whether and to what degree they function in that role.

But, surprisingly, after applying the paradigmatic approach to generalizations in biology, she *does not state any norms* that would enable us to determine what scientific laws are used in real science, what is their structure, and so on. In such a way stated norms need not be any more alienated from the practice of science and need not be reductive, that is, *need not have a uniform character in the sense that all scientific laws in all sciences should have the same structure* and would thus *correspond to only one norm*. On the contrary, a pragmatically rooted set of norms with respect to the concept of scientific laws could lead to a diversified approach to scientific laws; in chapters to follow I will make an attempt at this.

Fifth, in addition to the failure of the D-N model to reconstruct the explanation of a scientific law from other scientific laws, the following three deficits of this model are worth mentioning:[10]

---

8   For these approaches see (Armstrong 1978), (Armstrong 1983), (Tooley 1977), and (Pargetter 1984).
9   From this failure, Van Fraassen draws the conclusion that in the philosophy of science one should give up the concept of scientific law and replace it with the concept of symmetry. For a critique of this replacement see, e.g., (Earman 1993).
10 For a complete list of counterexamples to the D-N model see (Salmon 1990, 47–49).

(a) *The "causal deficit."* S. Bromberger shows in the following example that the D-N model cannot differentiate between valid causal explanations and non-valid pseudoexplanations; (Bromberger 1966, 92):

There is a point on Fifth Avenue, M feet away from the base of the Empire State Building at which a ray of light coming from the top of the building makes an angle of θ degrees with a line to the base of the building. From the laws of the geometric optic, together with the "antecedent' condition that the distance is M feet, the angle of θ degrees, it is possible to deduce that the empire State Building has a height of H feet ... By doing so, [one] would not, however, have explained why the Empire State Building has a height of H feet.

This type of deficit of the D-N model is also acknowledged by Hempel, who gives the following example (Hempel 1965a, 352):

The law for the simple pendulum makes it possible not only to infer the period of a pendulum from its length, but also conversely to infer its length from its period; in either case, the inference is of the form (D-N). Yet a sentence stating the length of a given pendulum, in conjunction with the law, will be much more readily regarded as explaining the pendulum's period than a sentence stating the period, in conjunction with the law, would be considered as explaining the pendulum's length.

It is worth noting here that Hempel views the law for the simple pendulum as a *noncausal type of scientific law*. I will analyze further and criticize this point below.

(b) *The "deficit of deduction."* Even if according to Hempel views the explanandum should be a logical consequence of the explanans, that is, scientific explanation based on universal laws is viewed by him exclusively as a *deductive argument*, this approach leads him again to the inability to distinguish genuine scientific explanations from pseudo-explanations; the following example, going back to H. Kyburg's article (Kyburg 1965) shows this clearly:

The question "Why this sample of table salt, after being placed into water, dissolved?" can be responded to in this way: "If placed into water, all samples of table salt subsequently dissolve" (the covering law), together with the introduction of the condition "This sample of table salt has been placed into water."

Yet the following "explanation" of why this sample of table salt dissolved seems to fit the D-N model as well:

"If placed into water and a magic spell is pronounced, all samples of table salt dissolve" and "This sample of table salt has been placed into water and a magic spell was pronounced over it."

Once we view scientific explanation based on universal laws as a logical argument, we then face one feature of it which is, from the point of view of the reconstruction of explanations, very unpleasant: if we add into the premises of a logical argument some – from the point of view of the very explanation – irrelevant or even nonsensical, additional true propositions, the validity of the very argument does not change but the coherence of explanation is destroyed. From this I draw the conclusion that there exist scientific explanations based on universal laws in which the explanandum is *not* a logical consequence of the explanans.

(c) *The (f)-deficit.* As shown above, because of his superficial analysis of scientific laws, Hempel ends up with the reconstruction of their structure given in /1/. This leads him to the claim that (Hempel 1958, 175)

The statement 'All crows are black' is a sentence of strictly universal form; and so is Newton's first law of motion, that any material body which is not acted upon by an external force persists in its state of rest or of rectilinear motion and constant speed.

But this view leads, in turn, to a view of scientific explanation that cannot differentiate between the following explanation:

All crows are black.
This is a crow.   /2/
This is black,

and the explanation, for example, of the acceleration of free fall of a body falling on Earth. Based on Newton's laws of motion and his law of gravitation one can derive for this acceleration the relation $a = G \times M/(R + h)^2$, and if one supposes that $R \gg h$ holds, that is, that the body falls on Earth from a height h that is much smaller than the radius R of the Earth, then it holds

$a = G \times M/R^2$ /3/,

where G stands for the constant of gravitation and M for the mass of the Earth.

But there is, according to J. Woodward, a profound difference between explanation /2/ and the explanation leading to /3/. The latter contains generalizations with (Woodward 1979, 46)

variables or parameters (mass, distance, acceleration ...) which are such that a whole range of different states or conditions can be characterised in terms of variations of their values ... [They formulate] a systematic relation between these variables. ... [and show] us how a range of different changes in certain of these variables will be linked to changes in others of these variables. In consequence, these generalisations are such that when the variables in them assume one set of values (when we make certain assumptions about boundary and initial conditions) the explananda ... are derivable, and when the variables ... assume other sets of values, a range of other explananda are derivable. For example, the second law of motion and the law of gravitation which occur in explanation [of /3/] ... are such that when the variables in them assume appropriate values (values for the mass and radius of the earth) Galileo's law is derivable. But these generalizations are also such that when the variables in them assume different values (*via* the combination of these generalizations with a different set of initial and boundary conditions) quite different explananda are derivable.

J. Woodward therefore imposes on proposed scientific explanations the *requirement of functional interdependence*, which is fulfilled neither by Hempel's reconstruction of deductive explanations based on universal laws nor by the example /2/ given above; it goes as follows (Woodward 1979, 46):

(*f*) The law occurring in the explanans of a scientific explanation of some explanandum *E* must be stated in terms of variables or parameters variations in the values of which will permit the derivation of other explananda which are appropriately different from *E*.

The conclusion that can be drawn from my analysis of the D-N model is at least two-fold. First, the problems and deficits stated above suggest that what is needed foremost is a reconstruction of the structure of scientific laws that approximates more closely to the structure of scientific laws used by science and that would at the same time overcome Hempel's oversimplified reconstruction given in /1/. Second, a reconstruction of scientific explanation based on such a reconstruction of the scientific laws should provide a more adequate model of explanation as compared to the D-N model and should overcome the deficits and problems of this model mentioned earlier.

1.2 Rudolf Carnap

The fact that the D-N model draws on a highly normative approach to the concept of scientific law and that deduction is one of its central concepts is not a whim of the history of philosophy of science but has its basis in the roots on which Hempel builds his own approach to laws and explanations. For their elucidation I now turn to the works of R. Carnap.

1.2.1 Carnap (and Popper) on Scientific Laws

Let me utilize Carnap's *Logische Syntax der Sprache* (*LSL*) of 1934[11] as a convenient starting point. He characterizes *Wissenschaftslogik* as the logical investigation of science, namely, the "the domain of all questions which are usually designated as pure and applied logic, as the logical analysis of the special sciences" (Carnap 1937a, §72, 280), and equates it with the logical syntax of language, for example, "the logical analysis of physics – as part of the logic of science – is the syntax of the physical language" (1937a, §82, 315). Based on such an understanding of the logic of science Carnap approaches the concept of scientific law in physics and states that (1937a, §82, 316)

[the] most general laws will be formulated as P-primitive sentences; we will call these *primitive laws* ... In the majority of cases, the primitive laws will have the form of universal sentences of implication or equivalence,

and where "P" stands for *physical*, understood by Carnap as a complement to "L" – the *logical*.[12]

Utilizing this approach and drawing on A. Einstein's *Zur Elektrodynamik bewegter Körper* of 1905, Carnap gives the following examples of physical laws (Carnap 1937a, §85, 328–330):

(1) If a magnet moves, then an electric field results.
(2) If an electric field arises, a current results.
(3) If the magnet does not move, then no field results, but an electromotive power results in the conductor which causes electric current.

In these examples, one can immediately identify a one-sided direction in the reconstruction of the logic of scientific laws: from an *a priori stated view on the structure of laws* – laws should have the character of a universal implication or equivalence – *to concrete examples of real laws of science, but never in the opposite direction – from concrete examples,* (for example laws (1), (2) and (3)), *to their logical reconstruction which would enable an adjustment of the initially normatively stated view to the real practice of science.*

In the *LSL*, Carnap treats the issue of scientific laws only very briefly. The situation changes when he shifts in his analysis of language from syntax to semantics, and views the logical analysis of science as focused on logical syntax *and* logical semantics (1942b, §39, 250). The most detailed analysis of the problem of lawlikeness in the semantic period is given by Carnap in (Carnap 1963b) and in

---

11 For the sake of convenience I quote from its English translation of 1937.
12 On this see also (Carnap 1934a).

(Carnap 1966). Here he proposes a four-step approach to the concepts of scientific law and causal modalities. *First*, one should, he claims, explicate the concept of sentence of a nomic form, that is, that of a law-like sentence (Carnap 1966, 211; Carnap 1963b, 951–952). *Second*, as a next step, one should define (with respect to a language L) the concept of fundamental/basic law so it holds that (Carnap 1966, 213; Carnap 1963b, 952):

A sentence S in L is a fundamental/basic law in $L =_{Df} S$ has nomic form and is true in L.

Third, one could give then the following definition (Carnap 1966, 214; Carnap 1963b, 952):

A sentence S in L is *causally valid/true* if it is a logical consequence of the class of all fundamental/basic laws in L.

Fourth, as the final step, this would enable one to obtain a logic of causal modalities; once S expresses in L the proposition p, then – once the former is causally true – p is causally *necessary*.

Carnap at the same time claims that all four steps can be *accomplished completely in the framework of logical semantics* because, in his view, (a) the very concept of a sentence of nomic form (of a law-like sentence) can be explicated in the framework of logical semantics (Carnap 1963b, 951; 1966, 211), (b) the concept of truth is a semantical concept, (c) the concept of logical consequence is a semantical concept, and (d) the differentiation between a sentence S of L and the proposition p it expresses in L can be explicated in the framework of logical semantics.

Unfortunately, Carnap's proposal to explicate the sequence of concepts *sentence of nomic form* → *basic/fundamental law* → *causal validity of a sentence* → *causal necessity* of a proposition completely in the framework of logical semantics, is a non-starter because already into the first step he introduces one *necessary* condition that is of a completely *non*semantical nature: "It seems clear that the following in a necessary ... condition for S having nomic form: S must not contain space-time coordinate constants, but only variables, as J. C, Maxwell has first pointed out" (Carnap 1963b, 952), and (Carnap 1966, 211):

The first condition for statement having nomic form was made clear by James Clark Maxwell. ... He pointed out that basic laws of physics do not speak of any particular position in space or point in time. They are entirely general with respect to space and time; they hold everywhere, at all times.

The requirement, that a fundamental scientific law should hold universally for all times and spaces has its roots *completely outside the framework of logical semantics*. It has, even if it is not acknowledged either by Hempel or Carnap, its roots in a view that can be here tentatively labeled as a certain type of *metaphysics/ontology* imposed on the philosophical reconstruction of scientific laws, but *because* it is imposed in an unacknowledged way, it is imposed *without any feedback control from the side of laws really stated in empirical sciences*, namely, *if they are in fact stated in such a way that they should hold for all times and spaces*.

The dominant position of logic in the attempt to frame the concept of scientific law is readily seen also in K. R. Popper's *Logik der Forschung* of 1934. Initially, Popper states the general claim that "my proposal is based upon an *asymmetry* between verifiability and falsifiability, an asymmetry which results from the logical form of universal statements" (Popper 1968, 41), without specifying here what he understands under "universal statements." Only when dealing with the issue of explanation and prediction does he characterize universal statements as "hypotheses of the character of natural laws" (1968, 60). Finally, when dealing with the very nature of these universal statements = hypotheses = scientific laws, he characterizes them as *strictly universal statements* – because they should hold for all times and spaces – and differentiates them from *numerically universal statements* holding only for a finite individual (or particular) spatio-temporal region (Popper 1968, 62). But even if Popper initially starts with the claim that the asymmetry between verifiability and falsifiability of universal synthetic statements has its roots in *logic*, when dividing the class of universal synthetic statements into two subclasses – strict universal statements and numerically universal statements – he startlingly claims that they are differentiated according to a criterion that *does not belong to logic*; "Formal logic (including symbolic logic) which is concerned only with the theory of deduction, treats these two [classes of] statements alike as universal statements ('formal' or 'general implications')" (Popper 1968, 62). Accordingly, this means that, contrary to his initial claim, *the characterization of the concept of scientific law has to draw on a criterion which does not stem from logic*. And therefore this criterion can and should be tested in confrontation with scientific laws as they are given in the practice of real science. But it is this confrontation that is, like in the approach of Hempel and Carnap, completely missing in Popper's *Logik der Forschung* of 1934.

It seems that Popper himself was not completely aware of this shift from a *declared logical differentiation* to a *de facto extra-logical differentiation*, because in his *Postscript to the Logic of Scientific Discovery* he returns to the "logical" differentiation and again treats it as being part exclusively of deductive logic (Popper 1983, 181–184).

What can we make of Carnap's and Popper's attempts to grasp the concept of scientific law by an approach grounded *exclusively in logic*, as well as of Hempel's classification of the adequacy condition (R2) that the explanans of a proposed explanation has to contain general laws as a *logical* condition? The development of Popper's approach to the concept of scientific law can be viewed as symptomatic of the fate of those attempts and this classification. In the late fifties, Popper, after a prolonged discussion with W. Kneale, gave up the view, initially stated in the 1934-edition of the *Logik der Forschung*, that scientific laws can be characterized as strictly universal statements (Popper 1968, 428), and in 1968 finally gave up all attempts to define the concept of scientific law (1968, 441). W. Stegmüller's claim given above can thus be made more precise: *All attempts to explicate the concept of scientific law by means of logical syntax and/or logical semantics have as yet failed*. I thus propose to view Hempel's condition (R2) as an *extra-logical* condition.

Let me now briefly analyze Hempel's classification of the condition (R3) that the explanans must have empirical content as a *logical* condition. Is this classification correct? It can be viewed as the conclusion of the following argument:

1. All statements of logic as a science about language can be constructed independently from any statements that are by their nature empiric (synthetic).
2. All statements outside the sphere of the analytic delineated by the rules of logic (and mathematics derived from logic), the latter being built in accordance with 1, are viewed as empiric (synthetic) by their nature.
3. (R1) The explanans should have empirical content.
4. Therefore (R1) is a logical condition.

But a closer look at premise 2[13] discloses, as a hidden premise, a supposition that runs as follows:[14]

2*. All statements of language can be exhaustively subdivided into two mutually exclusive classes: the class of all analytic statements and the class of all empiric (synthetic) statements.

But this supposition, as a classification of all statements of language, does not belong to logic. The conclusion 4 thus depends on at least one *extra-logical premise*; (R3) is therefore *not a logical condition*.

---

13 I do not analyze here the viability of premise 1.
14 On Carnap's views on the mutual relations of the realms of the logical, analytic, and the synthetic see (Carnap 1934a), (Carnap 1958), (Carnap 1975), and (Carnap 2000).

1.2.2 Carnap on Scientific Explanation

In the *LSL*, Carnap claims that (1937a, §82, 318–320)

> Exact rules for deduction can be laid down, the L-rules of the physical language. Thus the laws have the character of hypotheses in relation to the protocol-sentences; sentences of the form of protocol-sentences may be L-consequences of the laws. ... The *explanation* of a single known physical process, the *deduction* of an unknown process in the past or in the present, from one that is known, and the *prediction* of a future event, are all operations of the same logical character. In all three cases it is, namely, a matter of deducing the concrete sentence which describes the process from valid laws and other concrete sentences. To explain a law ... means to deduce it from more general laws.

The fact that in Carnap's, and then later on in Hempel's, approach[15] scientific explanation based on universal laws is viewed as a logical argument can be understood by a specific peculiar feature of Carnap's syntactical and semantical approach to language. Here I mean Carnap's reconstruction of the logical (*L*) and physical/factual (*P/F*) dimensions of language of science.

In his "syntactical" period Carnap differentiates between "all the logico-mathematical transformation rules of [language] S ... or *L*-rules; and all the remainder, physical or *P*-rules" (Carnap 1937a, §51, 180), where as an example of extra-logical rules he mentions the universal laws of physics (Carnap 1937a, §51, 180). By analyzing different cases of relations between a class of expressions $\mathfrak{K}_1$ and a sentence $\mathfrak{S}_2$, so that the latter is a consequence of the former (1937a, §51, 181), he classifies this relation of consequence either as *L-consequence* or as *P-consequence*. A closer look at the method he uses to introduce the concept of *P*-consequence reveals that he first introduces the concept of L-consequence and then views P-consequence as a complement of the former. That it is so is readily seen in §52 of the *LSL* in which he states: "$\mathfrak{S}_2$ is called a P-*consequence* of $\mathfrak{K}_1$, if $\mathfrak{S}_2$ is a consequence, but not an L-consequence, of $\mathfrak{K}_1$" (1937a, §52, 184).

Carnap holds to this "complement-approach" to the P-realm also in his "semantical" period; now he uses the F-symbol for the realm of the empirical (synthetic). So, for example, in the *Foundations of Logic and Mathematics* he states: "If a sentence is either L-true or L-false, it is called *L-determinate*, otherwise (L-indeterminate or) *factual*" (Carnap 1939b, §7, 13). In the *Introduction to Semantics* (*IS*) he uses the same approach.[16] He defines the concept of truth and falsity

---

15 Also K. R. Popper views in his *Logik der Forschung* scientific explanation as deduction, for example, (Popper 1968, 59–60).

16 One has to bear in mind here that according to Carnap the semantical term "F-implicate" from *IS* is the translation of the syntactical term "P-consequence" from *LSL* (1942, §39, 250–251).

for $\mathfrak{T}_i$ and $\mathfrak{T}_j$, the latter being signs of metalanguage designating sentences and classes of sentences, and then defines the concept of implication, symbolized as "$\rightarrow$", and L-implication, symbolized as "$-_L\rightarrow$", for the relation between $\mathfrak{T}_i$ and $\mathfrak{T}_j$. By means of these concepts he, finally, states "$\mathfrak{T}_j$ is an **F-implicate** of $\mathfrak{T}_i$ ($\mathfrak{T}_i$ F-implies $\mathfrak{T}_j$; $\mathfrak{T}_i$ "$-_F\rightarrow \mathfrak{T}_j$) (in S) $=_{Df}$ $\mathfrak{T}_i \rightarrow \mathfrak{T}_j$ but not "$\mathfrak{T}_i -_L\rightarrow \mathfrak{T}_j$" (1942b, §21, 143).

Finally, in *Meaning and Necessity* (*M&N*) (1947, §2, 10–12) Carnap proceeds by a similar method. He first, via a convention, gives the concept of L-true and then defines the concept L-implies as follows: $\mathfrak{S}_i$ ***L-implies*** $\mathfrak{S}_j$ (in S_1) $=_{Df}$ the sentence $\mathfrak{S}_i \supset \mathfrak{S}_j$ is L-true. Then – as a *complement* to the concept L-true – defines the concept F-true so that it holds: $\mathfrak{S}_i$ is ***F-true*** (in S_1) $=_{Df}$ $\mathfrak{S}_i$ is true but not L-true. And finally he provides the following definition: $\mathfrak{S}_i$ ***F-implies*** $\mathfrak{S}_j$ (in S_1) $=_{Df}$ $\mathfrak{S}_i \supset \mathfrak{S}_j$ is F-true. Given this chain of definitions, the concept F-implies is in fact defined as follows: $\mathfrak{S}_i$ ***F-implies*** $\mathfrak{S}_j$ (in S_1) $=_{Df}$ $\mathfrak{S}_i \supset \mathfrak{S}_j$ is true but not L-true, which can thus be restated as follows: $\mathfrak{S}_i$ ***F-implies*** $\mathfrak{S}_j$ (in S_1) $=_{Df}$ $\mathfrak{S}_i$ implies $\mathfrak{S}_j$ but it does not L-imply $\mathfrak{S}_j$.

Where does the importance of Carnap's method for the introduction of the L-terms and P/F-terms lie? This method serves *primarily* the needs of the construction of the language of logic and mathematics. Once these two sciences take center-stage one has to define the respective L-terms and *then* one can proceed further and define the respective P/F-terms. The whole P/F-realm is *from the point of view of logic and mathematics* only an "external" complement of the L-realm. But the situation changes fundamentally once the philosophy of empirical sciences, and, here, especially the issue of scientific explanation takes center-stage. To understand the profoundness of this change let us return to §52 of the *LSL*, where Carnap gives the following examples of sentences from a P-language (1937a, §52, 185):

$\mathfrak{S}_1$: "this body a is of iron";
$\mathfrak{S}_2$: "a is of metal";
$\mathfrak{S}_3$: "a cannot float on water."

According to Carnap, "$\mathfrak{S}_2$ and $\mathfrak{S}_3$ are consequences of $\mathfrak{S}_1$, and, specifically, $\mathfrak{S}_2$ is an L-consequence, but $\mathfrak{S}_3$ is not, and is therefore a P-consequence" of $\mathfrak{S}_1$ (1937a, §52, 185). If we take the reconstruction of the relation between $\mathfrak{S}_1$ and $\mathfrak{S}_3$ as the subject matter of philosophy of science and as pertaining also to the issue of scientific explanation, then it readily can be seen that Carnap's treatment of their relation is completely deficient. Carnap's logical syntax and logical semantics enable one to state, "$\mathfrak{S}_1$ P/F-implies $\mathfrak{S}_3$," but what he can state about this relation is – *from the point of view of the philosophy of science* – hardly satisfying: "$\mathfrak{S}_1$ implies $\mathfrak{S}_3$ but it does not L-imply $\mathfrak{S}_3$."

That we have here, from the point of view of the philosophy of science, a massive deficiency in Carnap's logical syntax and logical semantics can also be seen in the light of the following three "facts." *First*, Carnap's project of a syntax and (later) of a semantics of language aimed at reconstructing the language of science, including the empirical sciences. But once we begin dealing with the latter, then Carnap's approach to the L-terms and P/F-terms cannot serve the very ends he himself sets: the analysis of the language of these empirical sciences.

Second, it becomes clear why Hempel tried to reconstruct scientific explanation based on universal laws as logical arguments. The logical syntax and logical semantics he draws on in his attempts to reconstruct scientific explanation provided him *only* the L-terms as rich enough to be applied in the philosophical analysis of explanation in empirical sciences. The P/F-terms are treated by Carnap in the *LSL*, *IS*, and *M&N* only as a "left-over"; he therefore declares already in the *LSL*: "In what follows we shall make very little use of the P-terms" (1937a, §52, 185).

Third, the deficit in Carnap's treatment of the P/F-terms also becomes evident in the light of my conclusion in Part I of this paper, namely, that one has to revise Hempel's criterion of adequacy (R1), that scientific explanation based on universal laws has *exclusively* the character of a logical argument. Once we accept this conclusion and unify it with Carnap's definition of the concept of "P/F-implies," we end up with the claim that there exist explanations based on universal scientific laws where the *explanandum is P/F-implied by the explanans*. For philosophy of science, however, this claim is vacuous because in the framework of Carnap's logical syntax and logical semantics we can state nothing else about "being P/F implied" than that it is "being implied, but not L-implied."

1.3 The "Standard Conception" of scientific theories

1.3.1 What Is the "Standard Conception" About?

The "Standard Conception", sometimes also dubbed "the received view of scientific theories," was, as Carnap confesses, "influenced by two different factors: (a) the explicit development of the axiomatic method by Hilbert and his collaborators, and (b) the emphasis on the importance and function of hypotheses in science, especially in physics, by men like Poincaré and Duhem" (Carnap 1963a, 77).[17] At the same time, it should provide a logico-linguistic framework for the

---

17 On the Hilbertian roots of the " Standard Conception" see also (Carnap 1958, 79) and (Hempel 1973, 369).

reconstruction of the structure of scientific theories and an explication of what these theories are "about." The views stated in the Standard Conception take as their starting point Carnap's *Testability and Meaning* of 1936/37, in which he already clearly differentiates between observable and nonobservable predicates of a language as follows (Carnap 1936/37, 454–455):

A predicate 'P' of a language L is called *observable* for an organism (e.g. a person) N, if for suitable arguments, e.g. 'b', N is able under suitable circumstances to come to a decision with the help of few observations about a full sentence, say 'P(b)', i.e. to a confirmation of either 'P(b)' or '~P(b)' of such a high degree that he will either accept or reject 'P(b)'.

In a next step Carnap propounds in his *Foundations of Logic and Mathematics* (hereafter, *FLM*) the view that a scientific theory can be characterized by "two fundamentally different components, a factual and a logical" (Carnap 1939b, 37). While the latter can be characterized as a set of axioms plus its logical consequences, the former comes only via a factual/observational interpretation of the logical consequences of axioms.

Finally, in his *Methodological Character of Theoretical Concepts* of 1956, as well in his papers (1958), (1959/2000) and (1961), he divides the total language of science into an observational language L_O – as a language completely interpreted by means of its observational vocabulary V_O – and a theoretical language L_T, the descriptive vocabulary V_T of which consists of theoretical terms, that is, terms speaking about directly nonobservable entities. For L_O, but *not* for L_T, Carnap accepts the following six requirements (Carnap 1956, 41–42):

1. The requirement of *observability* for the primitive terms.
2. The requirement of explicit *definability* for the non-primitive descriptive terms.
3. The requirement of *nominalism*: the values of the variables must be concrete, observable entities (e.g. observable events, things, or thing-moments).
4. The requirement of *finitism*: the rules of the language L_O do not state or imply that the basic domain (the range of values of the individual variables) is infinite (i.e., L_O has at least one finite model).
5. The requirement of *constructivism*: every value of any variable of L_O is expressed by an expression in L_O.
6. The requirement of *extensionality*: the language L_O contains only truth-functional connectives.

For L_T holds (Carnap 1975, 76; 1958, 237):

that it contains a type-theoretic logic with an infinite sequence of domains D^0, D^1, D^2, etc. ... The elements of D^0 are the members of the infinite sequence 0, 0',0", etc. Officially, no meaning is given to these logical expressions. Their use follows from the rules of the language. We shall, however, in order to relate these expressions to familiar concepts, unofficially regard D^0 as the domain of natural numbers with '0' denoting the number 0, '0'' the number 1, etc.

In addition, a theory is built in the extra-logical part of L_T on the basis of theoretical postulates, that is, on extra-logical axioms, and contains also other theoretical statements derived as theorems from these axioms. What comes in also – according to Carnap – are the so-called correspondence-rules/postulates (hereafter, C-rules), that is, statements containing both theoretical and observational terms. These rules/postulates give to the theoretical postulates and terms from V_T a certain *interpretation*.[18]

A succinct characterization of the "Standard Conception" is given by Hempel as follows (Hempel 1952, 36):

A scientific theory might be linked to a complex spatial network. Its terms are represented by the knots, while the threads connecting the latter correspond, in part, to the definitions and, in part, to the fundamental and derivative hypotheses included in the theory. The whole system floats, as it were, above the plane of observation and is anchored to it by the rules of correspondence. These might be viewed as strings which link certain points [of the network] with specific places in the plane of observation.

This approach to scientific theories can be schematically represented as shown in Figure 1.14 (here $\alpha, \beta, \chi, \ldots$ stand for undefined terms while the lines connecting them stand for axioms; a, b, c, ... stand for defined terms and the lines connecting them stand for theorems; $o_1, o_2, \ldots$ stand for observational terms).[19]

The Standard Conception is at the same time a vehicle for certain specific philosophical views on the nature of scientific theories and scientific knowledge in general, namely, that *only observational terms* from V_O have a *direct reference*; they are about *something*, while the terms from V_T have *no direct reference*, but at the same time are not just part of a purely computational device. This approach should enable, Carnap claims, to escape both the instrumentalist view of scientific theories – theories serve us just as instruments for computation – and the anti-nominalist (realist) view on the nature of theoretical terms, namely, that they refer to entities in the extra-linguistic realm. So, already in his *FLM*, Carnap claims that (1939b, 68):[20]

---

18 On this see (Carnap 1956, 46).
19 For a different scheme see (Feigl 1970, 6).
20 On this see also (Carnap 1966, 254–256).

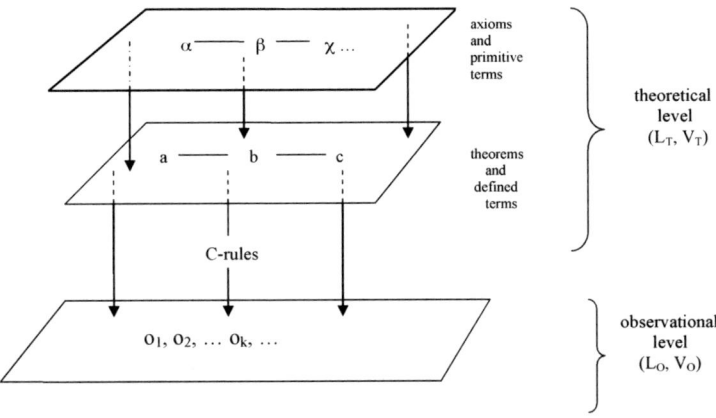

Figure 1.1 A spatial representation of the "Standard Conception"

philosophers ... content that ... modern theories ... are not at all theories about nature but "mere formalistic constructions," "mere calculi." But this is a fundamental misunderstanding of the function of a physical theory. It is true that a theory must not be a "mere calculus" but possess an interpretation, on the basis of which it can be applied to facts of nature. But it is sufficient ... to make this interpretation explicit for elementary terms; the interpretation of the other terms is then indirectly determined by the formulas of the calculus, either definitions or laws, connecting them with elementary terms.

And in his papers (Carnap 1956; 1958), Carnap underscores his nominalism by claiming that the variables of L_T do *not* range over nonobservable entities but over *mathematical* entities. The example he gives for this is as follows: "Let the constant 'n_p' be defined as 'the cardinal number of planets.' This constant is descriptive, to be sure, but the thing described by it is a natural number which belongs to the domain of D^0. The number n_p is identical with 9, but the identity '$n_p = 9$' is synthetic" (1975, 80–81; 1958, 243). In general this means that according to Carnap we have two designators "f" and "g", where the former is a *descriptive* constant and the latter a *mathematical* constant, but it holds that they have the *same extensions* for the same arguments, that is, they are *extensionally identical*. And in order to escape any possible charge of realism with respect to "g"'s ranging over natural numbers, he states also that this "should not be taken literally but merely as didactic help by attaching familiar labels to certain kind of entities or, to say it in a still more cautious way, to certain kinds of entities of expressions in L_T" (Carnap 1956, 45–46).

1.3.2 Negative Consequences of the Nominalism of Standard Conception

The negative consequences of the nominalism of the Standard Conception becomes readily seen when one analyzes Carnap's views in *FLM* on the possible relations between the elementary (i.e., observational) terms and the abstract (i.e., theoretical) terms during the process of theory construction. As examples of the former he mentions terms like "yellow," "bright," "dark" and, of the latter, terms like "electric field," "frequency of oscillations," "wave function." He then goes on as follows (Carnap 1939b, 62–66):

> Suppose we intend to construct an interpreted system of physics – or of the whole science. We shall first lay down a calculus. Then we have to state the semantical rules … for the specific signs, i.e., for the physical terms. … Since the physical terms form a system, i.e., are connected with one another, obviously we need not state a semantical rule for each of them. For which terms, then, must we give rules, for the elementary or for the abstract ones? … Either procedure is … possible. … The *first* method consists in taking elementary terms as primitive and then introducing on their basis further terms step by step, up to those of highest abstraction. … The first method has the advantage of exhibiting clearly the connection between the system and observation. … However, when we shift our attention from the terms and the methods of empirical confirmation to the laws, i.e., universal theorems of the system, we get a different picture Would it be possible to formulate all laws of physics in elementary terms, admitting more abstract terms only as abbreviations? … But it turns out … that it is not possible to arrive in this way at a powerful and efficacious system of laws. To be sure, historically, science started with laws formulated in terms of a low level of abstractness. But for any law of this kind, one nearly always later found some exceptions and thus had to confine it to a narrower realm of validity. The higher the physicist went in the scale of terms, the better did they succeed in formulating laws applying to a wide range of phenomena. Hence we understand that they are inclined to choose the *second method*. This method begins at the top of the system, so to speak, and then goes down to lower and lower levels. It consists in taking a few abstract terms as primitive signs and a few fundamental laws of great generality as axioms. Then further terms, less and less abstract, and finally elementary ones, are to be introduced by definitions. … At least this is the direction in which physicists have been striving with remarkable success, especially in the past few decades. … Now let us examine the result of the interpretation if the first or the second method for the construction of the calculus is chosen. In both cases the semantical rules concern the elementary signs. In the first method these signs are taken as primitive. Hence the semantical rules give a complete interpretation for these signs and those explicitly defined on their bases. … If, on the other hand, abstract terms are taken as primitive – according to the second method, the one used in scientific physics – then the semantical rules have no direct relation to the primitive terms of the system but refer to terms introduced by long chains of definitions. The calculus is first constructed floating in the air, so to speak; the construction begins at the top and then adds lower and lower levels. Finally, by the semantical rules, the lowest level is anchored at the solid ground of observable facts.

Carnap can thus provide the following schematic representation of these two methods which he views as *alternative* methods of theory construction (Carnap 1939b, 63):

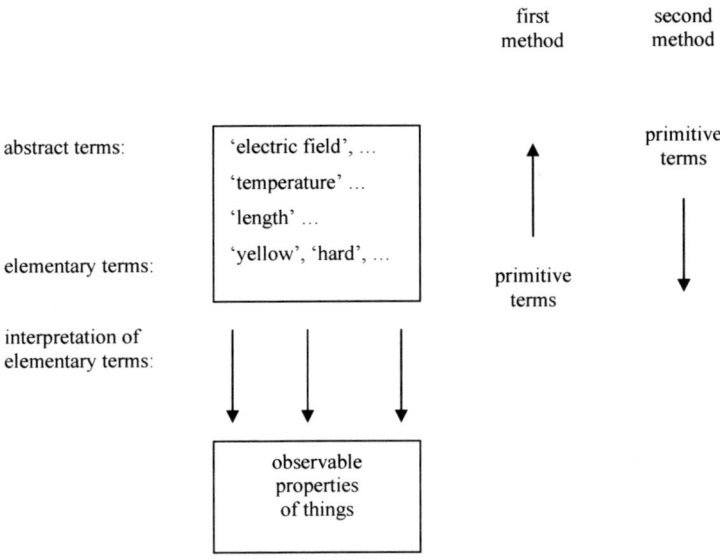

Figure 1.2 Carnap on the alternative methods of theory construction

What has to be emphasized here is, first, that already in his *FLM*, like in his later papers (1956), (1958), (1959/2000) and (1961), Carnap holds to the view that abstract terms like "electric field," "wave function," (hereafter "Ψ") lacks any reference. So, for example, he claims (1939b, 68):

If we demand from the modern physicist an answer to the question what he means by the symbol 'Ψ' of his calculus, and are astonished that he cannot give an answer, we ought to realize that the situation was already the same in classical physics. There the physicist could not tell us what he meant by the symbol 'E' in Maxwell's equations. Perhaps, in order not to refuse an answer, he would tell us that 'E' designates the electric field vector. To be sure, this statement has the form of a semantic rule, but it would not help us a bit to understand the theory. It simply refers from a symbol in a symbolic calculus to a corresponding word expression in a calculus of words.

What has to be emphasized here, second, is that Carnap in *FLM*, contrary to his papers (1956), (1958), (1959/2000) and (1961), understands under semantical interpretation of a term *only* the assignment of its *designatum*. As I will show below in section 1.4, the meaning/intension of terms is in the *FLM* as yet *not* taken into account because in 1939 Carnap in his semantics did not as yet arrive at the distinction between the intension and extension of language expressions.

The first of these two features of Carnap's *FLM* leads to the following negative consequence for the very reconstruction of the structure of scientific theories. So, as the abstract terms are in the sequence Carnap labels as the "first method" viewed as the *end*-product, and at the same time are viewed as the *point of departure* in the sequence he labels as the "second method" – but at the same time are viewed by him as mere *syntactical entities lacking any extra-linguistic designata*, and, thus, their reference cannot have any causal determinations related to the causal determinations of the designata of the less-abstract terms – *the whole sense and importance of the movement from the elementary terms as primitive terms "back" to the elementary terms as derived terms gets lost*. Hempel expressed that sense and importance as follows (Hempel 1970, 142):

Theories are normally constructed only when prior research in a given field has yielded a body of knowledge that includes empirical generalizations or putative laws concerning phenomena under study. A theory then aims at providing a deeper understanding by construing those phenomena as manifestations of certain underlying processes governed by laws which account for uniformities previously studied.

The first of the earlier mentioned features of *FLM* leads also to the inability to realize that scientific theories are in fact *not* built by *either* the first *or* the second method, but via a *unity* of them, that is, using *cyclical method*. In chapter 3 I will reconstruct the cyclical method of theory construction as it is given in Newton's *Principia*.

The second of the already mentioned features of *FLM* enables to understand the other negative consequences for the very reconstruction of the structure of scientific theories. Even if Carnap starting from 1942/43 on substantially changed his semantics by bringing in the differentiation between the intension and extension of language expressions, and even if he explicitly provided the semantics for them in his *Meaning and Necessity* (hereafter, *M&N*) of 1947, at the level of the very philosophy of science one finds a *surprising absence of the application of these semantics for a detailed analysis of the intension/meaning of the terms of scientific theories*. Even if Carnap views the meaning of the terms of the observational language as fixed by analytic/meaning postulates, and even if he deals with the *latter* in a separate paper (Carnap 1952), *neither this paper nor any other work published by him contains any analyses of the meanings of those terms that he labels as "elementary/observational terms."*

Let me now turn to the terms which Carnap labels as "abstract/theoretical terms." Worth quoting here is Carnap's letter to H. Feigl of August 4, 1958, in which he refers to previously mentioned example "The cardinal number of planets is 9" which appears in his article (Carnap 1958) in square brackets (RC 102-07-05):[21]

The entities to which the variables in the Ramsey-sentence refer are characterized not purely logically, but in a descriptive way; and this is the essential point. These entities are identical with mathematical entities only in the customary extensional way of speaking; see my example in square brackets ... In an intensional language (in my own thinking I use mostly one of this kind) there is an important difference between the intension 9 and the intension n_p. The former is *L*-determinate ... the latter is not. Thus, if by 'logical' or 'mathematical' we mean '*L*-determinate', then the entities to which the variables in the Ramsey-sentence refer are not logical.

From the point of view of the semantics of *M& N*[22] this means that while the extension of the expression "9" can be found out by using only the rules of language of mathematics determining its intension, that is, without turning to extra-linguistics facts, the finding of the extension of the expression "n_p" requires, in addition to the use of rules fixing its intension, some investigation into the extra-linguistic facts. In addition, according to Carnap's views stated in this letter, for "$n_p = 9$" it holds that while the *extensions* of the expressions flanking the sign of identity are *identical*, their *intensions are different*. But if one turns to Carnap's papers (1956), (1958), (1959/2000) and (1961) written *after* he has already accomplished the shift to the semantics of extension and intension, one finds out that what Carnap investigates are *not the intensions of theoretical terms of the language of science*, but the ways how to differentiate between the analytic and the synthetic components of the language L_T and how (via the Ramsey-sentences) to eliminate theoretical terms in favor of variables bound by existential quantifiers.[23]

Such a lack of any investigation into the intensions of the terms of the language of science then leads to yet another negative consequence. Even if Carnap takes into account that there exist *terms of the language of science which are both the point of departure and the "end"-point* in the process of theory construction (see Figure 1.2),, still he does not pose the following crucial question. *Does the intension of these terms change once they are transformed from the point of departure to the "end"-point of theory construction?*

---

21 The symbol "RC" stands for the catalog number of Carnap's manuscripts in the archives at the University of Pittsburgh.
22 "... an L-determinate intension is such that it conveys to us its extension" (Carnap 1947, §22, 88).
23 Carnap's aim in his paper (1959/2000) is to define explicitly theoretical terms in a language that contains both Hilbert's ε-operator and the whole logic and mathematics.

What is behind such a possible change of meaning of terms was spelled out, at least at the general level, by G. Schlesinger in (Schlesinger 1964). Let us suppose that M is a term to which we should grant admission into our language of science. In order to do so, we should be able to construct a proposition S_M containing M, so that S_M entails an observation sentence S_O which has to be tested. If S_K stands for some other propositions, T for a theoretical system and C for correspondence rules, we should require no more than the following two requirements (Schlesinger 1964, 395–396):

(A) S_O is logically implied by the conjunction of S_M, S_K, T, and C.
(B) S_O is not logically implied by the conjunction of S_K, T, and C alone.

Once S_O is confirmed, then M acquires its meaning relative to T and C and becomes a member of the class of terms in our language. From this he draws the conclusion (Schlesinger 1964, 402, 404, 405):

that a word which is endowed with significance and is made to stand for a given concept may lose this significance upon the change of context and cease to stand for that concept. After it has lost its original significance it may assume some new signification and stand for another concept or it may not do so. In the latter case, it becomes an empty combination of letters. ... On each occasion ... [a term] is applied anew – in the context of the same theory but in different setting – one has to re-examine that collection of propositions which originally bestowed meaning on it and see whether it is still relevant to this situation. If it is not, one has to find out whether some other set of S_O is logically implied by the conjunction of T, C, S_K and S_O may not lend significance to it. If such a set is found then we have to probe into the question whether we ought to regard the concepts determined by the two sets as being essentially identical concepts and hence should employ the same term in both settings. We have to compare the two groups of T, C, S_K and S_O and use our judgment to decide whether there are sufficient connecting elements between them to render the use of the same term in both cases a reasonable and useful practice. ... a concept which belongs to a highly developed science is molded by a very complex set consisting of an enormous number of T, C, S_K and S_O sentences. When a new sentence joins this set or an old leaves it the character of the concept is altered to some degree. Its complete loss of identity comes about by a gradual process.

In chapter 3, I will analyze the change of meaning of Kepler's third law in Newton's *Principia*.

The approach of the so-called Standard Conception to the method of theory construction can thus be characterized by the following three catch-phrases: *no meaning-change*; *one-directionality of thought-movement*; *and movement from the nonobservable to the observable*.

1.4 Carnap and Language: From Vienna to Santa Fe

1.4.1 Between the Zirkelprotokolle of June 1931 and the Logische Syntax der Sprache of 1934

A convenient starting point when reconstructing Carnap's views on language are the so-called *Zirkelprotokolle* of June 1931, in which he presents his investigations into language. He understands language as composed of objects that have the character of sentence-signs (*Satzeichen*) and introduces special (Gothic) signs to denote the objects of the object-language, stating that "in a metalogical sentence about a formula we are not allowed to write the formula itself but only its metalog[ical] description" ([RC 081-07-17], 11. Juni 1931, 4).

At the same time he characterizes metalogic as follows: "What is in question here are not sentences about sentences but sentences ... about physical structures" ([RC 081-07-17], 25. Juni 1931, 4).[24] From this it follows, Carnap stresses, that in his approach and project *any hierarchy of languages disappears* and, thus, there *exists in fact just one language*; "... there exist indeed sentences of very different kinds ... but all, including the metalogical ones, are in one language" ([RC 081-07-19], 25. Juni 1931, 8). Responding to H. Hahn's question "Do we need at all the technical term 'metalogic,' i.e. are the metalogical sentences principally different from the other" with a clear "No," Carnap adds that "only because of reasons of expedience does one sum up a certain class of sentences under the name 'metalogical sentences'" ([RC 081-07-19], 25. Juni 1931, 8), while responding to O. Neurath's question "Is the metalogic of metalogic again expressible in the original language" ([RC 081-07-19], 25. Juni 1931, 8), by "Yes, one can arrange it so that this is the case" ([RC 081-07-19], 25. Juni 1931, 8). Carnap characterizes this approach to metalogic as "the theory of the forms which appear in a language, i.e. the representation of the syntax of language" ([RC 081-07-19], 11. Juni 1931, 1).

This syntactical approach to language is also present in Carnap's paper (1932b), in which he views all statements of philosophy, as long as they are not nonsensical, as syntactical statements dealing with the forms of the object language. This approach culminates in the *LSS*, in which he puts the question about the differentiation of the syntax-language and the object-language as follows: "Are these necessarily two separate languages?" (Carnap 1937a, §18, 53). To this he gives the following answer: "... it is possible to manage with one language only; not, however, by renouncing syntax, but by demonstrating that ... [the] syntax of this language can be formulated within this language itself" (Carnap 1937a, §18, 53). Carnap starts the *LSS* by programmatically characterizing his project as follows (Carnap 1937a, §1, 1, 4):

---

24 On this see (Hempel 1935).

By the **logical syntax** of a language, we mean the formal theory of the linguistic form of that language – the systematic statement of the formal rules which govern it together with development of the consequences which follow from these rules. A theory, a rule, a definition, or the like is to be called *formal* when no reference is made in it either to the meaning of the symbols (for example, the words) or to the sense of the expressions (e.g. the sentences), but simply and solely to the kinds and order of the symbols from which the expressions are constructed. ... we are concerned with two languages: in the first place with the language which is the object of our investigation – we shall call this the **object-language** – and, secondly, with the language in which we speak *about* the syntactical forms of the object-language – we shall call this the **syntax-language**.

Carnap accomplishes the project of a syntax of language initially for the language of elementary arithmetic of natural numbers or, as he says, for Language I. For this language he introduces five categories of symbols (Carnap 1937a, §4, 16): (numerical) variables, individual symbols like "≡", "∃", "∨", etc., numerals, predicates like P, Q, R, etc, and functors, while for the metalanguage he introduces Gothic symbols, which are syntactical names representing symbol-categories: expressions are designated by "𝔄" (*Ausdruck*), numerical expressions by "𝔷" (*Zahlenausdruck*), sentences by "𝔖" (*Satz*).

From the point of view of this development of Carnap's approach to language, a crucial role is played by the terms *analytic, contradictory, synthetic, equipollent* and *synonymy* which he initially defines for Language I (Carnap 1937a, §14, 38–43). After dealing with Language I, Carnap moves on to Language II, which contains Language I as a sublanguage as well as the whole classical mathematics, including expressions that make it possible to state the sentences of physics. For the sentences of Language II, he then defines concepts like L-determinate, synthetic, compatibility, and incompatibility of sentences (Carnap 1937a, §34e, §34f, §34g, 115–120). Based on such an approach to Language II, Carnap constructs a syntax for language in general of which the central importance is assigned to terms like *consequence, consequence class, validity* and *contravalidity* of a class of sentences, *content, null content*, and *total content* of a class of sentences (Carnap 1937a, §§48–49, 172–186).

Based on these terms Carnap introduces the term "range" as follows (Carnap 1937a, §48, 172, §56, 199):[25]

[1.] If every $\mathfrak{R}$ (and consequently every $\mathfrak{S}$ of S) is dependent on $\mathfrak{R}_1$, then $\mathfrak{R}_1$ is *complete* and, thus, every sentence is either affirmed or denied; a complete $\mathfrak{R}$ leaves no questions open. And if $\mathfrak{R}_1$ is contravalid, then $\mathfrak{R}_1$ is complete in a trivial sense: every sentence is at the same time affirmed and denied.

[2.] $\mathfrak{R}_1$ is called a premise-class if $\mathfrak{R}_1$ is complete but not contravalid, and if there exists no complete class which is a proper sub-class of $\mathfrak{R}_1$.

---

25 "$\mathfrak{R}$" stands here for the designation (of the syntax-language) of classes of expressions (sentences).

[3.] $\Re_1$ is called a premise-class of $\Re_2$, – in the sense of a correlate of 'consequence-class' – if $\Re_1$ is a premise-class and $\Re_2$ a consequence-class of $\Re_1$.
[4.] By a **range** (*Spielraum*) we understand a class $\mathfrak{M}_1$ of premise-classes such that each class which is equipollent to a premise-class belonging to $\mathfrak{M}_1$, belongs also to $\mathfrak{M}_1$.
[5.] By the range of $\Re_1$ one understands the class of premise-classes of $\Re_1$.
[6.] If one understands under a *material interpretation* of a sentence $\mathfrak{S}_1$ its interpretation from the point of view of its representation of an object-property attributed to the object designated by the expression $\mathfrak{U}_1$ appearing in $\mathfrak{S}_1$, then in such an interpretation
   a) every premise-class represents one of the possible states of the object-domain with which language S is concerned.
   b) That $\mathfrak{M}_1$ belongs to the range of $\mathfrak{S}_1$ means that the former is the class of all possible cases in which the latter is true, i.e., it is the domain of the possibilities left open by $\mathfrak{S}_1$, while the term "range" goes back to Wittgenstein's *Tractatus*: "The truth-conditions determine the range which is left open to the facts by the proposition" (4.463).

Carnap's 1931–1935 approach can thus be summarized by the following thesis:
Thesis 1: At the theoretical level, Carnap views language as a self-contained entity, the signs and expressions of which are not related to any extra-linguistic entities.

Such a *one-level approach* to language can be represented as follows:

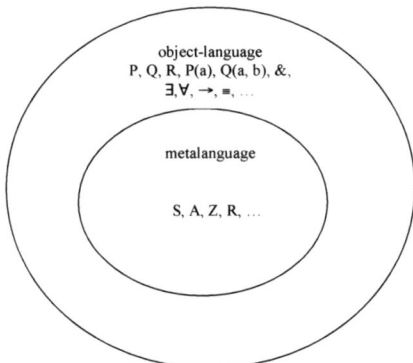

Figure 1.3 Carnap's 1931–1935 one-level approach to language

That this is so can be seen from the fact that in the years 1931–1935, Carnap's aim was to get rid of the *designata* and the relation of *designation*. For example, at the very beginning of the *LSS* he states that "it will not be assumed that … a symbol (*Zeichen*) … designates anything (Carnap 1937a, §2, 5). And the *LSS*, finally, culminates in Carnap's effort to *unmask* those sentences that on the surface seem to refer to *object-sentences*, that is, seemingly have to do with objects, but, in fact, refer only to syntactical forms, (Carnap 1937a, §74, 285):

41

to the forms of designations of those objects with which they appear to deal. Thus these sentences are syntactical sentences in virtue of their content, though they are disguised as object-sentences. We will call them *pseudo-object-sentences*.

He then proves that they are *quasi-syntactical sentences of the material mode of speech* which can be translated into the correlated syntactical sentences and, thus, into the formal mode. Carnap gives the following examples of such a translation (Carnap 1937a, §75, 289):

Material mode of speech (quasi-syntactical sentences)	*Formal mode of speech* (the correlated syntactical sentences)
1a. Yesterday's lecture treated of Babylon.	1b. In yesterday's lecture the word 'Babylon' (or a synonymous designation) occurred.
2a. The word 'daystar' *designates* (or: *means*; or: *is a name for*) the sun.	2b. The word 'daystar' is synonymous with 'sun'.
4a. The word 'luna' in the Latin language *designates* the moon.	4b. There is an equipollent expressional translation of the Latin into the English language in which the word 'moon' is the correlate of the word 'luna'.

Table 1.1 *Carnap's 1934 examples for the elimination of the relation of designation*

1.4.2. The Tarskian Turn and Carnap's Semantics

A. Tarski's presentation of the paper "Grundlegung der wissenschftlichen Semantik" at the Paris conference in September 1935 had a profound impact on Carnap's works in the field of philosophy of language which lasted until 1943.

A. A Semantician in Paris

In connection to Carnap's works from the years 1931–1935 Tarski's paper (1936) brings in the following innovations:
1) It conceives the endeavor of semantics as (Tarski 1936, 1; Tarski 1956, 401)

> ... the totality of considerations concerning those concepts in which, roughly speaking, find their expression certain connections between the expressions of a language and the objects and states of affairs referred to by these expressions.

2) In order to realize the enterprise of semantics one has to build a language that is different from the language *about which the former speaks*, and it holds that "the semantical concepts simply have no place in the framework of the language to which they relate" (Tarski 1936, 2; Tarski 1956, 402).
3 The constitution of a sufficiently exact semantics of a language consists of two steps: first, the construction of a *formalized object-language* and, second, the construction of the corresponding metalanguage containing semantical terms like *denotation* (*Bezeichnen*), *satisfaction* (*Erfülllen*). and *definition* (*Definieren*).
4) Such semantics (Tarski 1936, 3; Tarski 1956, 403)

is exact and clear only if it carries a purely structural character, that is to say, if one employs in it only those concepts which relate to the form and the arrangement of the signs and other expressions of the language,

i.e., it should have in fact the character of a *syntax-language*. In the concepts of such a semantics (Tarski 1936, 3; Tarski 1956, 403)

are expressed certain relations between objects and states of affairs 'about which one speaks' in the investigated language. Hence the sentences which establish the essential properties of semantical concepts must contain both designation of the objects referred to (thus the expressions of the language itself), and the terms which are used in the structural description of the language. The latter terms belong to the domain of the so-called *morphology of language* and are the designations of individual expressions of the language, of structural properties of expressions, of structural relations between expressions, and so on. The metalanguage which is to form the basis for semantical investigations must thus contain both kinds of expression: the expressions of the original language, and the expressions of the morphology of language.

Tarski's approach to semantics in his Paris-conference paper can thus be represented as follows:

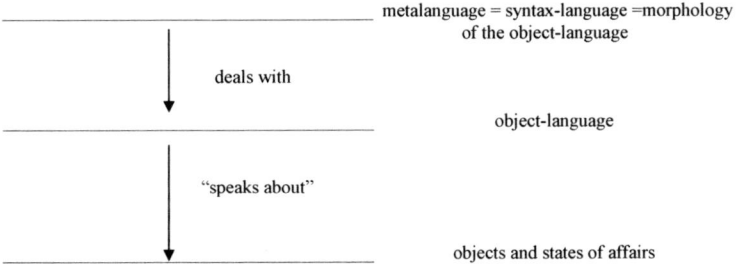

Figure 1.4 Tarski's 1936 approach to semantics

B. Carnap's incomplete semantics of 1935–1942

What is the relation of Carnap's approach to language in the years 1931–1935 to that of Tarski in his 1936 paper? An answer to this question is given by M. Kokoszyńska in her paper (1936a),[26] in which she distinguishes the following possible approaches to the construction of the syntax of language:

a) one can limit himself or herself either to *logical* signs or one can allow in the construction also *descriptive* signs; the former approach is labeled by Carnap as *pure syntax*, the latter as *descriptive syntax*;
b) if one opts for the latter option, one can still choose between an approach in which only *designations of expressions and of their relations are permitted*, or *also designations of other objects*; the first approach is that of a *syntax proper (eigentliche Syntax)*, via the second approach one obtains an *enlarged* or an *non-proper (uneigentliche) syntax*.

Carnap's approach in the years 1931–1935 can then be characterized as that of a *proper syntax*, while Tarski's approach in the Paris-conference paper as that of an *enlarged syntax*.[27]

How did Carnap himself react to Tarski's and Kokoszyńska's appearance at the Paris conference? It seems that Carnap initially was not willing to give up his proper-syntax-approach from the years 1931–1935. In the discussion that followed the presentations of Tarski and Kokoszyńska, Paul Bernays (Neurath 1935, 399–400)

suggested that this approach opens up a new region, what CARNAP refused with the hint that via semantics no third sphere is added to the "languages" and "facts." Semantics sets up only a peculiar relation between these two spheres.

To illustrate his point Carnap used the following example: "If the zoologist investigates into the animals and the botanist into the plants, then no new object-region is opened up if somebody investigates into the relation between animals and plants" (Neurath 1935, 400).

Carnap's denial of the existence of a "third sphere" (*dritter Bereich*) reappears in his paper "Von der Erkenntnistheorie zur Wissenschaftslogik," which he presented at the Paris conference. In this paper he clearly differentiates between (Carnap 1936a, 40):

---

26 See also her papers (1936b), (1936c), (1937), and (1937/38).
27 As a third criterion M. Kokoszyńska introduces the relation of the richness of the syntax-language to that of the object-language Both Carnap's and Tarski's syntax is logically richer than the object-language.

two kinds of objects of scientific investigation: on the one side, the things, processes, facts etc., and on the other, the linguistic forms. The investigation into the facts is the task of the real-scientific, empirical research; the investigation into the linguistic form is the task of the logical, syntactical analysis. We do not find any third object-sphere besides the empirical and the logical.

And in a footnote he adds (Carnap 1936a, 40):

The semantics which is treated in the presentations of Tarski (Part III) and Mrs. Lutman [Kokoszyńska] (Part III), deals with relations between objects and language-expressions. In it we thus also do not find any third sphere.

However, at the same conference Carnap presented another paper, "Wahrheit und Bewährung" (Carnap 1936b), in which he explicitly draws upon Tarski's paper (1936), which he evaluates positively (Carnap 1936b, 18), and clearly differentiates between the concept of truth and the concept of confirmation, and, in addition, demands that the question of the definition of the term *true* be distinguished from the question of a criterion of *justification* (Carnap 1936b, 23). Nevertheless, Carnap does not explicitly accept Tarski's view that the definition of "truth" is part of the enterprise of semantics.

A thoroughly accomplished shift to semantics is clearly recognizable in the works of Carnap written after the Paris conference, namely, his (1937b; 1938; 1939a; 1939b). In (1937b) Carnap gives the following survey of kinds of symbols used in symbolic logic (Carnap 1937b, 1):

Kinds of symbols	Level	Letters Used As Constants	As Variables	Designate
sentential symbols	--	A, B, C ...	p, q, r	-- (states of affairs)
individual symbols	0	a, b, c...	x, y, y, u ...	objects
predicate 1-place	1	P, Q ...	F, G ...	properties
" 2-place	1	R, S ...	H, K ...	relations
" (see #9)	2	^2P ...	^2F ...	properties of properties
	:			
	n	^nP ...	^2F ...	
functors (see #10)		k, l	f, g ...	

Table 1.2 Carnap's 1937 survey of kinds of symbols used in logic

In the paper "Logical Foundations of the Unity of Science" (Carnap 1938) Carnap differentiates between two types or parts of the analysis of linguistic expressions of science: *logical syntax* and *semantics*. The former is "restricted to the forms of the linguistic expressions involved, i.e., to the way in which they are constructed out of elementary parts (e.g. words) without referring to anything outside language" (Carnap 1938, 408), while the latter, he writes (Carnap 1938, 409):

goes beyond this boundary and studies linguistic expressions in their relation to objects outside of language ... This investigation ... takes into consideration one important relation between linguistic expressions and other objects – that of designation. An investigation of this kind is called *semantics*. ... What is designated by a certain expression may be called its *designatum*.

In a similar vein Carnap states in his paper (1939a) that in *"semantics* we study the relation between expressions in a language and their designata" (Carnap 1939a, 222) and as its possible theme he mentions the case that (1939a, 223):

... a semantical analysis would show that to describe a certain thing in a certain respect, e.g. its weight, we can use quite different forms, among which we may distinguish three chief kinds: first, one place predicates designating properties (e.g. "this thing is heavy"); second, two place predicates designating relations (e.g. "this thing is heavier than that"); and third, functors designating magnitudes (e.g. "the weight of this thing is 5 pounds").

The relation between an expression of a language and its designatum he now views as the *proper subject matter of semantical analysis*, and states (Carnap 1939a, 223–224):

We might equally well regard it as exhibiting the truth-conditions of the sentence of the language in question. This is merely a different formulation of the aim of semantics. Suppose a sentence of the simplest form is given, consisting of a proper name combined with a one-place predicate (e.g., "Switzerland is small"); if, now, we know which object is designated by the name and which property is designated by the predicate then we know also the truth-condition of the sentence: it is true if the object designated by the name has the property designated by the predicate. Thus the concept of *truth* turns out to be one of the fundamental concepts of semantics. We may say that the result of the semantical analysis of a sentence is the understanding of the sentence. To understand a sentence is to know what is designated by the terms occurring in the sentence, and hence, to know under what conditions it will be true.

In addition, such an approach to semantics finds its continuation in Carnap's *Foundations of Logic and Mathematics* of 1939, where semantics is again understood as dealing "with the expressions of a language and their relation to the designata" (Carnap 1939b, 4). In this work, Carnap, first, introduces an object-language B and English as its metalanguage and, then, constructs a semantical system B-S corresponding to language B, in which the former should consist of rules establishing the relation between the expressions of B and their designata. As a preparation for the construction of rules of B-S it is presupposed that the expressions of B are composed of signs that can be divided into *logical constants* and *logical variables*, as well as into *descriptive names* and *descriptive predicates*. He then provides the rules (the so-called B-SD-rules giving the designata of descriptive signs) for the system B-S, so that names designate things and predicates designate properties of things, as well as truth-conditions for sentences of B-S (the so-called B-SL-rules). So as "the semantical rules give an *interpretation* of the language system" (Carnap 1939b, 11), the interpretation of a language is here understood as assigning designata to its descriptive signs and assigning truth-conditions to its sentences. Finally, Carnap explicates a variety of important concepts, among them semantical synonymity, L-truth, F-truth (factual truth), L-equivalence, and L-implication.

The ultimate stage of the development of Carnap's semantics in the Tarskian framework is articulated in his *Introduction to Semantics* (hereafter, *IS*) of 1942. Its roots go back to early August 1939, when Carnap initiated an extensive correspondence with philosophers, logicians, and mathematicians (e.g., W. V. O. Quine, C. G. Hempel, A. Tarski and A. Church) in which he tried to identify a unified English terminology for semantics and, what is important from the point of view of this paper, to determine the terms for designata of various language expressions. His choice is as follows (Quine and Carnap 1990, 270):[28]

expression	designatum
predicate or functor	function (wider sense)
predicate	attribute
one-place predicate	property
more-place predicate	relation
functor	function (narrower sense)
sentence	proposition

Table 1.3 Carnap's 1939 proposal for expressions and their designata

This semantics then reappears at the very beginning of the *IS* in the following form (Carnap 1942b, §6, 18):

---

28 For a first critique of Carnap's approach to viewing propositions both as the designation of sentences and as the meanings of sentences see (Nagel 1942, 471).

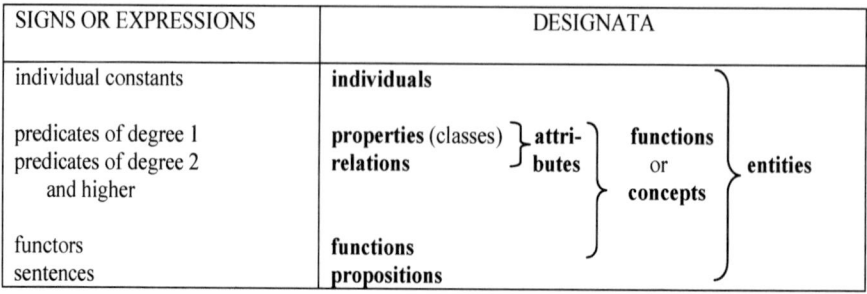

*Figure 1.5 Signs/expressions and their designata in **Introduction to Semantics***

It is worth noting that Carnap's semantics as presented in the *IS* of 1942 not only draws on Tarski's views on semantics, but also unifies them with Wittgenstein's views, especially with the latter's approach to the so-called *state of affairs*,[29] which Carnap labeled in the *LSS* of 1934 as the *material interpretation* of sentence $\mathfrak{S}_i$. The center stage in Carnap's effort to construct a semantical (or interpreted) system as a system of rules stated in a metalanguage referring to an object-language, so that these rules determine the truth-conditions for every sentence of this object-language (Carnap 1942b, §7, 22), goes in the direction of explicating the L-concepts and, here, especially the concepts L-truth and L-implication, the first being understood in the sense of "analytic (truth)" and the latter as "(analytic) consequence."[30]

Carnap proceeds in *IS* by, first, introducing (§14) the terms "L-truth" and "L-implication" as elementary terms together with postulates for other L-terms like "L-determinate," "L-exclusive," "L-complete," "L-comprehensive," "L-interchangeable" and "L-perfect." He, then, defines the first two terms, deriving from their definitions these postulates as theorems (§20). Here he already uses the term "L-range" introduced (§18) as an elementary term, which, in turn, is informally introduced by means of the concept of state of affairs (Carnap 1942b, §18, 95):[31]

---

29 Wittgenstein presents his views on state of affairs in the *Tractatus* as follows (4.25–4.26): "If an elementary proposition is true, the state of affairs exists, if an elementary proposition is false, the state of affairs does not exist. If all true elementary propositions are given, the result is a complete description of the world."
30 According to Carnap a sentence in a semantical system S_i is L-true "if and only if it is true … in such a way that its truth follows from the semantical rules" (Carnap 1942b, §15, 79–80) of this system alone.
31 "$\mathfrak{S}_i$" stands here for a syntactical name of a sentence from an object-language.

A possible state of affairs of all objects dealt with in system *S* with respect to all properties and relations dealt with in *S* is called an *L-state* with respect to *S* ... A given L-state leaves no question in *S* open; every sentence in *S* either admits or excludes that L-state. The class of the L-states admitted for $\mathfrak{S}_i$ is called the *L-range* of $\mathfrak{S}_i$.

At the same time Carnap claims that once (Carnap 1942b, §18, 96)

... we understand a sentence we know what possibilities it admits. The semantical rules determine under what conditions the sentence is true; and that is just the same as determining what possible cases are admitted by it. Therefore, the L-range of a sentence is known if we understand it – in other words, if the semantical rules are given; factual knowledge is not required.

Then, by drawing upon the claim that "L-states are propositions" (Carnap 1942b, §18, 95), and upon Wittgenstein's view that what "propositions" stand for can be labeled as a "state of affairs" (Carnap 1942b, §37, 235; i.e., that L-ranges are classes of propositions) Carnap in §20, finally, explicates the terms "L-true" and "L-implies." He proceeds in such a way that he considers a class of all L-states with respect to a semantical system S which is the *universal L-range* in S. A sentence $\mathfrak{S}_i$ is L-true in S if, and only if, its range is the universal L-range in S, and $\mathfrak{S}_i$ L-implies $\mathfrak{S}_j$ if the L-range of the former is included in the L-range of the latter. So, in *IS*, *proposition is the type of entity which is denoted by a declarative sentence* (Carnap 1942b, §37, 235).

The influence of Alonzo Church is evident in Carnap's choice between various approaches to the designatum of a declarative sentence. While Carnap initially suggested in his set of questions of early August 1939 (Quine and Carnap 1990, 270) two options, (a) propositions or (b) state of affairs, Church's answer in a letter to Carnap of August 15, 1939 was as follows ([RC 089-08-02], 4):

The proposal that the <u>designatum</u> of a (declarative) sentence be a state of affairs is <u>unsatisfactory,</u> because it provides no designata for false sentences. It must be amended to read that the designatum of a sentence is to be an <u>alleged state of affairs</u>. But the underscored phrase is merely a complicated substitute for the <u>familiar term</u> *proposition.*

C. LSS and Carnap's semantics of 1935–1942

Let me now compare Carnap's 1931–1935 approach to language with that of the years 1935–1942. The fundamental difference between these two approaches can be located at the *theoretical* and *metatheoretical* levels and can be stated via the following two theses.

Thesis 2: At the level of a theory of language Carnap relates the signs/expressions of a language to their designata.

From the point of view of Carnap's 1935-1942 semantics the sentences appearing in both columns in Table 1.1 can now be understood as follows:

i. The sentence 1a is part of an object-language and the word 'Babylon' appearing in it is a word of an object-language referring to the extra-linguistic entity Babylon. The sentence 1b is part of an object-language, while the word "Babylon" appearing in it is a metalanguage expression referring to the object-language-word 'Babylon' from sentence 1a.
ii. The word "daystar" in 2a is a metalanguage expression, like the whole sentence 2a, while 'sun' is an object-language-word embedded in a sentence of this metalanguage. The same holds for the sentence 2b, where "sun" is a metalanguage-word referring the object-language-word 'sun' from 2a.
iii. The word "luna" in 4a is a metalanguage expression like the whole sentence 4a, while 'moon' is an object-language-word embedded in this metalanguage expression. Sentence 4b is a sentence from a language which is a metalanguage with respect to the object-languages Latin and English, while the words "moon" and "luna" are object-language-words integrated into this metalanguage.

At the *metatheoretical* (metasemantical) level the fundamental difference between the 1931–1935 and the 1935–1942 approach becomes apparent when one compares Figures 1.3 and 1.4. The move from the syntactical to the semantical approach takes place also at the level of *reflection on the status of these theories.* *Thesis 3: Once Carnap differentiates between metalanguage, object-language, and its objects, he explicitly acknowledges that he has to give up his one-level approach to language in favor of a more advanced two-level approach.*

Nevertheless, we still face the following question. Why did I, even with this type of theoretical and metatheoretical advance in Carnap's approach to language, label his 1935–1942 semantics in the caption of B. as an "incomplete" semantics? It is because Carnap inherited from his pre-semantical period one important feature – *the abstraction from meanings of language expressions as the third logical entity* in addition to the very *expressions* and their *designata* – which lacks the explication of the term "meaning." One can, from the point of view of such an abstraction from meaning, trace a *continuity* between the presemantical and the semantical, 1935–1942 period in Carnap's works.

First, at the very beginning of the *LSS* Carnap already states that he will not assume that a symbol possesses a meaning (*Bedeutung*) (Carnap 1937a, §2, 5), and he acknowledges that meanings are the objects of the investigation of semantics (Carnap 1937a, §2, 9). It is worth noting here that the *LSS* culminates in the attempt to replace sentences about meaning by certain syntactical sentences. Carnap gives the following examples (Carnap 1937a, §75, 289–290):

Material mode of speech (quasi-syntactical sentences)	Formal mode of speech (the correlated syntactical sentences)
3a. The sentence $\mathfrak{S}_1$ *means* (or: *asserts*; or: has the *content*; or: has the *meaning*) that the moon is spherical.	3b. $\mathfrak{S}_1$ is equipollent to the sentence 'The moon is spherical.'
6a. The expression 'merle' and 'blackbird' have the same meaning (*Sinn*) (or: *mean* the same; or have the same *intensional object*).	6b. 'Merle' and 'blackbird' are L-synonymous.

Table 1.4 Carnap's 1934 examples for the elimination of the meaning of an expression

Second, in his paper (1936a) from the Paris-conference Carnap (correctly) emphasizes that neither his own syntactical approach from 1931–1935 nor Tarski's and Kokoszyńska's semantical approach – *and thus not his own 1935–1942 semantical approach, either* – introduces a third logical realm, in addition to expressions and their designata.

Third, in the *IS*, the fact that meanings are not taken into account is readily seen in Carnap's approach to the term *proposition*. Among its various uses he mentions that it is "that which is expressed (signified, formulated, represented, designated) by a (declarative) sentence" (Carnap 1942b, §37, 235).

1.4.3 Back to Frege: Alonzo Church's Rediscovery of Frege

Even if Carnap already acknowledges (in the *LSS*) that words have meanings and that they are the proper subject matter of semantics, it is only from 1942 on that he starts to reflect on meanings, thus inaugurating yet another important shift in his semantical endeavor.

The first trace of a shift in Carnap's semantics can be found in his excerpt-notes of May 1, 1942 from C. J. Ducasse's reflections on two possible meanings when using a predicate-term. C. J. Ducasse differentiates there between the following two possible cases (Ducasse 1941, 225–226):

we may be referring either to the applications made of it – i.e., to the variety of subjects of which we predicate it – or to the implications of it, i.e., to the set of attributes we assert of anything by means of the given predicate. ... The first may be called the indicative or monstrative use of a term, and the second its quiddative use. The distinction is ultimately that between *where* and *what*, between the place of something and the properties of it. ...

The distinction is customarily described as one between the denotation and the connotation, or the extension and the intension of a term.

The main parts of Carnap's excerpt-notes are as follows [RC 089-08-04, 1]:

I	II
application (indicative use) where? (place) denotation extension	implication (quiddative use) what? (properties) connotation intension
(class of) <u>denotata</u> <u>extension</u>	<u>designatum</u> –

Table 1.5 *Carnap's excerpt-notes on extension/intension of a predicate from May 1942*

It is important to note here that Carnap differentiates between the *denotata* and the *designatum* of a term. This clearly deviates from his own understanding in his (1942a),[32] in which he characterizes semantics as a "[t]heory of the relation between signs and what they refer to (their 'designata' or 'denotata')" (Carnap 1942a, 288).

A. Alonzo Church's rediscovery of Frege

Another impulse for the conceptual shift in Carnap's semantics in the years 1943– 1944 came from Alonzo Church, whose initial works in the field of mathematical logic and especially on theory of function led him ultimately to Frege's semantical works.

Before I turn to the works of Church, let me briefly and elementarily explain Frege's views[33] on function and object. Frege, starting from and generalizing the

---

32 I am forced here to quote from the published contribution of Carnap to D. D. Runes' *Dictionary of Philosophy* because the manuscript of this contribution is not available. The same holds for the contributions of A. Church to this dictionary from which I quote below. Even if both of them in the *Journal of Philosophy* (1942, Vol. 39, 139) publicly disavowed their contributions to Runes' *Dictionary of Philosophy*, the passages quoted here do not deviated from the views to which Carnap and Church held during the analyzed period.

33 For a more detailed analysis of Frege's approach to functions and objects see, e.g., (Martin 1982), (Kreiser 1986, 45–62), and (Klement 2002, 28–43).

mathematical understanding of function, differentiates between function and object (*Gegenstand*), so that the former, but not the latter, is by itself principally incomplete, unsaturated, and in need of supplementation (*ergänzungsbedürftig*) (Frege 1967a, 128; Frege 1984a, 140). At the same time function is the reference of a function-name, while the object is the reference of a proper name. Frege also differentiates between the function with one variable, which he labels as *concept*, always assigning to an object as its argument a truth-value, and the function with two (or more) variables, labeled as *relation*, always assigning to two (or more) objects as its arguments a truth-value. In addition, Frege introduces the term "course of values" (*Werthverlauf*) of a function, understood by him as the course of values of the function for certain arguments and which, in the case of a concept, he equates with its range *(Umfang)*. Let me take as an example the sentence "Caesar conquered Gaul." In Frege's semantics this sentence is the name of the object True and is composed of the proper name "Caesar" and of the name of the function or concept "conquered Gaul". Schematically the situation in Frege's semantics can be represented as follows:[34]

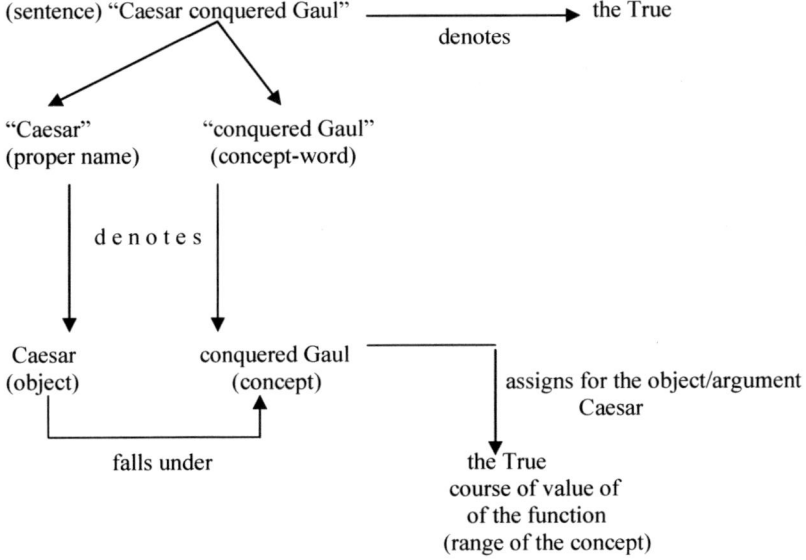

Figure 1.6 Frege on object and concept

---

34 For the sake of simplicity I do no represent here the sense (*Sinn*) of the expressions.

Church's turn to Frege's semantics has its roots in the work of the former in the field of mathematical logic and initially in his reflections on the nature of functions. Already in (Church 1936), independently of Frege, Church clearly differentiates between a *function* and its *values*, where a function, say, $3x^2 + 5x + 1$ "denotes an ambiguous real number" (Church 1936, 2), and only by fixing x can one obtain an unambiguous result, the value of that function. In order to differentiate between function and its values, Church introduces the λ-notation, so that for the example given, the notation for the function is $\lambda x[3x^2 + 5x + 1]$. By means of such a notation he can differentiate, like Frege, between the identity of two functions, for example,

$\lambda x[3x^2 + 5x + 1] = \lambda y[3y^2 + 5y + 1]$,

and the identity of values of two functions, for example,

$\{\lambda x[3x^2 + 5x + 1] (2)\} = \{\lambda y[3y^2 + 5y + 1] (2)\}$.

Church, also like Frege, *generalizes* the concept of mathematical function by introducing the term *propositional functions*, understood by him as "expressions which become propositions or truth-values when values were assigned to all variables involved" (Church 1936, 1).

In a manner similar to that of Frege he states that "'truth' ... and 'falsehood' ... denote different things" (Church 1936, 13), but not being fully acquainted with Frege's semantics, Church deviates from him with respect to the understanding of the relation between *sentence, proposition*, and *truth-value*. He characterizes propositional function, abbreviated as ppfn, as follows (Church 1936, 13):

In a usage employed elsewhere, a ppfn is a function the range of whose dependent variable is included in the set of propositions. The distinctions may be illustrated: "x is a man" is an ambiguous proposition, $\lambda x[x$ is a man$]$ is a function whose values are propositions, and $\lambda x[\text{TrV}(x$ is a man$)]$, where TrV(*) means "truth-value of *", is a ppfn in our sense. Without attempting to answer the question of whether there is a tenable distinction between $\lambda x[x$ is a man$]$ and $\lambda x[\text{TrV}(x$ is a man$)]$, a question which is irrelevant to our present purpose, we shall merely agree to use the latter function and to speak of it as the ppffn "to be a man."

The first important shift in Church's views on functions can be found in (Church 1941). Here, he characterizes function as "a rule of correspondence by which when anything is given (as <u>argument</u>) another thing (the <u>value</u> of the function for that argument) may be obtained" (Church 1941, 1) and, then, differentiates between the *function in extension* and the *function in intension*, where the latter is the *meaning* of the function, as follows (Church 1941, 2–3):[35]

> The foregoing discussion leaves it undetermined under what circumstances two functions shall be considered the same. The most immediate and, from some points of view, best way to settle this question is to specify that two functions f and g are the same if they have the same range of arguments and, for every element a that belongs to this range, (fa) is the same as (ga). When this is done we shall say that we are dealing with <u>functions in extension</u>. It is possible, however, to allow two functions to be different on the ground that the rule of correspondence is different in meaning in the two cases although always yielding the same result when applied to any particular argument. When this is done we shall say that we are dealing with <u>functions in intension</u>. ... We ... shall speak of functions in intension in any case where a more severe criterion of identity is adopted than for functions in extension.

The turning point in Church's oeuvre appears to occur with his presentation of a short paper (on December 31, 1941 at the meeting of the *Association of Symbolic Logic*) with the title "On sense and denotation,"[36] in which he presents Frege's semantics of proper names and sentences as given in "Über Sinn und Bedeutung." The fact that Church became acquainted with Frege's works and realized the latter's importance is readily seen in the articles he wrote for the D. D. Runes' *Dictionary of Philosophy*, the most important of which are the following:

i. *Abstraction*. If A is a formula containing a free variable x, then the process of forming a corresponding monadic function[37] is called *abstraction* or *functional abstraction* and can be understood as an operation upon formula A yielding a function, and is relative to a particular system of interpretation for the notations given in the formula, and to a particular *variable*, as x. The notation used by Church, for the function obtained from A by abstraction relative to x is $\lambda x[A]$. Church also relates the term "abstraction" to Frege as follows (Church 1942b, 3):

> Frege ... uses a Greek vowel, say ε, as the variable relative to which abstraction is made, and employs the notation $\acute{\varepsilon}(A)$ to denote what is essentially the function in extension (the 'Werthverlauf' in his terminology) obtained from A by abstraction relative to ε.

---

35 *(fa)* and *(ga)* stand here for the values of functions f and g for the argument a, respectively.
36 Unfortunately the manuscript of this lecture is not preserved in the archives of Princeton University; only a short abstract of it appeared in print as (Church 1942a).
37 "Monadic function" is understood by Church as "a law of correspondence between an argument (or *value of the independent variable*) and a *value* of the function (or *value of the dependent variable*)" (Church 1942c, 113).

ii) *Proposition*: From various meanings of the term "proposition" Church chooses the following one (1942d, 256):

> ... the content of meaning of a declarative sentence, i.e., a postulated abstract object common not only to different occurrences of the same declarative sentence but also to different sentences (whether of the same language or not) which are synonymous or, as we say, mean the same *thing*.

iii. *Propositional function*. Church understands it as the function for which the range of the dependent variable is composed of propositions each of which stands for the value of the propositional function for the given arguments. At the same time he understands it as the *propositional function in intension*, that is different from the *propositional function in extension*; for the latter holds that "the assumption is made that two propositional functions are identical if corresponding values are materially equivalent ... The values of a propositional function in extension are truth-values ... rather than propositions" (Church 1942e, 257). From this Church draws the conclusion that a *monadic propositional function in intension* is a property (of things belonging to the range of the independent variable), while a *monadic propositional function in extension* is a class (or set or aggregate).

iv. *Signification*. Here Church under its various meanings understands also the case that "it may be used to indicate the intensional rather than the extensional meaning of a word" (Church 1942f, 292).

v. *Truth, semantical*. Church characterizes it as "the property of a propositional formula (sentence) that it expresses a true proposition" (Church 1942g, 322).

So, in Church's approach, beginning in 1942, the sentence's *intensional meaning* is the *proposition* expressed by the sentence, while for the propositional function, as he understood it in 1942, the following scheme can be used:

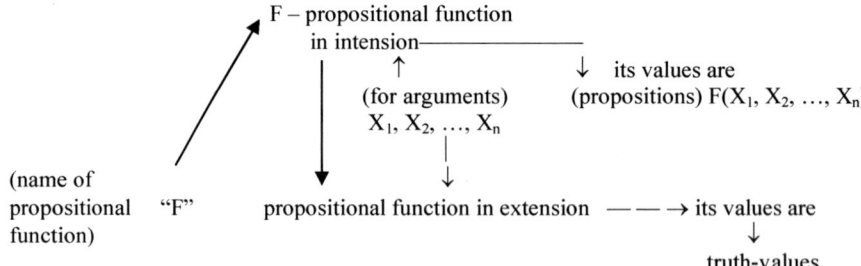

Figure 1.7 Alonzo Church on propositional function in 1942

It is important to notice here that Church does not as yet differentiate in his 1942 articles from the *Dictionary of Philosophy* between the semantical "characteristics" of the variables $X_1, X_2, \ldots, X_n$ with respect to the propositional function in *intension* and their semantical "characteristics" with respect to the propositional function in *extension*. Neither are the full arrows in Figure 1.7 labeled by Church in the 1942 articles.[38]

Church's new approach also can be discerned in his 1942 lectures, "Elementary Topics in Mathematical Logic", where he differentiates between a sentence and the proposition expressed by it as follows (Church 1942h, 27):

> By a proposition we shall mean the <u>content of meaning of a declarative sentence</u>. (It would be a first approximation to say that a proposition <u>is</u> a declarative sentence, but this is not quite adequate because we wish to say that different sentences which are synonymous – and likewise intertranslatable sentences from different languages – express the same proposition.)

In these lectures he also differentiates between the proposition expressed by a sentence and the truth-value denoted by it (Church 1942h, 52–53):

> If A is any sentence ... containing a (free, object) variable, say x, then the notation xA is used to mean <u>the class of things x such that</u> A is true. ... the notation (x)A is used to express the proposition that A is true for all values of x ...; and the notation (Ex)A is used to express the proposition that A is true for at least one value of x ... The class denoted by xA and the propositions expressed by (x)A and (Ex)A are said to be obtained from sentence A by <u>abstraction</u> with respect to x, <u>universal quantification</u> with respect to x, and <u>existential quantification</u> with respect to x, respectively.

Church's new approach from 1942 to propositions and (propositional) functions had a profound impact on Carnap's semantical endeavor in the years 1943–1944 via their mutual correspondence and via Church's review of the *IS* (1943b). In this review he proposes two fundamental changes in Carnap's semantics. First, he proposes to take as designata of sentences not propositions but truth-values. This proposal represents a clear departure from his own views stated in the 1939-correspondence with Carnap analyzed above, and on which he reflects in the review as follows: "On this point the reviewer confesses to have changed his own former opinion ..." (Church 1943b, 299–300). Second, he proposes, while directly referring to Frege, yet another change in Carnap's semantics, (Church 1943b, 303):

---

38 In (1956) Church already differentiates for the range of the variables $X_1, X_2, \ldots, X_n$ – now understood as syntactical entities (i.e., as "X_1," "X_2," ... "X_n") – between their respective *concepts* and *denotations*, the former being relevant for the function in intension, the latter for the function in extension. He also labels in his book (1956) these "arrows" so that a name of a function *expresses* a function-concept, while the latter, in turn, *determines* the function in extension.

namely, that the notion of sense not only should receive treatment but should be taken into account throughout as of equal importance with that of designation. In particular, the notion of sense should be prominent in connection with propositions, truth-conditions, absolute L-concepts, extensionality, L-synonymy.

Once Church proposed these two changes, he arrives, on the metasemantical level, at a new understanding of the goals of semantics, which he characterizes in another review as follows (Church 1943a, 47):

There remains the important task, which has never been approached, of constructing a formalized semantical system which shall take into account both kinds of meaning, the relation between a name and its denotation, and the relation between a name and its sense. It would be a desideratum for such a system that the object language should contain for every name in it a name of the associated sense, and should be able of expressing the relation between a sense and the denotation which it determines.

And, at the same time, at the very level of Carnap's semantical theory he proposes (Church 1943b, 303):

... the statement about 'meaning' ... on page 75 probably should be replaced by an explicit postulate about sense, that two expressions have the same sense if and only if they are L-synonymous. The assumption on page 92 that L-equivalent sentences have the same designatum, and are therefore synonymous, should then be replaced by the assumption that L-equivalent sentences have the same sense, and are therefore L-synonymous.

Later on, Carnap accepted both of Church's proposals for changes from this review together with the just-quoted proposals for replacements and integrated them into his semantics, which I will explain shortly.

After the publication of *IS* in 1942 and *Formalization of Logic*[39] in 1943 Carnap continued his endeavor in the field of semantics, which included a letter to Quine of January 21, 1943, together with a questionnaire, distributed in May 1943, which he believed would enable him to clarify the meaning of the central semantical concepts. In this letter, in a manner that can be viewed as a generalization of his differentiation between designata and denotata of predicate given in his excerpt notes from (Ducasse 1941) analyzed above, Carnap gives the following semantical classification of kinds of expressions (Quine and Carnap 1990, 306):

---

39 Church in his review of Carnap's *Formalization of Logic* shows where his views deviate from those of Carnap. While for the latter "an interpretation of a calculus is merely an assignment of truth-conditions for sentences" (Church 1944b, 493), Church understands the interpretation of a calculus as "assigning to each sentence a meaning as expressing a certain proposition" (Church 1944b, 493).

Designation and denotation

(1) Kinds of expressions	(2) Designatum	(3) Denotatum (?)
a. Predicate (degree 1)	Property	Classes
b. Sentence	Proposition	Truth-Value
c. Individual constant	Individual concept (?)	Object

Table 1.6 Carnap's semantics of expressions from January 1943

Here Carnap clearly deviates – with respect to individual constants – from his own views given in the *IS* – by claiming that "it would be better to say that the designatum of an individual constant is a concept of an individual type (I call them tentatively 'individual concepts')" (Quine and Carnap 1990, 306). But he claims that he still holds to his views from the *IS* with respect to sentences and predicates (degree 1); "As before I regard properties as designata of predicates and propositions as designata of sentences" (Quine and Carnap 1990, 306). In an explication of this table he states the following (Quine and Carnap 1990, 307):

It is true ... that individuals designate objects, predicates designate classes, and sentences designate truth-values (as Frege said). However, we might consider even here to call the entities (3) not designata but to use another word, perhaps 'denotata'. If so, we would say ... a predicate designates a property and denotes a class. Then we should also say: 'Pegasus' designates something but does not denote anything.

The semantics of expressions given in the Quine-letter of January 1943 represents an important "in-between stage" in the development of Carnap's semantics between, on the one hand, the *IS* and, on the other hand, the manuscript "Extension and Intension" of 1943. To evaluate this "in-between stage" one has to take into account his *Questionnaire on Terminology for Expressions and Meaning* of April 1943 (Quine and Carnap 1990, 314–322) and here especially the following passage (Quine and Carnap 1990, 317–318):

The word 'green' has a certain semantical relation R_1 to a certain property, viz. the color or greenness and another relation R_2 to a certain class of green things. In general, R_1 is a relation between an expression and what is usually called its meaning; R_2 is a relation between an expression and what sometimes is called an extension. Then there is a third relation R_3 between a meaning and the corresponding extension, e.g. between a property and the corresponding class. R_3 is not a semantical relation since no expression is involved.

It is, then, obvious that when conceptually differentiating between the *designatum* (i.e., the *meaning*) and the *denotatum* (i.e., the *extension* of an expression) Carnap wrongly claims that the semantics of expressions proposed in the letter to Quine of January 1943 still conforms to the semantics of predicates and sentences from the *IS*. But, as shown above, in the semantics of the *IS meaning* – as the third logical entity – *is not existing*. The semantics from the Quine letter of January 1943 represents a profound shift as compared to the semantics of the *IS*, even if Carnap was not immediately aware of the profoundness of this shift.

It was again Alonzo Church who brought to Carnap's attention the existence of this shift as well as its importance and profoundness. In a four-page letter to Carnap of May 10, 1943, Church writes the following:

The meaning of the word "designatum" proposed on the first page of your letter to Quine seems to me to be a very definite departure from the usage of your Introduction to Semantics. The passage at the top of page 9[40] of the book seems to make it clear that your designatum is the same as Frege's Bedeutung oder Bezeichnung ... and this is reinforced by the passage on pages 53–55.[41] Moreover, the very writing (as true) of such things as "Des ('Pferd', horse)" and "Des ('gross', large)" indicates that your relation of designation is the same as Frege's denoting – since "Pferd" and "horse" certainly denote the same thing, whether the thing be a class or a property. And "designatum" by its etymology, as well as by your explanation on page 9, means the designated object.

It is worth noting here also that in the very questionnaire of April 1943 mentioned above, Carnap, under the heading of the *meaning* of expressions, mentions already *propositions and attributes* – of any degree (Quine and Carnap 1990, 318) – thus, clearly deviating from his own semantics from the *IS*, in which the latter two were classified as *designata* of expressions or signs.

1.4.5. The Manuscript "Extension and Intension": The Missing Link

During his stay in 1943 in Santa Fe, New Mexico Carnap completed the manuscript "Extension and Intension", about which he states the following in his intellectual autobiography (Carnap 1963a, 63):

In 1943, I wrote a book manuscript, called "Extension and Intension." With both Quine and Alonzo Church, who read copies of it, I had detailed discussions by correspondence which

---

40 According to Carnap "we may distinguish ... the expression, and what is referred to, which we shall call the designatum of the expression. (We say, e.g., that in German 'Rhein' designates the Rhine, and that the Rhine is the designatum of 'Rhein' ...)" (Carnap 1942b, §4).
41 Carnap 1942b, §12.

greatly helped to clarify my conceptions. Later, I worked out a considerably changed and extended version which appeared under the title *Meaning and Necessity*.

Unfortunately, this manuscript is available neither in the Carnap archives of the University of Pittsburgh nor in the Church archives of Princeton University or W. V. O. Quine's archives at Harvard University. It is thus impossible at present to find out the form of semantics to which Carnap holds in this mediating link between his *Introduction to Semantics* and his *Meaning and Necessity*. In the latter he holds to the following semantics of language expressions:

Sign	Extension of the sign	Intension of the sign
individual expression	individual	individual concept
predicate of degree 1	class	property
predicate of degree 2 (or higher)	class of ordered pairs (or *n*-tuples) of things	relation
sentence	truth-value	proposition

*Table 1.7 Carnap's semantics in **Meaning and Necessity** of 1947*

It is thus readily seen that Carnap finally accepted Church's suggestions for revisions of his semantics as proposed in (Church 1943b). I can thus state the following thesis:

Thesis 4: Carnap, finally, views the expressions and signs of language as related to two different extra-linguistic entities: extensions and intensions.

When one compares the semantics from *M&N* with that of the *IS* as expressed in Figure 1.5, as well with the approach to language from the *LSS* given in Part 1 above, one can realize the profoundness of the change to which Carnap's approach to language was subject in the years 1931–1947, from a *purely syntactical approach*, via an *incomplete semantical* approach, to a more complete, *intension/extension* approach. This latter approach is the one to which Carnap holds to, at least for the next twelve years after the publication of *Meaning and Necessity* in 1947. That it is so can be seen from his *Notes on Semantics* of 1955/1959, in which he gives the following terminology of intensions and extension ([RC 086-17-01], 8):

Designator	Intension	Extension
individuator	individual concept	individual
one-place predicate	property	class
n-place "	n-adic relation	class of n-tuples
sentence	proposition	truth-value (T, F; or 0, 1)

Table 1.8 *Carnap's semantics from* **Notes on semantics** *of 1955/1959*

Worth mentioning here, by way of conclusion, is the fact that Church, in the course of the written correspondence with Carnap, also changed his views in important ways.[42] In a letter of February 9, 1944, while commenting on Carnap's manuscript "Extension and Intension," he refers to his own paper (Church 1940), which he *now* understands as a starting point for "a treatment of extension and intension, or of denotation and sense, along the lines of Frege's ideas" (Church letter Carnap, February 9, 1944, 1–2). Initially, the paper was considered merely a formulation of the mathematico-logical (simple) theory of types, in which Church introduced the Greek symbols ι and o so that (Church 1940, 56–57):

… the type symbols enter our formal theory only as subscripts upon variables and constants. In the interpretation of the theory it is intended that the subscript should indicate the type of variable or constant, o being the type of proposition, ι the type of individuals.

Church, thus, did not in 1940 as yet differentiate between, on the one hand, the *very type-symbols* "ι" and "o" and, on the other hand what they *express* and what they *refer to*. But, in the letter to Carnap of February 9, 1944 Church already considers his paper (1940) as a convenient basis for a logic of sense and denotation.[43] Now, he

---

42 This development can be readily seen when one compares Church's views on functions, propositions and propositional functions, as given in his (1935), (1941), (1944), and (1956). From the point of view of this paper one of the most important aspects of this development is that he regards the *sentence as expressing a proposition* (Church 1944a, 34), while a *propositional variable represents truth-values rather than propositions* (Church 1944a, 112). An important place in this development is played also by his letter to Carnap of May 17, 1943 ([RC 089-07-04]), where he comments on Carnap's terminological questionnaire; he even implicitly refers to this letter in the second edition of *Introduction to Mathematical Logic* (Church 1956, 6).

43 For a continuation of Church's work on the logic of sense and denotation see his (1946), (1951), (1973), (1974), and (1993) works. For an analysis of this logic see (Potts 1979), (Klement 2002), and (Parsons 2001).

replaces the basic types ι and o by two sets of basic types $\iota_0, \iota_1, \iota_2, \ldots$ and $o_0, o_1, o_2, \ldots$ and at the same time starts to speak about symbols of the type ι and o which should *denote extra-linguistic entities* (Church to Carnap, February 9, 1944, 2):

Names of type o_0 ... would be construed as names of truth-values, names of type ι_0 would be construed as names of individuals, names of type o_{n+1} would be construed as names of senses of names of type o_n, names of type ι_{n+1} would be construed as names of senses of names of type ι_n ....

1.5 Transparent Intensional Logic

So, as neither Frege's nor Carnap's semantics provides a satisfactory solution to the issue of identity of meanings[44] and, thus, of changes of meaning, I employ in this book *Transparent Intensional Logic*[45] (hereafter, TIL). In chapters 3, 5, and 6 I will use TIL in order to prove that a shift of meaning takes place in the process of theory construction both in the natural and the social sciences.

The latter provides a significant extension of intensional logic, such as R. Montague's, in whose framework it is possible to deal with the language of scientific theories. Because TIL is not widely known, even among logicians, and has not, to best of my knowledge, until now been used in the framework of the philosophy of science, I will state briefly the main principles of TIL.

In addition to the widely used semantical entities – language expression and meaning/intension and extension – TIL brings in a new semantic entity, namely, construction and at the same time reorganizes the whole semantics for language expressions. TIL claims that each meaningful expression *represents* something, while meaning is understood as an instruction, or procedure, known as *construction*. In any state of affairs, the procedure can be executed yielding the entity denoted by an expression or, in well-defined cases, failing to yield anything. In the case of empirical expressions the denoted entity (if any) is a intension of possible-world semantics. Thus, TIL modifies the Frege-Church semantic schema by shifting intension down to denotation and explicating "sense" as a construction. Thus shift makes it possible to distinguish between denotation and reference of empirical expressions. According to P. Materna (2005, 31)

In the case of nonempirical expressions there is no reason to make the distinction [between denotation and reference]. In the case of *empirical expressions* it holds that they *denote intensions*; their *reference* is the value of the respective denotation in the actual

---

44 See Hanzel 2006.
45 This term was initially coined by P. Tichý in (Tichý 1988). Here I draw on (Materna 1995), (Materna 1998), and (Materna 2005).

world+time ... we can speak about their reference with respect to *the couple ⟨possible world, time point⟩*.

This means that the very knowledge of (the meaning of) the language expression suffices to determine its denotation while, in order to determine its reference, the language user has to turn to the empirical world. The relation of these semantical concepts in the framework of TIL can be expressed in general as follows (Materna 1998, 9):

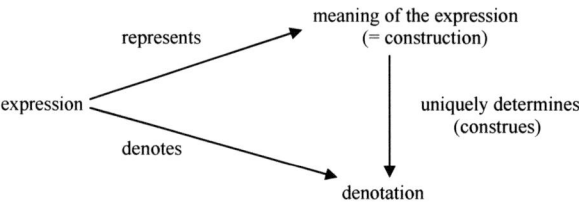

Figure 1.8 The semantics of TIL

As possible denotations TIL lists (a) no object; (b) a "nonempirical" object (intension); (c) an empirical object (extension); and (d) a higher-order object (a function with constructions as arguments or values). For the extension and intension the following classification and examples can be given (Materna 1998, 19–22):

Example of a language expression	Its intension	Its extension
"Aristotle"	X	individual
"the teacher of Alexander the Great"	individual role/office	the individual who is the holder of the particular role/office
"9"	(number)	9
"mountain"	property of being a mountain	classes of objects with that property
"the number of planets of the Solar system"	magnitude	(number) 9

"is identical with"	the relation of identity	a relation-in-extension
"is larger than"	a relation in intension	a relation-in-extension
"The Morning Star is the Evening Star"	proposition	truth-value

Table 1.9 Examples of intension and extension of expressions in TIL

In Table 1.9 it can be readily seen that in the framework of TIL intension should mediate between an expression and its respective extension thus, it views them as a type of function that to each possible world assigns a "chronology" of a certain type α, where the latter is in turn a function that to each moment of time τ assigns at most one element of a certain type α, that is, a certain *chronology of values*. In order to model the fine-grained typology of extensions and intensions given in Figure 1.8, TIL, by drawing upon Alonzo Church, introduces the following atomic (elementary) types:

i. o-type understood as the set {True, False}
ii. τ-type understood as the type of real numbers/time-points
iii. ι-type understood as the universe (individuals)
iv. ω-type understood as the class of possible worlds

In the framework of TIL, then, the following claims hold (Materna 1998, 30–32):

1. If α is an arbitrary type, then the above mentioned chronologies are $(\alpha\tau)$-objects. For example proposition is a type of function that associates each possible world with a chronology of truth-values by assigning to particular τ-points at most one truth-value.
2. In general, α-intensions are functions of type $((\alpha\tau)\omega)$, that is, $((\alpha\tau)\omega)$-objects, abbreviated as $\alpha_{\tau\omega}$-objects.
3. First-order objects are α-objects, where α is a simple type.
4. Extensions are first-order objects that are not intensions.
5. Nontrivial intensions are intensions the values of which are distinct in at least two possible worlds.
6. Any empirical expression denotes a nontrivial intension.

Based on these claims, the typology of extension/intensions of expressions in Table 1.9 can be assigned the respective types as shown in Table 1.10 (Materna 1998, 31):

Intension	The corresponding type	Extension	The corresponding type
proposition	$o_{\tau\omega}$	truth-values	o
individual roles	$\iota_{\tau\omega}$	individuals	ι
properties of ι-objects	$(o\iota)_{\tau\omega}$	classes of ι-objects	$(o\iota)$
relations-in-intensions	$(o\beta_1...\beta_n)_{\tau\omega}$	relations-in-extensions	$(o\beta_1...\beta_n)$
magnitudes	$\tau_{\tau\omega}$	numbers	τ

Table 1.10 Intension/extension of expressions and their corresponding types in TIL

Yet another claim of TIL is the following one:

7. First-order objects are *not* meanings, because they are not structured (they are "flat"), while *meanings are structured entities.*

In order to take into account this claim, TIL brings in the concept of construction via something that goes well beyond the standard intensional semantics. Constructions are viewed in TIL as ideal entities having the character of an instruction (possibly containing subinstructions) leading to certain results because of their "calculating" (in the broadest sense) nature. TIL introduces, first, *simple* (*elementary*) constructions that are variables, understood *objectually*, constructing a certain element from their range depending on the valuation, that is, if x_i is a variable and v any of its valuations, then x_i v-construes that entity which is assigned to it via v. For example, the expression "5 + x" represents such a construction that for each valuation v, if x v-construes 9, then this construction v-construes the number 14.

TIL also differentiates between the following three types of *composed* constructions. The first, called *trivialization*, is of such a type that when an object comes in as an "input," it comes out without any change, so that if X is a first-order object in the sense given above, then ^0X is trivialization of X. The second, *composition*, is defined as follows (Materna 1998, 42):

Let X be construction that v-constructs a function F (type $(\alpha\beta_1...\beta_n)$) and let $X_1, ..., X_n$ be constructions that v-construct $\beta_1...\beta_n$-objects $b_1, ..., b_n$ respectively. If F is defined on $\langle b_1, ..., b_n\rangle$, then the construction $[X, X_1, ..., X_n]$, called *composition*, v-constructs the value of F on that tuple. Otherwise it v-constructs nothing: it is v-*improper*.

As an example I take the division 18 : 2. In the framework of TIL, it is reconstructed as the composition $[^0: {}^0 18\ {}^0 2]$, where the $^0:$ construes the function of division which is of the type $(\tau\tau\tau)$.[46]

Finally, the construction labeled *closure* is defined as follows (Materna 1998, 43):

Let $x_1, ..., x_m$ be arbitrary pairwise distinct $\beta_1...\beta m$-variables and X an α-*construction*. Then $[\lambda x_1, ..., x_m X]$ is an $(\alpha\beta_1...\beta m)$-*construction* called *closure*. It v-constructs the following function F: Let the tuple $\langle b_1, ..., b_m\rangle$ of $\beta_1...\beta m$-objects, respectively, be an argument of F. Let v' associates with $x_1, ..., x_m$ the respective members of the above tuple and be otherwise identical with v. Then the value of F on that tuple is the $(\alpha$-$)$-object v'-constructed by X; if X is v'-improper, then F is not defined on $\langle b_1, ..., b_m\rangle$.

For example (Materna 1998, 44) $[\lambda x_1 x_2\ [^0: x_1 x_3]]$ v-constructs, when the valuation v assigns to the variables $x_1, x_2,$ and x_3 the values 1, 5 and 8 respectively, the function (of type $(\tau\tau\tau)$) which associates each pair $\langle j,k \rangle$ with the result of dividing j by 8.

Being aware of the highly unusual character of the concept of construction and of its fine-grained distinctions, I give the following examples of constructions.

Example 1: "the largest city." As shown above, the denotation of the expression "city" is an entity C of the type $(o\iota)_{\tau\omega}$, while the denotation of the expression "largest" is an intension L that picks up at the respective world-time couple one individual from the set of individuals, namely, the largest one, that is, it is of the type $(\iota(o\iota))_{\tau\omega}$. By their composition the individual office/role of the type $\iota_{\tau\omega}$ can be obtained. This intension is constructed according to TIL by the following construction:

$\lambda w \lambda t\ [^0 L_{wt}\ {}^0 C_{wt}]$,

which at the same time is the meaning of the expression "the largest city."

Example 2: "the husband of the American president." Here "the husband of" denotes a function H of the type $(\iota\iota)_{\tau\omega}$, while "the American president" denotes a function Ap of the type $\iota_{\tau\omega}$. The whole expression denotes a function of the type $\iota_{\tau\omega}$ and the corresponding construction is $\lambda w \lambda t\ [^0 H_{wt}\ {}^0 Ap_{wt}]$.

---

46 TIL views also the operation of addition as a function of the type $(\tau\tau\tau)$, while τ-identity is viewed a relation of the type $(o\tau\tau)$.

Example 3: "New York is a megacity." Here "New York" is a name of an individual, NY, while "megacity" is a name of a property, MC; the corresponding construction is $\lambda w \lambda t \ [^0 MC_{wt} \ ^0 NY]$.

Chapter 2: Empirical Knowledge, Measurement, and Experimentation

The aim of this chapter is chapter is to give a reconstruction of the structure of empirical knowledge, experimentation, and measurement. I start with an analyses of Carnap's and Hempel's approach to measurement-statements as well as of Quine's attempt to discard – *via* reduction – named, or "impure," numbers in favor of pure numbers. Then I deal with the cyclical nature of empirical knowledge and provide a detailed typology of measurement at the level of empirical knowledge, together with the central categories enabling us to deal with the differentiation between the already observed and the as yet not observed.

2.1 Carnap and Hempel on Measurement Statements

2.1.1 R. Carnap on measurement

A convenient starting point here is Carnap's characterization of the physical language, which runs as follows (Carnap 1963c, 404, 406):

statements of the simplest form (e.g. the temperature of such and such place at a specified time is so much), attach to a specific set of co-ordinates (three space, one time co-ordinates) a definite value or a range of values of a coefficient of physical state. ... The concepts of physics are quantitative concepts, having numerical values. ... Another peculiarity of physical concepts ... consists in their abstractness and the absence of qualities from their enunciation.

Carnap relates then two types of languages. He claims that it is possible to state (Carnap 1963c, 407–408)

which physical term (or class of physical terms) corresponds to a definite qualitative term in ... [the] protocol language ... That determinations of that kind are theoretically always possible due to the fortunate circumstance that the protocol has certain ordinal properties. This emerges in the fact of the successful construction of the physical language in such a fashion that qualitative determinations in protocol language are uniquely determined by the numerical distribution of coefficients of physical states.

But Carnap does not restrict himself to the claim that once the quantitative physical language is given, the qualitative protocol language is determined as well,

69

but claims also that every protocol sentence of a subject *S* can be translated into a physical statement and that these two languages are mutually *isomorphic* (Carnap 1963c, 418).

The approach of Carnap to the issue of measurement is at the same time characterized by a strong ambiguity. On the one hand, his views on measurement statements underwent some important shifts in the years 1929 to 1958 due to changes in his views about the language in which measurement results are stated. On the other hand, his approach to the differentiation between pure and named numbers displays in these years a surprising conservatism. Let me analyze the first aspect.

In his paper "Die physikalische Sprache als Universalsprache der Wissenschaft" of 1932 Carnap puts the protocol language, as containing only *qualitative* determinations, into a strict opposition to the physical language, containing quantitative determinations. In his *Logische Syntax der Sprache* of 1934, he further specifies how objects are described by numerical coordinates stating their mutual positions. Based on this numerical language Carnap then introduces functors enabling us to express properties or relations of positions by means of numbers (Carnap 1934, §3, 13). As an example he gives the temperature-functor "te" so that, for example, "te(3) = 5" means "the temperature at position 3 is 5."

While in the *Logische Syntax der Sprache* Carnap, as mentioned earlier, in chapter 1, discards the notion of designation of the language expressions and, thus, also of functors, in his first semantical period lasting from 1935 until 1942/1943, he changes his approach and states that *functors appearing in the language of physics designate physical magnitudes* (Carnap 1940, 222), for example, the functor "'*temp*(*x*)' designates the temperature of the body *x*" (Carnap 1942, §6, 17), and, in general, *functors designate functions* (Carnap 1942, §6, 18).

Finally, during his second semantic period, starting approximately in 1943, functors are treated just as one of several types of expressions to which semantical analysis can be applied, all labeled by the term "designator" (Carnap 1947, 6).

Let me now turn to Carnap's second, more *conservative* approach to the issue of measurement. In his *Abriss der Logistik* of 1929, he presents his approach to measurement-numbers (the so-called *Maßzahlen*), that is, numbers given in measurement as follows: "*Measurement-numbers* are to be understood logically as relations between numbers and objects (things, states, world-points, etc.). These relations are of the type one-many, therefore they lead to the formation of descriptions" (1929, §42, 94). He then gives the following examples (where "⟨⟩" stands for a two-place relation, "" stands for a description, and "M" and "fr" stand for the currencies named "Mark" and "Franc") (1929, §42, 94):

1. The price of this book is 3.50.
2. 3.50 ⟨price⟩ of this book.
3. ⟨Price⟩'of this book = 3.50.
   The *unit of measurement* belongs not to the number, but to the relation! The arrangement "3.50 M" has no sense; but "⟨price in M⟩'of this book" is the description of a pure number, namely, of the number 3½. ⟨price in M⟩, ⟨price in fr⟩ are different relations.

He also gives the following examples (1929, §42, 94):

1. The amount of gas has in this moment the volume 100.
2. ⟨Volume⟩'⟨present state⟩'⟨this amount of gas⟩ = 100.

To this view, namely, that numbers obtained in measurement (*Maßzahlen*) are pure numbers, Carnap holds also in his *Introduction to Symbolic Logic*, where he declares (1958, §41a, 168)

Quantitative concepts, e.g. length, weight, temperature, price, degree of attention, etc., are also called "measurable quantities" because the procedure for establishing their value is that of measurement. Such concepts are most conveniently designated by means of functors; their value expressions are considered of greatest general usefulness when they are real number expressions.

He then faces the following question: "Values of a measurable magnitude are expressed in terms of some *unit* of measure (e. g. a centimeter or an inch, a second or a day, a shilling or a dollar); where and how should this unit be specified?" (1958, §41b, 169), and gives his answer as follows (1958, §41b, 169):

Strictly speaking ... the specification of the unit is part of the definition of the functor; the value of the functor is always a pure number. Should an explicit indication of the unit be wanted in the symbolization of the measurable magnitude ..., this indication must be achieved by way of an inseparable part of the functor sign, e.g. by a subscript. For example, in the matter of length we might write "$lg_{ccm}(a) = 5$" or "$lg_{inch}(a) = 2$"; each of 'lg_{cm}' and 'lg_{inch}' is to be regarded as one sign, and each designates a different magnitude.

Finally, the same view on "pure" numbers given in measurement reappears in Carnap's 1966 work, in which he characterizes measurement by "rules for the process of measuring. These rules are nothing other than rules that tell us how to assign a certain number to a certain body or process, so we can say that this number represents the value of the magnitude for that body" (Carnap 1966, 62).

What is behind these views of Carnap is spelled out in his paper "Beobachtungssprache und theoretische Sprache" of 1958. In order to escape the problem of reference of the terms of the theoretical language L_T of science, he claims that its variables do *not* range over nonobservable entities but over *mathematical* entities. The example he gives for this is as follows:[47] "Let the constant 'n_p' be defined as 'the cardinal number of planets'. This constant is descriptive, to be sure, but the thing described by it is a natural number which belongs to the domain of D^0 [of natural numbers]. The number n_p is identical with 9, but the identity sentence 'n_p = 9' is synthetic" (1975, 80–81; 1958, 243). In general, this means (Carnap 1975, 80; Carnap 1958, 242–243):

> We shall assume the ... observable objects are numbered in an arbitrary way (for example, let body k have the number 17). Let us ... consider a physical magnitude whose value for every observable body is a real number (e.g., the mass of the body). This magnitude can then be construed ... as a function f whose numerical value is not ascribed to the body itself rather to its number (e.g., the proposition that the body k has mass 5 would be expressed by the sentence '$f(17)=5$'). The function symbol 'f' is here, to be sure, a descriptive constant but the function f is identical with a mathematical function g (i.e., for every n $f(n) = g(n)$).

It is worth mentioning here that even if Carnap holds to the pure-number-approach to measurement, very often when dealing with concrete examples of measurement, he speaks not about pure numbers, but (at least implicitly) about *specified numbers* or *impure (named) numbers*, so, for example, he speaks in the case of measurement of temperature about "temperature-numbers" (*Temperaturzahlen*), in the case of measurement of length about "length-numbers" (*Längezahlen*), and in the case of measurement of time about "time-numbers" (*Zeitzahlen*) (Carnap 1929, 27, 27, 38).

2.1.2 C. G. Hempel

Hempel's approach to measurement-statements moves in the same direction as Carnap's approach. Hempel introduces the relations of coincidence, C, and precedence, P, and then defines the term "metricizing" as follows (Hempel 1952, 63):

(12.1) Let C and P be two relations which determine a quasi-serial order for a class D. We will say that this order has been metricized if criteria have been specified which assign to each element x of D exactly one real number, $s(x)$, in such a manner that the following conditions are satisfied for all elements x, y of D:
(12.1a) If xCy, then $s(x) = s(y)$
(12.1b) If xPy, then $s(x) < s(y)$.

Finally, he states: "Any function s which assigns to every element x of D exactly one real-number value, $s(x)$, will be said to constitute a quantitative or metrical

---

47 D^0 is the domain of natural numbers.

concept, or briefly a quantity (with the domain of application D)" (Hempel 1952, 63). As an example of a metrical concept Hempel gives in his paper (1958) the concept of weight, w, of a body x, represented by the expression "$w(x)$" (Hempel 1958, 45), while the results of measuring w for a certain body b and a certain liquid l are represented by him as "$w(b) = w_1$," and "$w(l) = w_2$," where w_1 and w_2 are according to Hempel just *positive real numbers* (Hempel 1958, 45).

One might also mention here that Hempel, like Carnap, when dealing with concrete examples of measurement-statements, deviates from the pure-numbers-approach in favor of the named-numbers-approach. He gives the following two examples (Hempel 1952, 29):

"the length of the distance between points u and v is r cm"
"the mass of physical body x is s grams"

But then he gives the following brief representation of these statements (Hempel 1952, 29):

length $(u, v) = r$,
mass $(x) = s$,

where r and s should be just nonnegative numbers, thus espousing the pure-numbers-approach.

2.2 Quine's attempt at reduction

One of Quine's ambitions in his 1964 paper, drawing on Carnap's views explained earlier, was to purify expressions containing impure numbers in such a way that they are disposed of in favor of pure numbers. He accomplishes such a disposal as follows. He uses the example of the two-place predicate "H" of temperature so that "$H(x, \alpha)$" stands for "The temperature of x is α," where "α" can be written as "$n°C$," so that "$H(x, n°C)$" should be explained as "$H_c(x, n)$" and where "H_c" is a two-place predicate of temperature in degrees Centigrade.

He then states the following three conditions that should be fulfilled, if one wants to move from a theory whose objects include place-times x, impure numbers α, and whose primitive predicates include "H" to a theory whose objects include place-times x, pure numbers n, and whose primitive predicates include "H_c" (Quine 1961, 24):

Condition 1. "$H_c(x, n)$" under the intended interpretation agrees in truth-value with "$H(x, n°C)$" under its originally intended interpretation for all values of x and n.

Condition 2. In order to eliminate all references to impure numbers, α is confined (confinable) to the specific form of context "$H(x, \alpha)$"

Condition 3. An impure number α can always be referred to in terms of a pure number n and a unit, for example, $n°C$, $n\mathrm{kg}$, and so forth.

Quine then specifies the third condition via the so-called *proxy function*, which he characterizes as a "function whose values exhaust the old things, in this example the impure numbers of temperature, as their argument range over new things, the pure numbers. Our proxy function is expressed in the notation '°C' which brings $n°C$ into correspondence with n" (Quine 1964, 214).

Based on the temperature example, Quine gives the following general standard of reduction of theory θ to a theory θ' (Quine 1961, 215–216):

> We specify a function ... whose values exhaust the universe of θ for arguments in the universe of θ'. This is the proxy function. Then to each n-place primitive predicate of θ, for each n, we effectively associate an open sentence of θ' in n free variables, in such a way that the predicate is fulfilled by an n-tuple of values of the proxy function always and only when the open sentence is fulfilled by the corresponding n-tuple of arguments. ... Reduction of a theory θ to natural number ... means determining a proxy function that actually enumerates the objects of θ and maps the predicates of θ into open sentences of the numerical model. Where this can be done, with preservation of truth values of closed sentences, we may well speak of reduction to natural number.

Against such an attempt at reduction at least the following three objections can be stated. First, on the metascientific level pertaining to the very inspiration of the attempt at reduction, Quine's attempt is completely *ad hoc* and *parasitic* in its very nature. Only *after* expressions are stated which contain predicates of the type "$H(x, n°C)$", "$H(x, n\mathrm{kg})$," does Quine's reduction-endeavor "descend" upon the already accomplished work of scientists yielding such predicates. That his attempt at reduction is in its very nature *ad hoc* and parasitic with respect to the results of empirical science is readily seen in what Quine himself says about the notation of the proxy function (Quine 1964, 214):

> It is not required that such a notation be available in the original theory θ to which "H" belonged, much less that it be available in the final theory θ' to which "H_c" belongs. It is required rather of *us*, out in the metatheory where we are explaining and justifying the discontinuance of θ in favor of θ', that we have some means of expressing a proxy function. Only upon us, who explain "$H(x, \alpha)$" away by "$H_c(x, n)$", does it devolve to show how every α that was intended in the old θ can be determined by an n of the new θ'; and "$n°C$" is *our* way.

Second, the following critique of the result of Quine's attempt at reduction can also be given. With respect to his example for the H-predicate, the proxy function f, together with its arguments and values can be represented as shown in Table 2.1:

θ'	−273	....	0	100	...
f		°C			
θ	−273°C	....	0°C	100°C	...

Table 2.1 Quine's proxy function f used in reduction of theories

Here it is readily seen that even if Quine claims to be reducing theory θ containing the predicate "H" to theory θ' containing the predicate "H_c", he is in fact proceeding in the *opposite* direction; by means of f he enlarges the vocabulary of θ' into that of θ by moving from n to n°C. But in order to move from n to n°C he has to know in advance for which values of n there exist values of the function (say, for $n = -280$ there does not exist the value -280°C of the function °C). This means that in order to state the proxy function for the respective pair of theories θ and θ', one not only has to know in advance the theory θ, but the whole reduction can never dispose of the theory θ in favor of θ'. Theory θ is, then, the "eternal" basis and framework for θ', hence, my second objection. That this is so can also be seen in Quine's requirement that the truth-value of "H_c" agree with the truth value of "H". In other words, when one wants to find out the truth-value of a statement from predicate "H_c", one always has to know in advance the truth-value of the corresponding statement from predicate "H".

Let me try to solve this problem by formulating for theories θ and θ' an "inverse" proxy function, namely, 1/°C. Again, this is a no-starter simply because in order to compute the respective values of n we always have to know in advance the values of the argument n°C; once again, theory θ is the indispensable basis and framework for theory θ'. Stated otherwise, Quine's attempt, as given in his 1964 paper, at reducing a scientific theory containing impure numbers to a theory containing only pure numbers – and thus eliminating the former in favor of the latter – is not only an *ad hoc* attempt but also one that is doomed to failure.

The third and final objection to Quine's attempt is as follows. Even if Quine claims that "n°C" is *our* way of a meta-notation for "α", what he has in fact done is just a simple replacement of one symbol for another (e.g., instead of "9 degrees Celsius" he writes "9°C," instead of "5 kilograms" he writes "5kg."). He, thus, has simply moved from one type of notation for impure numbers to another type of notation without truly disposing of them. In fact, Quine acknowledges this when he states the following for the function fx: "I must admit that my formulation suffers from a conspicuous element of make-believe. Thus in the Carnap case I had to talk as there were such things as x°C, much though I applaud Carnap's repudiation of them" (1964, 215).

2.3 Kyburg Jr. on basic magnitudes

Let me take, as an example, the basic magnitude of length (of something).[48] Let "$L(x, y)$" stands for "x is longer than y" and where x and y obtain their values from the set of *rigid bodies*, that is, bodies that do not change their shape. Following H. E. Kyburg, I view

$$\neg L(x, y) \wedge \neg L(y, x)$$

to be an equivalence relation labeled as "is the same length as" and establishing equivalence classes that serve us as the actual world-time values of the magnitude of length. As the next step one can choose an equivalence class of objects neither shorter nor longer than the platinum bar in the International Bureau of Weights and Measures in Paris, which I denote as "l" and which should serve us as the standard for the measurement of the magnitude of length. Finally, one determines the way the objects of the equivalence class should be mutually concatenated; here we choose the procedure of putting ends to ends. One could, of course, obtain another construction if she or he would choose another type of concatenation, say, B. Ellis's famous "right-angled" concatenation,[49] where the objects from the chosen equivalence class are viewed as the perpendicular sides of a triangle and where the value of their concatenation is computed by means of the Pythagorean theorem.

In the case of the magnitude of length, the chosen standard l enables one not only to pick up a specific equivalence class of objects but also – via the latter – to determine the unit, $[l]$, for its measurement; the unit is here the equivalence class picked out by l. Let us suppose[50] that $[l]$ contains at least N instances $l_1, l_2, ..., l_N$. By means of $1 \times [l]$,[51] I denote the equivalence class to which belongs any one-fold concatenation of the unit object l, and by $n \times [l]$ (where $n < N$), I denote the equivalence class to which belongs any n-fold concatenation of l. One can also introduce "fractional" lengths by means of the equivalence class $(m/n) \times [l]$. Its elements are objects equal in length to the concatenation of m objects equal in length and where n of them are equal in length to l. The equivalence classes $n \times [l]$ and $(m/n) \times [l]$, for which holds that their elements are either longer or shorter than l, can be denoted by the expression with the structure "$r \times [l]$," where "r" denotes a numerical variable whose values are positive real numbers.

What has to be emphasized here is, first, that the previously introduced equivalence classes are viewed in this approach as the extensions of the magnitude

---

48 Here I draw upon the paper (Kyburg 1997).
49 On this see (Ellis 1960).
50 Here I draw again on (Kyburg 1997).
51 The symbol "×" stands for multiplication.

denoted by the name "length (of something)" which exist once there are equivalence classes based on the respective relation. Second, they are viewed as really existing before they are at all numerically represented. And, *third*, 1 × [*l*], 2.54 × [*l*], and so on are viewed here as different from pure numbers 1, 2.54, etc. The expression "*r* × [*l*]" does *not* stand for a number. The numeral "*r*" is here just a part of complex symbol standing for a certain equivalence class of objects under the relation expressed by "is the same length as."

2.4 Experimentation, cycles of knowledge and the differentiation between the already observed and the not as yet observed

Let me now place the issue of measurement into the broader context of experimentation and empirical knowledge. In order to be as close as possible to the practice of real science I first analyze two periods in early modern science: the early stages of thermodynamics, namely, the attempts at the construction of the thermoscope and the early stage of modern mechanics as given in the endeavor of C. Huygens to measure the distance covered by a freely falling body in the first second of free fall, that is, what we today label as "gravitational acceleration." My analyses draw here primarily on the excellent reconstructions of these periods of early modern science as given in (Chang 2004) and (Yoder 1988).

2.4.1 Early thermodynamics and the construction of the thermoscope

The study of heat was centered around the problem of constructing a device we today label as "thermometer," and the crucial issue here was how to find stable, fixed points based on which one could built a scale and then attach them to the "thermometer" so that the latter could function as a reliable measuring device. But, as H. Chang shows convincingly, in order to find such points in the thermic behavior of, say, air, mercury and so on, one needs *in advance a measuring device* that would enable us to find certain points on the scale for certain materials as fixed points. In this paradoxical situation the problem initially focused on the issue what stuff, or "medium," should be used at all in the process of finding the fixed points. R. Boyle proposed as such stuff congealing oil of aniseed *or* freezing distilled water, Huygens, boiling water *or* freezing water, and Isaac Newton melting snow *and* the temperature of the human body.[52] What followed after such a plethora of

---

52 Here "*or*" stands for a proposed system with one fixed point, and "*and*" for a proposed system with two fixed points.

approaches to construct a reliable measuring device was a "weeding out" of the unreliable "media" for the construction of the measuring device. The boiling point of water was chosen despite its vagaries based on (Chang 2004, 23–29):

(i) the realization that there is a difference between the temperature (up to 112 °C) that water can withstand without boiling (the so-called effect of superheating) and the temperature that water maintains while boiling (the maximum is 103 °C), so that the first state and its fixed point can be discarded;
(ii) the discovery of factors allowing the stabilization of the boiling point and to escape the phenomenon of superheating (e.g., the choice of the vessel for water; the adding of certain ingredients into the water);
(iii) the observation that the pressure of saturated vapor is equal to the normal atmospheric pressure when the temperature is 100°C. This means that one obtains the fixed point of boiling of water at 100°C under the condition that the external pressure is that of the normal atmospheric pressure;
(iv) on the choice of inserting the measuring device not into the liquid itself but into the steam in order to reduce the variations of the boiling point due to miscellaneous factors/causes.

It was thus possible to use (in the very beginning of the attempts to find the fixed points, in the situation of absence of a pre-established measuring device providing a standard of fixity) just a simple device – a tube filled with some liquid – based on the common sense knowledge that once the tube is subject to an external change of heat, the liquid in the tube either expands or shrinks. Based on such an analysis it is possible to differentiate between two types of measuring device: one built before the "fixing" and discovery of fixed points, and then used in the search for the fixed points, and the second built once those points were "fixed" and discovered. Chang labels the former device as "thermoscope," while for the latter he uses the term "thermometer." For thermoscopes it holds that "they are not graduated by any principle that would give systematic meaning to their readings" (Chang 2004, 41). They provide the user only with an *ordinal scale* of temperature, whose numbers (and the numerals, as symbols of numbers, placed on the scale of the thermoscope) "only indicate a definite ordering, and the arithmetic operations such as addition do not apply meaningfully to them. In contrast, a proper thermometer is meant to give numbers for which some arithmetic operations yield meaningful results ... proper thermometers give us a *numerical* temperature scale" (Chang 2004, 41).

Thermoscopes, as stated above, are based on the common sense knowledge that when they are subject to external changes of heat, then their liquids change their volume. This knowledge is *not* grounded on results obtained via the use of the thermoscope; the latter cannot be used in the proof of that knowledge. Instead, it finds its *initial* grounding in the unaided and not quantified human sensations.

According to Chang "We get the idea that liquids expand with heat because that is what we observe in the cases that are most obvious to the senses. For example, we put a warm hand on the thermoscope that feels quite cold to the touch and see it gradually rise … What we see here is that human sensation serves as the prior standard for the thermoscope" (Chang 2004, 42).

2.4.2 Ch. Huygens' kinematics

What makes, from the point of view of the philosophy of science, Huygens endeavor to find out the value of acceleration on Earth similar to that of the construction of thermoscopes is the fact that in order to find the "fixed" value of that acceleration, that is, the path covered by a freely falling body in the first second of this fall, one needs *in advance* a timepiece capable of measuring that fixed time-interval, namely, a one-second interval.

This problem was encountered already by Marin Mersenne upon whose works Huygens initially drew. The former built a pendulum, based on Galileo's idea of the isochronism of the pendulum (i.e., that any pendulum has a fixed period of swings depending only on the length of its suspension cord), so that it would have one-second beats, and where the swings of this seconds-pendulum were calibrated by means of a mechanical clock.[53] By a series of experiments Mersenne obtained a value of 3.5 (Parisian Royal) feet for this seconds-pendulum which he then via experiments decided to shorten to 3 (Parisian Royal) feet. By means of such a pendulum he then measured the distance covered by a freely falling body in such a way that he fixed the 3-foot pendulum to a wall and placed its bob and the body into one of his hands. Then he determined the height from which they should be released so that the sound of the bob hitting the wall and the sound of the freely falling body hitting the floor were heard simultaneously. He found out that the falling body in half a second covered the distance of 3 feet, thus – via Galileo's law of free fall covered – the distance covered in the whole second would be 12 feet. Unfortunately, Mersenne obtained not only that value for the distance covered by a freely falling body in half a second, but values ranging from 3 feet up to 6 feet (Yoder 1988, 13).

Huygens in 1659 repeated Mersenne's experiments, accepting initially his value of 3 (Parisian Royal) feet, that is, 3.5 (Rheinisch) feet for the seconds-pendulum. But even if he used Mersenne's method of synchronizing the free fall

---

53 M. Mersenne states that he used different clocks (*horologes communes*) (Mersenne 1636, 220). It remains, however, an open question as to how these clocks themselves were calibrated. According to the view of J. Yoder, communicated to me in an e-mail, eventually all timepieces of that time had to be calibrated by means of the sidereal day.

of the body with the swings of the seconds-pendulum, he obtained for the first second of the free fall the value of 14 feet, which deviated substantially from Mersenne's value of 12 feet.

In a next step Huygens embarked on a strategy that represented an important widening – as compared to the method used by Mersenne – of the method to find the value of the distance covered by the falling body in the first second. He viewed both the free fall of a body *and* the swing of the pendulum as based on the same type of a physical process, namely, free fall (even if in the case of the swings of the pendulum the free fall is restricted by the cord) and where the free fall has weight as its *cause*. He superimposes a parabola – representing the path of a falling body projected horizontally – on a circle. Next, he approximates a circle with a parabola whose *latus rectum* equals the diameter d of the circle; the diagram is then as shown in Figure 2.1:

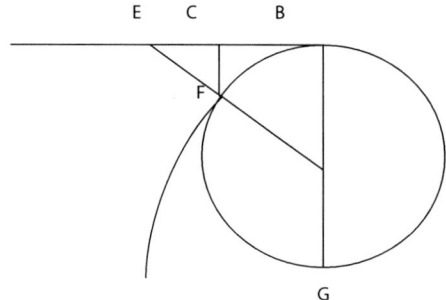

Figure 2.1 Huygens' formula $EB^2 = EF \times BG$

For the *latus rectum* holds the equation $CB^2 = CF \times BG$. So as F is considered to be infinitesimally close to B (the point of release of the body), it holds that $CB = EB$ and $CF = EF$. From the previous equation, we then obtain $EB^2 = EF \times BG$. By making a generalization of this equation to the full circle given in the 2.1, and at the same time putting weight equal to the centrifugal force acting on the body moving on that circle, he finds out that in the time a body moves around a circumference πd of a full circle with a diameter of d, the body moves through the distance of $(\pi d)^2 \times \frac{1}{d}$ $(= \pi^2 d)$ due to the centrifugal force and, thus, equivalently, it falls under the influence of weight through the same distance of $\pi^2 d$.

Next, using the expression $\pi^2 d$ and drawing upon the value of 14 feet obtained before from his first free-fall experiments, he calculated the radius of the circle whose circumference is circuited by a body in one second, again supposing that its weight equals its centrifugal force, and obtained the value of 8.5 (Rheinisch) inches.[54] Then, in order to test this value of r, he supposed that the body circuits not in one second but in 24 hours and, thus, he should obtain the value of the Earth-radius. His calculation yielded the value of well over 5 billions of (Rheinisch) feet. In a next step he compared this calculated value with the value calculated by W. Snell, which was less than 20 millions feet – a miss by a factor of 265. Once Huygens realized that the value of 14 feet for the first second of free fall leads to a value of the radius of Earth which is completely off-beam, he repeated his experiments in such a way that he still accepted Mersenne's value of the length of the seconds-pendulum but increased its length so that the free fall would take not one but 1.5 seconds so that he could get a more distinct sound-differentiation between the striking of the bob and of the ball. Based on these experiments he found out that the value of the covered distance in the first second equaled to 13 feet 8 inches. Huygens applied, at the same time, the method of equilibrating weight and centrifugal force together with this value of "gravitational acceleration," also in the very construction of a pendulum-clock, which should have enabled him to measure seconds-intervals in such a way that its cord sweeps out a surface of a cone while the bob is suspended from the cord at a fixed angle (45°). So as it now holds 13 feet 8 inches $= d \times \pi^2$ and if we suppose a circle around which a body moves so that we have an equality of weight and centrifugal force, we have for the diameter $d = $ 13 feet 8 inches $\times (\frac{7}{22})^2$ and obtain $d = 16\frac{6}{10}$ inches and, thus, a radius $r = \frac{d}{2} = 8\frac{3}{10}$ inches. Based on the latter value he considers the case when the bob suspended from a cord at a fixed angle is moved in a circular path so that the cord sweeps out the surface of a cone. For this type of a clock – the so-called *conical clock* – he calculates the length of its cord so that its bob transits the circle in one second; it is $8\frac{3}{10}$ inches $\times (2)^{\frac{1}{2}} = 11\frac{8}{11}$ inches, assuming that the cord is kept at 45°.[55]

This enabled Huygens to calculate the length of the cord of a conical pendulum completing 4320 and 5040 revolutions per hour so that the former can rotate in 0.72 of a second and the latter in 0.83 of a second. Once we have the length $11\frac{8}{11}$ inches for a conical pendulum rotating in one second, we obtain for 4320 revolutions per hour (i.e., 3600 seconds) $(3600/4320)^2 \times 11\frac{8}{11}$ inches $= 8\frac{1}{7}$ inches and for 5040 revolutions per hour $(3600/5040)^2 \times 11\frac{8}{11}$ inches $= 6$ inches. Based on these calculations he turned again to the construction of a pendulum-clock and based on this clock obtained by careful observation the value of 8 feet and 9.5 inches for the distance covered in three quarters of a second of free fall, thus arriving at a value of $15\frac{6}{10}$ feet for the first second of free fall. Then he computed again

---

54 It holds $14 = \pi^2 d$, while for Huygens $\pi = 22/7$, thus $d=17$ inches and $r = 8.5$ inches.
55 For details and the respective illustrations see (Yoder 1988, 27).

the radius of the circle circuited in one second and, by using for the value of the value of $\frac{355}{113}$ arrives, finally, at a value of 15 feet and 7.5 inches for the "constant of gravitational acceleration" corresponding, if expressed in units and notation used nowadays, to the value of $g = 9.81$ ms^{-2}.[56]

2.4.3 The practico-cognitive cycle

I can now state the general features of Mersenne's and Huygens' endeavor to find the value of the distance covered by a freely falling body in the first second. It holds, first, that their endeavor goes hand in hand with the endeavor to determine the length of the seconds-pendulum and enabling, thus in turn, to construct a timepiece which is more precise than that one from which they initially proceeded, namely, the pendulum calibrated by means of sidereal day.

Second, the measurements performed by Mersenne and Huygens were based on *knowledge* and *simultaneous conscious use* of two *scientific laws*: Galileo's law of free fall, which states that the covered distance of freely falling body is proportional to the square of the time taken – $s \propto t^2$, for short – and his law for the pendulum stating that the square of the period of its swings is proportional to its length, $t^2 \propto l$, for short. The measurement was here based on an *interchaining* of the following two procedures: once we fix the length of a pendulum and simultaneously calibrate it in such a way that it beats at one-second intervals, we have a mean for measuring the distance covered by a freely falling body in the first second. We have here a case when, by *measuring magnitudes given in a scientific law we can measure magnitudes given in another law*, thus a case of measurement we label from now on as *nomic measurement*. The latter has to be *differentiated* from two other types of measurement. *One type* is exemplified by the case of the construction of the thermoscope as already discussed. Its construction was *not* based on the explicit knowledge and use of a scientific law but only on the *common sense* knowledge that under the impact of an external change of heat, materials like water, mercury, and so on change their volume. To state it in even stronger terms: in order to state a precise scientific law that would relate the change of the temperature to a change of, say, length or volume of some material, we have to have in advance a measuring device capable of measuring temperature. Thus, the moral of Chang's reconstruction of the road leading to the construction of the thermoscopes is that it is a case of a construction based on *pretheoretical knowledge* relying on human sensations, where the latter are, of course, as *human* sensations, *conceptualized* sensations because they are *ori-*

---

[56] On this see (Yoder 1988, 31–32; 64–65).

ented *towards the observations of certain chosen entities, where this choice is framed by pregiven practical and cognitive interests.*

The *second type* of measurement, which I will reconstruct in all its details in chapter 3, is a nomic measurement but, in *addition* and *contrary* to the nomic measurement exemplified by Mersenne's and Huygens' measurement of "gravitational acceleration," simultaneously a measurement consciously employed to *measure a cause by means of its effect*; I label it from now on as the *nomic measurement of the cause by means of its effect*. Once Huygens was able to develop a method for measuring "gravitational acceleration," a next step immediately suggested itself: the *measurement of* the *cause* – of the force of weight, W, via one of its *effects*, gravitational acceleration, g for short, based on the knowledge and use of the scientific law $W \propto g$. This type of measurement is already part of the part of mechanics labeled *dynamics* and is, as I will show in chapter 3, at the very basis of Newton's *Principia*.

As we have seen, Huygens' aim was to measure *length*, then *period*, and, finally, what we today label as "*gravitational acceleration*," while simultaneously *discarding the underlying cause*, namely, weight, by putting it equal to the centrifugal force. This enabled him to remain in the conceptual space determined by the concepts of distance, time, and acceleration, that is, to remain in the framework of what we nowadays label as *kinematics*. So as the type of measurement consciously employed by Huygens remains at the level of the *phenomena-effects before passing to their cause*, we label from now this type of measurement as the *nomic measurement of phenomena-appearances*. A more detailed explication of typology of the phenomena-effects will be given in chapter 3. There a more precise characterization of *nomic measurement of phenomena-appearances* will be given.

Third, Huygens' endeavor draws, not only on geometry, but also on *experimentation* serving him as an independent source of information input and, also on an information yielded by methods which are – as in the case of Snell's determination of the radius of the Earth by trigonometrical methods – independent both from the experimental determination of the value of "gravitational acceleration" and the physics of free fall.

Fourth, for Mersenne's and Huygens' endeavors it holds that their inputs and outputs are part of a *repeating* sequence of both experiments trying to find out the value of g – the "gravitational acceleration" and calculations of the length of the seconds-pendulum and a repeating sequence of attempts to build a more and more precise timepiece based on the isochronism of the pendulum. Their endeavors were thus of a *cyclical* nature, that is, of a sequence of cycles, where these elements continuously change the mutual positions. If E_i stands for the experimental determinations of g, C_i stands for the computational determinations of g, Cl_l stands for the computational determinations of the length l of a seconds-pendulum *and* its construction, while C_s stands for Snell's computational determination of the radius of the Earth, then one such a cycle can be schematically represented in Figure 2.2:

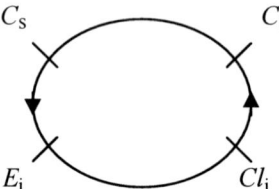

Figure 2.2 One cycle for the relations between experiments, calculations and additional information in the determination of g

The whole development of Mersenne's and Huygens' combined endeavors can then be represented by a sequence of cycles as follows (here "M" stands for the cycle of Marin Mersenne, "H" for the cycles of Christian Huygens, "C_s" gives the value of Earth's radius as computed by W. Snell, "$l =$" gives the respective value of the length of the seconds-pendulum, while "$g =$" gives the respective value of g). See Figure 2.3.

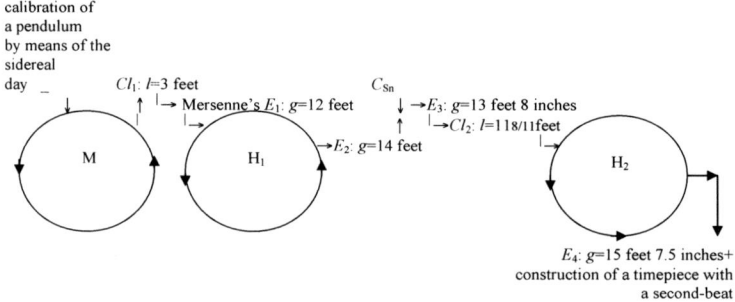

Figure 2.3 Cycles for Mersenne's and Huygens' determination of g

If "CS_i" and "O_1" stand for common sense and unaided observation, respectively; "D" for the realization of the existence of a difference between superheating and the temperature water maintains while boiling; "S" for factors stabilizing the boiling point; "I" for the immersion of the measuring device into the steam; "F" for fixing of the boiling point of water at 100°C at normal atmospheric pressure; "th_i" for the construction of the thermometer; and "PR" for the principle of respect to what is given via unaided senses, while the dotted arrows stand for a feed back, then the sequence of cycles shown in Figure 2.4 can tentatively be discerned in the case of the construction of the thermoscope.

Figure 2.4 Cycles in the construction of the thermoscope

This sequence of cycles as well as the sequence of cycles in the Mersenne-Huygens case can generally be characterized as follows:[57]

1. Measuring instruments for basic/fundamental, that is, *nonderived* magnitudes in a certain empirical science, like the thermoscope for the measurement of temperature in physics, are initially justified by and built drawing on their conformity to human sensations simply because initially there is no other alternative to them, but, on the other hand, once they are built, they enable one to, in turn, *augment* and *correct human sensations*.
2. We have, as the case of the construction and improvement of the thermoscope shows, here, a case of the measuring device, initially built upon sensations and common sense, *plus* the discovery of factors determining the fixity of the searched-for "fixed" points, leading to a progress in the construction of the thermoscope, and where (Chang 2004, 44–45)

    progress comes to mean a spiral of self-improvement if it is achieved while observing the principle of respect. Investigations based on prior standards can result in the creation of a new standard that improves upon the prior standard. Self-improvement is possible only because the principle of respect does not demand that old standards should determine everything. Liberality in respect creates the breathing space for progress.

3. So as this self-improvement is a unity of epistemic thought-operations – via which cognition takes its course – *and* practical operations in experimentation and instrument-construction, it can thus be labeled as a process of a *practico-epistemic iteration*.
4. The process of the practico-epistemic iteration is characterized by its open-endness; certain suppositions functioning initially just as "supporting" inputs, unforeseeably, become the much improved outputs of the cycles through which the practico-epistemic iteration takes its course. So, for example, Mer-

---

57 Here I draw partially on (Chang 2004, 43–45, 220).

senne in order to measure *g* had to use, initially, a pendulum calibrated by the sidereal day, that is, an inherently imprecise timepiece and, then, through the sequence of practico-epistemic cycles Huygens was able, not only to determine the value of *g* more and more precisely, but also to construct a much more precise timepiece.
5. As shown above in the case of the thermoscope, the construction of the most elementary measuring devices is initially based on accumulated common-sense knowledge and observation and, as shown in the case of the search for the value of the "gravitational acceleration," one can perform measurements of phenomena as effects of certain causes while *bypassing* the causes underlying these phenomena. If one understands, as I do in this book, under the terms "theory" and "theory construction" the *endeavor to obtain certain information about the causes underlying phenomena as effects*, then I have to coin yet another term that would correspond to and be a methodological generalization of Huygens' ability to remain at the level of kinematics by abstracting from weight as the cause of free fall of a body and of the swings of a pendulum. I, therefore, introduce the term "prototheory" (*PT*, hereafter), and, then, the cycle through which the practico-epistemic iteration takes its course can be schematically represented as follows (here *PE* stands for the practice of measurement together with the construction and employment of measuring instruments):

$PE \rightarrow PT \rightarrow PE' \rightarrow PT^* \rightarrow ...,$

the sign "*" stands increase in the precision of the measuring instruments and the growth of knowledge about the phenomena.

2.4.4 The lifeworld, the already observed and the as yet not-observed

As already mentioned, the early thermoscopes were initially based both on bodily sensations of hot/cold and the accumulated experience about the expansion/contraction of water and other liquids under the impact of warming/cooling. Then the very construction of the thermoscope enabled to improve observation and determination of fixed points on the nominal scale and at the same time led to changes of previously accumulated knowledge about the boiling behavior of liquids in various conditions. This means that in the cycle of cognition where *PT* and *PE* continually change their mutual positions from a (relative) presupposition to a (relative) result, *the very boundary* between *what is already observed* and *what is as yet not observed continually shifts. Where* the boundary is given at a certain time depends on the boundary of the practical operation in the external world at that time. This operation involves always a *body* of human being together with his or her *senses*. In addition, it can – and in modern science it has to – involve also the *construction and use of instruments* and, finally, it involves a certain amount of, even if sometimes only common-sense, *information/ knowledge* about the external world. The *human body/human senses* plus their *extensions via instruments* and the corresponding *information/knowledge* stand for two, mutually inseparable dimensions of observability. First, the givenness of a certain *chain of interactions* so that on one "end" of the chain the human body with its senses is given and on the other one the object to be observed, changed and manipulated. Second, via the chain, *to* the human being are being conveyed *information* – always with respect to the knowledge/information already given to this human being. This means[58] that with regard to an object x and a property P of x, one can say that the ordered pair <x, P> is observable if there can be a chain of interactions between x and the human senses so that the information that x is P is conveyed from x to the human senses. These senses are, of course, senses fixed on "something" that is framed and picked out by the observing and experimenting human being.

Based on this analysis, I can now list the *central categories* of the practico-empirical aspect of natural science: *information/knowledge about the objects of the external nature*; *the human body and its senses*; and, finally, *the interaction chain between the latter and the former*. I will label the unity of the conveyance of information from the external object to the human being and its senses with the interaction between bodies/senses of human beings with the object of the external nature as "work." These categories enable me to delineate the infrastructure of the *lifeworld of natural science*.

With respect to what has been said up to this point, it is readily seen that all the structural components of the lifeworld are, and thus the lifeworld itself is, *thoroughly historical*. Such a characterization of lifeworld enables me to approach the issue of measurement both from a *structural* and a *historical*, that is, *developmental*, viewpoint. Based on my differentiation of the various types of measurement given above, we can now reconstruct the following typology of measurements, where each type can be viewed as a structural element of the lifeworld.

1. *Type 1 of measurement*: the measurement of types of magnitudes that are viewed as basic/fundamental in the sense that they can be measured by *themselves*, that is, not via the conscious use of other types of magnitudes related to the former via some form of a geometric relation or scientific laws of physics.

---

58 Here I draw on (Koss 1988); see also (Koss 1989).

As such type of measurement one can view, for example, the measurement of the *length* when it is based only on the measurement of length, the measurement of the duration of processes when it is measured by the duration of other processes, and the measurement of weight when it is viewed not as a concept of dynamics, but only as yet as a common-sense property of an object to be "heavy" and when it is measured by a comparison with the weight of other objects, say, on a beam-balance. This type of measurement is based on common sense and accumulated knowledge and originates in the need to master certain practical tasks in the external world, like the construction of buildings and the advance planning of the permanently reoccurring need of crop-harvesting, for example.

It remains here, however, the task of a detailed historical analysis in the framework of the *historical metrology* to find out if the ancient astronomical measurements of time were really based *only* on the measurement of time *or* were based already on the measurement of other magnitudes, say, the apparent distances covered in the sky by the moving heavenly bodies. If the latter be the case, then we have here already the following type of measurement.

2. *Type 2 of measurement*: the measurement of types of magnitudes measured via the conscious use of other types of magnitudes associated to the former via some relations, say, the measurement of area and volume by means of relations given in geometry. Again, the measurement of this type originates in the need to master certain practical tasks in the external world, like the construction of buildings and the need to measure the space needed to store a certain amount of crop.[59]

3. *Type 3 of measurement*: the measurement of types of magnitudes measured by the conscious use of scientific laws when these magnitudes are basic/fundamental magnitudes (e.g., distance and time), that is, they can be measured independently of each other and, in addition, have with respect to each other the character of *phenomena-effects of a common cause* and, thus, do *not* stand to each other in the relation of cause and effect. This type of measurement was labeled above as that of *nomic measurement at the level of the phenomena-appearances*.

4. *Type 4 of measurement*: the measurement of types of magnitudes measured by the conscious use of scientific laws when these magnitudes stand to each other in the relation of cause and effect; this type was measurement was labeled above as of *nomic measurement of the cause via its effect*.

I view this sequence of types of measurement both as expressing a *hierarchical structure*, in the sense that each type incorporates the previous type(s), and at the same time as expressing the *historical sequence* of types of measurement in the sense that types 2 through 4 of measurement each developed from the preceding one.

---

59 On this see, e.g., (Schmitt 2005).

From the point of view of the *historical sequence* of types of measurement, two issues are worth mentioning here. First, types 3 and 4 are *culturally specific* in the sense that they were initially not given universally in all human societies, but only in certain ones, initially in Italy and, later, in countries on the Atlantic rim, primarily, France, the Netherlands, and England.

Second, it seems that types 1 and 2 are *culturally universal* in the sense that they were generated *independently from each other in different human societies*. For example not only the measurement of distances[60] but also the measurement of areas was practiced not only in ancient Greece but also in pre-Columbus Central America.[61] This means that the lifeworld is Janus-faced: it displays certain *historically limited* features and at the same time certain universal, *non*historical[62] features which can be viewed as its *a priori*. It was E. Husserl who in Part 3 of his *Crisis of the European Sciences* was confronted with these two types of features of the lifeworld. Ludwig Landgrebe's characterization and explication of Husserl's views on this is as follows (1991, 126–127):

Husserl's concept of the life-world ... does *not* refer to the empirical description of the multiplicity of worlds in which various groups of human beings live, although ... Husserl *also* uses the term in this sense. "Life-world" names the most basic dimension of experience in which all higher dimensions are grounded ... It is in this sense *arché*, the ground of being, ground of becoming, ground of knowledge in Aristotle's sense: life-world, where every wherefrom and whereto has its beginning. Translated into the language of phenomenology, "life-world" is thus the sum total of elementary functions of consciousness which all human beings, regardless of the group or epoch to which they may belong, have in common. So the life-world is the ground of this a priori in terms of which there is even the possibility of comparing these groups and cultures. It is an a priori which all investigations and sciences of man make use of, without, as empirical sciences, having to give an account of this life-worldly ground.

We thus face the following question. What is the basis of these universal features of lifeworld? One answer suggests itself immediately: because all humans share the same biological outfit based, say, the same DNA, the culturally specific lifeworlds are rooted in a lifeworld *a priori* that is *non*social in its very nature. The other possible answer, to which I hold in this book is that the lifeworld *a priori* is *social* in its very nature. So, for example, I explain the fact that various societies independently of each other developed the types of measurement 1 and 2 by the fact that all of them not only faced but at the same time *had to* face the same *social practical tasks*: the construction of buildings, the permanently reoccurring

---

60 On this see (Brinton 1885) and (O'Brien – Christiansen 1986).
61 On this see (Harvey – Williams 1980).
62 "Nonhistorical" means here, as long as there existed, exists, and will exist a species engaged systematically in agriculture.

need of crop-harvesting, the need to measure the space for the storage of the crop, and so forth.

The view that the lifeworld is both *historical* and at the same time *socially based* even in its ahistorical, *a priori* features enables also to evaluate the views of B. Van Fraassen incorporating a widely accepted approach to the nature of observable. The following three claims of Van Fraassen are worth quoting here:

[1] (Van Fraassen 1980, 16–17):

> [1] ... the moons of the Jupiter can be seen through a telescope; but they can also be seen without a telescope if you are close enough. ... A look through a telescope at the moons of the Jupiter seems to me a clear case of observation, some astronauts will no doubt be able to see them as well as from close up. But the purported observation of micro-particles in a cloud chamber seems to me a clearly different case ... while the particle is detected by means of the cloud chamber, and the detection is based on observation, it is clearly not a case of the particle's being observed.

[2] "I regard what is observable as a ... function of facts about us *qua* organisms in the world" (Van Fraassen 1980, 57–58).

[3] (Van Fraassen 1980, 81):

... what is the world in which I live, breathe and have my being, and which my ancestors of two centuries ago could not enter? It is the intensional correlate of the conceptual framework through which I perceive and conceive the world. But our conceptual framework changes – but the real world is the same world.

So, while according to Van Fraassen what is/is not observable depends solely on our biological outfit, which is the result of our past biological evolution (say, of the genetic code) as a species, in my view what one should differentiate between is only what is *already* observed and what is *as yet not* observed, and where this difference *depends solely on our practical operation in the external natural world by means of our bodies/senses plus our equipments. Our present senses are thus the result of the history of the human practical operation, including experimentation and the construction of instruments and devices, in the external natural world*, that is, not only of our past *natural* history as a species, but also, and primarily, of our past *social* history. And in this social history changes, not only the intensional correlate of our conceptual framework but also the external world which we continually populate with new and new artifacts. Thus there is no difference *in principle* between the observability of the moons of Jupiter and of the particles in a cloud chamber. In order to see both of them we have to construct in the external world special equipments and, thus, to change this world: in the case of the moons of Jupiter, more and more precise telescopes based on the use of light, ultrasound, X-rays, and so-forth, and possibly even a spaceship; in the case of particles, a cloud chamber.

Chapter 3: The Cyclical Method of Theory Construction I: The Retreat into and Coming Out from the Formal Ground

The aim of this chapter is to reconstruct the first type of thought-movement that goes beyond the realm of empirical knowledge, experimentation and direct measurement analyzed in chapter 2.

I start with a detailed reconstruction of the method of theory construction as given in Book I and Book III of Newton's *Principia*,[63] which I regard as a paradigmatic case of a cyclical method of theory construction and I show how it deviates from the "Standard Conception" analyzed in chapter 1. By drawing partially upon the works of W. L. Harper, I draw certain conclusions pertaining to the internal structure of the *Principia*. After showing the limitations of Harper's reconstruction of Newton's method of theory construction, I improve it by bringing in Hegel's categories of theory construction like *appearance, external (apparent) measure, (formal and real) ground* and *manifestation* from his *Science of Logic* (Hegel 1923; Hegel 1969). These categories enable me to show that the method of *Principia* is a specific type of cyclical method of theory construction. I then reconstruct the specific type of scientific law as well as specific types of scientific explanations corresponding to the structure of scientific laws and scientific explanations given in Newton's *Principia* and in Newtonian mechanics, in general, and, which clearly deviate from Hempel's and Carnap's view on scientific laws, scientific explanations and theory-construction as presented in chapter 1. I compare then this reconstruction with the approaches of B. Ellis and J. Bigelow (Ellis 1992; Bigelow and Ellis and Lierse 1992), of J. Woodward and C. Hitchcock (Woodward 1979; Hitchcock and Woodward 2003), of C. Liu (Liu 2004a, Liu 2004b) to scientific laws, scientific explanations, and idealizations as well as with the views of M. Thalos (Thalos 1999) on analytic mechanics.

I then apply Transparent Intensional Logic to the harmonic law of planetary motion in order to deal with the problem of meaning change of the terms, where this change takes place in the course of theory construction. Then I reconstruct the structure of the nomic, sometimes called "derived" form of measurement. Finally, I deal with the futile attempts of C. G. Hempel, B. Ellis and E. Mach to dispose of "metaphysically biased" concepts of science.

---

63 I reconstruct here, from the point of view of the applied method of theory construction, neither the relation of Book II of the *Principia* nor the methods used in its internal construction. Books I and III deal with movements of bodies in nonresistant spaces (media) while Book II with their movements in resistant spaces.

3.1 The cyclical method of theory construction: From appearances to the formal ground and "back"

3.1.1 Definitions and Laws in the Principia

Newton begins the construction of the *Principia* by relying on notions like "time, space, place and motion," which "are very familiar to everyone" (Newton 1999, 408)[64] and then tries "to explain the senses in which less familiar words are to be taken in this treatise" (1999, 408). From the point of view of *order* in the process of theory construction, the sequence of concepts at the very beginning of the *Principia* is as follows: *space* and *time* (explicitly discussed in the scholium after the definitions), *velocity* and *acceleration* (not explicitly discussed), *inherent force* corresponding to inertia (Definition 3), *impressed force* (Definition 4), *centripetal force* (Definition 5), *accelerative measure of centripetal force* (Definition 7). Definitions 2, 6, and 8 have a very special position in the conceptual hierarchy of the *Principia* because they are related to the term *quantity of matter* (body, mass). While the latter appears as a "measure of matter" (Newton 1999, 403) already in Definition 1, because it is, as I. B. Cohen labels it, "a primary quantity" (Cohen 1999, 92), it is *not*, from the point of conceptual understanding and grasping, defined by means of the terms "density" and "volume" appearing in that definition. The concept of force has, with respect to the concept of mass, in the *Principia*, an initial priority. Mass is understood first as what we discover as a property of a body, when an external force acts on it (i.e., as a measure of the body's resistance to the action of that force). This priority of the concept of force with respect to that of mass can be readily seen in corollary 4 of proposition 4, Book III, where Newton states that "particles have the same density when their respective forces of inertia are as their sizes" (1999, 810), while "a body exerts this force only during a change of its state, caused by another force impressed upon it" (1999, 404). The concept of mass becomes theoretically operational in the *Principia* as far as Book III by means of reflections pertaining to gravity, namely, by the procedures of finding out the mass of concrete bodies. Only here, thus, the absolute measure of centripetal force as a notion, first introduced in Definition 8, acquires its operational status.

This means that *at the very beginning* of the *Principia*, Newton *bypasses* the sequence of concepts mass, motion, change of motion, motive quantity of cen-

---

[64] But even with respect to the former he makes the remark that "these quantities are popularly conceived solely with reference to the objects of sense perception. And this is the source of certain preconceptions; to eliminate them it is useful to distinguish these quantities into absolute and relative, true and apparent, mathematical and common" (Newton 1999, 408).

tripetal force, absolute quantity of centripetal force, and draws exclusively on the following chain of concepts: space and time, velocity, acceleration, accelerative quantity of centripetal force. This explains the *prima facie* surprising fact that the concept of accelerative measure of force clearly dominates in the *Principia* as a whole and, in fact, is the only type of force appearing in Book I of the *Principia* up to proposition 11. In part 3.1.2.C I will analyze the place of the notion of mass in the conceptual hierarchy of the *Principia*, and in part 3.3.4 I will reconstruct Newton's method of the measurement of the masses of planets as given in Book III of the *Principia*. The concept of accelerative force is thus the real starting point for any further thought-movements in the *Principia* after the definition-section.

Let me now compare definitions 4 and 8 with the respective laws (axioms)

Definition 4	Law 1
Impressed force is the action exerted on a body to change its state either of resting or of moving uniformly straight forward (1999, 405).	*Every body preserves in its state of being at rest or of moving uniformly straight forward, except insofar as it is compelled to change its state by forces impressed* (1999, 416)
Definition 8	Law 2
The motive quantity of centripetal force is the measure of this force that is proportional to the motion which it generates in a given time (1999, 407).	*A change of motion is proportional to the motive force impressed and takes place along the straight line in which that force is impressed* (1999, 416).

What is readily seen is the fact that each of these two laws is a converse of its respective definition, or, to be more precise, a special type of "converse." In the definitions we identify (*discover*) the cause (impressed force on a body, centripetal force) by means of its effects (change of state of this body, the motive quantity generated proportional to the motion generated in a given time), that is to say, *we proceed in thought from the effects of forces to the very forces*. In both laws, *we proceed from the forces* (their non-action on a body; motive force*) to their respective effects* (the preservation of the state of the body; change of motion in time). I view each of the two pairs of definitions/laws as a case of *a cyclical thought-movement* about which W. L. Harper states "Newton's use of inferences from phenomena is a part of a process of theory construction that is like ... an information feedback process," (1993, 156) and which Newton himself characterized in general physical terms in the 1687 preface of the *Principia* as follows: "The basic problem of philosophy seems to be to discover the forces of nature from the phenomena of motion and then to demonstrate the other phenomena from these forces" (Newton 1999, 382).

In sections 3.1.2 and 3.1.3 I will use the phenomena of motion → forces of nature → phenomena of motion-type of theory construction as key to the analysis of Books I and III.[65] in such a way that I will try to find in it, wherever possible, first, thought-movements from certain types of phenomena-effects to their causes; second, thought-movements from those causes to the *same* types of phenomena, that is to say, from which the causes were initially derived; and, third, the thought-movements from causes to *other* phenomena, that is to say, to phenomena-effects *different* from those from which the causes were initially derived. Exempted from my reconstruction will be sections 4 through 6 of Book I, because they represent an exercise in pure geometry.[66]

3.1.2 Book I of the Principia

A. From the Phenomena of Motion to the Forces of Nature

Section 2[67] as a whole can be viewed as that part of the *Principia* where Newton accomplishes thought-movements from phenomena-effects to their causes; the whole section being symptomatically labeled "*To find centripetal forces*" (1999, 444). In proposition 2 Newton states (1999, 446):

Every body that moves in some curved line described in a plane and, by a radius drawn to a point, either unmoving or moving uniformly forward with a rectilinear motion, describes areas proportional to the times, is urged by a centripetal force tending toward the same point.

By proving it he, thus, proves that the effect-phenomenon of motion of a body according to the law of areas has as its cause the force acting in a centripetal manner on that body. Proposition 3 then generalizes proposition 2 to the case when the body *L* (the Moon) orbits a second body *T* (the Earth) under the impact of a centripetal force to that second body, so that both are subject to yet another force acting on them along parallel lines (e.g., from the Sun), proving that "the difference of the forces tends ... toward the second body as a center" (1999, 448).

---

65 It remains the task of a future investigation to find out if such a cyclical form of relation is given also between Book II of the *Principia* and its two other books, and whether such a cyclical form of relation can be discerned also inside Book II.
66 This view on sections 4 through 6 of Book I was communicated to me in written correspondence by the late Professor J. B. Brackenridge.
67 Section 1 provides with its "first and ultimate ratios" the mathematical apparatus for the *Principia*.

But even if section 2 as a whole should deal with finding centripetal forces from their effects, proposition 1 in this section is in reversed order. It states that "the areas which bodies made to move in orbits describe by radii drawn to an unmoving center of forces lie in unmoving planes and are proportional to the times," (1999, 444) that is to say, that this proposition moves from the centripetal force, acting on a body, to one of the *effects* of this force: the character of areas described by the body moved by that force in orbits. This shows what we already have seen already in the relation of Laws 1 and 2 to Definitions 4 and 8, namely, that Newton permanently accomplishes a *bi-directional thought-movement*: from the effects to their cause and the other way round.

Proposition 4 states (1999, 449):

The centripetal forces of bodies that describe different circles with uniform motion tend toward the centers of those circles and are to one another as the squares of the arcs described in the same time divided by the radii of the circles.

It is readily seen from this that Newton determines here the ratio of forces acting on bodies moving uniformly in circular orbits via the characteristics of these orbits (arc and radius) or, as in corollaries 1 and 2 of proposition 4, via the characteristics of the movement of these bodies (speed and period).

To get a better understanding of the method of determining the cause via its effects, let us go through the proof, not of proposition 4 of the *Principia*, but of the corresponding to it second theorem in the manuscript *De Motu Corporum in Gyrum*.[68] It runs as follows: "The centripetal forces of bodies revolving uniformly in the circumferences of circles are as the squares of the arc described in the same time divided by the radii of the circles" (1965a, 278). The problem Newton faced was how to *measure* the centripetal force. The solution, upon which his thought-movement from an effect of a cause to the very cause is based, is as follows. Centripetal forces causes the deviation of bodies from their rectilinear motion and, thus, the former are proportional to the latter, or "It is the centripetal forces that perpetually draw the bodies back from the tangent to the circumferences, and hence they are to each other as the distances ... gained by their bodies" (1965a, 278). Based on this idea of how to measure the forces, the proof can be accomplished. We have two circles that can be represented as follows (S is the center of the smaller circle and s the center of the larger circle):

---

[68] The correspondence of Theorem 2 to Proposition 4 is, of course, not complete. In the former, Newton implicitly presupposes that the forces are centripetal; in the latter he provides a proof of that. A detailed comparison of *De Motu* with the *Principia* is given in Chapter 7 of (Brackenridge 1995).

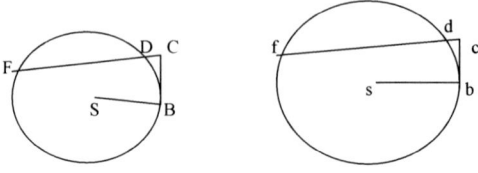

Figure 3.1 Newton's measurement of forces by means of the deflection they cause

On each circle a body circulates, so that $SB \neq sb$ and $bd \neq BD$ holds. It can be proved[69] that ("×" stands for multiplication) $cd \times cf = cb^2$ and $CD \times CF = CB^2$ and, thus, $CD = CB^2/CF$ and $cd = cb^2/cf$ holds. So as the traversed arcs are very small, D is close to B, d is close to b, cf and CF are close to be the diameter of the respective circle, that is, $2BS = CF$ and $2bs = cf$ holds and therefore $CD = DB^2/2BS$ and $cd = db^2/2bs$ holds as well. Finally, one obtains the proportion between the forces f and F to their respective deflection cd and CD as $f \propto db^2/bs$ and $F \propto DB^2/BS$ ("$\propto$" stands here for "is directly proportional to"). From this result, it is then possible to prove the respective corollaries in *De Motu* and the *Principia*. So as $db = v \times t_1$ and $DB = V \times t_2$ holds, where v, V are the respective speeds; t_1, t_2 the time (and $t_1 = t_2$), from the ratio $F : f = DB^2/BS : db^2/bs$ it follows $F : f = V^2/BS : v^2/bs$, what is stated in corollary 1. Because for the periods t and T of the movement of the bodies along the whole circle holds that $v \times t = 2\pi \times bs$ and $V \times T = 2\pi \times BS$, one obtains $F : f = BS/T^2 : bs/t^2$, as stated by corollary 2. If $T = t$, then $F : f = BS : bs$ as stated in corollary 3 of proposition 4, Book I, in the *Principia*. If the ratio on the right side of this equation is equal to one, then $F = f$ as stated in corollary 3 of the *De Motu* and corollary 4 of proposition 4. If the periods T and t are as the radii the ratio is $F : f = bs : BS$ as stated by corollary 4 of *De Motu* and corollary 5 of proposition 4.

Corollary 6 of the *Principia* (corollary 5 of *De Motu*) has a special place in proposition 4 of the *Principia* (theorem 2 of *De Motu*). Newton brings in here the relation between the period and the radius of the orbit typical for celestial bodies, namely, that the square of the periodic time is as the cube of the radius (i.e., what we today label as "Kepler's third law"). He then proves that $F : f = db^2 : DB^2$, that is to say, the centripetal force diminishes with the square of the distance.

Proposition 6 states the following (1999, 453–454):

If in a nonresisting space a body revolves in any orbit about an immobile center and describes any just-nascent arc in a minimally small time, and if the sagitta of the arc is understood to be drawn so as to bisect the chord and, when produced, to pass through the center of forces, the centripetal force in the middle of the arc will be as the sagitta directly and as the time twice inversely.

---

69 By means of proposition 36 (Book 3) of Euclid's *Elements*.

In order to explain Newton's proof,[70] I partially draw upon Figure 3.2 drawn not by Newton but by I. B. Cohen (1999, 319).[71]

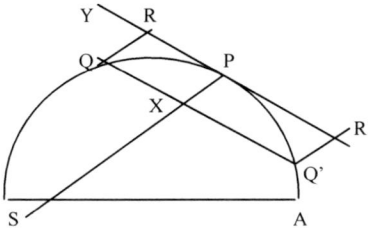

Figure 3.2 Proof for proposition 6 of Book I

Here XP is the sagitta (i.e., the line that when drawn bisects the chord and passes through the center S of force), P approaches Q, and Newton proves[72] that $F \propto PX/\Delta t^2$. The "problematic" component in this expression is the magnitude of time. We are in need of an "Ersatz"-measure for it. Newton, drawing upon proposition 1, uses what we today call "Kepler's area law," and substitutes for Δt the area of the sector swept out by the line SP. The surface of the triangle SPQ can serve as the measure of Δt. If QT stands for the perpendicular dropped from Q on SP and SPR can, under the "nascent" considerations, be viewed as a triangle whose surface is given by $\frac{1}{2}(SP \times QT)$, then we end up with the proportion[73] $F \propto QR/(SP^2 \propto QT^2)$.

Propositions 7, 8, 9 and 10 (Newton 1999, 455–461) are viewed by Newton as examples or as applications of the proportion $F \propto QR/(SP^2 \times QT^2)$ from proposition 6 to different types of situations. In proposition 7, Newton presupposes that a body P revolves in the circumference of a circle and the force tends toward any point S. The solution yields the result $F \propto 1/(VP^3 \times SP^2)$, and in corollary 1 he proves that if S is located in the circumference of the circle, so that $V = S$, it holds that $F \propto 1/SP^5$. In proposition 8 Newton presupposes that the center S of force is very distant from the body P, while the lines PN and RM are parallel (P, R are the points given in Figure 3.2, while N and M are the points of intersection of the line PS and RS with the diameter). For the force then it holds that $F \propto 1/PM^3$. In proposition 9, the body revolves in a spiral that intersects all its radii. For the force

---

70 For analysis see also (Pourciau 2007)
71 The figure used by Newton (1999, 454) contains neither the sagitta nor the chord.
72 The proof is based on corollary 4 of proposition 1 and corollaries 2 and 3 of Lemma 11.
73 $PX=QR$ and $\frac{1}{2}$ can be omitted because I deal here not with equations but only with proportions.

that tends to the center of this spiral $F \propto 1/SP^3$ holds. Proposition 10, concluding section 2, supposes a body P revolving in an elliptical orbit. For a force tending towards its center (not focus) C it holds that $F \propto PC$.

Section 3 as a whole is not devoted exclusively to the derivation of proportions for the force given certain specific trajectories. It is titled as "*The motion of bodies in eccentric conic sections.*"

It starts with proposition 11, which formulates the following problem to be solved: "*Let a body revolve in an ellipse; it is required to find the law of the centripetal force tending toward the focus of the ellipse*" (1999, 462). Newton provides two proofs: one longer and one shorter.[74] Figure 3.3 shows the latter (*CO* is parallel with *SP*):

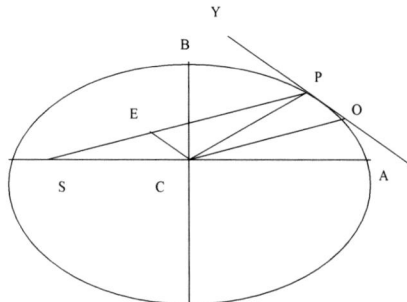

Figure 3.3 The figure for the proof of proposition 11 of Book I

It can be shown (by corollary 3 of proposition 7) that the ratio of the force F, by which the body P revolves in an orbit around the center of force C, to the force f, by which the same body P can revolve in the same obit and in the same periodic time around any other center, is $F : f = CP \times SP^2 : OC^3$. Because *ECOP* is a parallelogram (i.e., *OC=PE*) we obtain $F : f = CP \times SP^2 : PE^3$. And because by proposition 10 it holds that $F \propto CP$, we have $f = F \times PE^3/(CP \times SP^2) = PE^3/SP^2$. But since *PE* is for a given ellipse a constant magnitude (equal to its semimajor axis *AC*), we end up with the proportion $f \propto 1/SP^2$.

Propositions 12 and 13 conclude that part of section 3, where Newton accomplishes his thought-move from the phenomena as the effects of forces to the forces. The former states the following problem: "*Let a body move in a hyperbola; it is required to find the laws of the centripetal force tending toward the focus of the figure,*" (1999, 463) and then Newton proves that $F \propto 1/SP^2$. The latter proposition states "*Let a body move in the perimeter of a parabola; it is required to find the*

---

[74] The former proof is given in all its details in (Brackenridge 1995, 107–117) and in (De Gandt 1995, 39–40).

laws of the centripetal force tending toward a focus of the figure" (1999, 466), and then he proves that $F \propto 1/SP^2$ holds again.

What is worth noting is the fact that while section 2 and section 3, up to corollary 1 of proposition 13, represents thought-movements from the phenomena of motion to their forces, and while the latter sections of Book I represent thought-movements from forces to their effects (either new, i.e., not previously derived, or already previously derived), still we can find scattered in Book I several propositions, whose subject matter is the discovery of forces from their phenomena-effects. Here I mean propositions 43 and 44, which deal with the case of finding the centripetal force causing a moving trajectory.

Proposition 53 is also symptomatic for Newton's derivation of the cause by means of one of its effects. It states, "*Granting the quadratures of curvilinear figures, it is required to find the forces by whose action bodies moving in given curved lines will make oscillations that are always isochronous*" (Newton 1999, 556). The pendulum with cycloidal cheeks can be represented in Figure 3.4 (the point *T* stands for the oscillating body, *STRQ* is the curved line in which the body oscillates, *AR* is the axis of *STRQ* which passes through the center of force *C*):

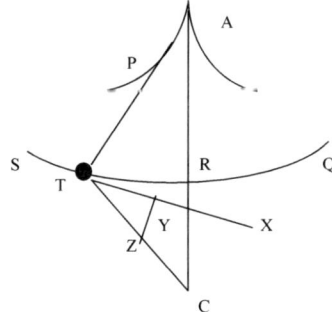

Figure 3.4 Finding forces causing isochronous oscillations of bodies

Newton takes the line *TX* which is tangential to the path *STRQ* at *T* because this is the only direction, at the instantaneous position of the body, in which the attracting force can *cause* a change of its speed, that is, accelerate or decelerate it in the curve *STRQ*. He uses the path-*effect TZ* of the action of the force of attraction on the bob and decomposes it into the components *TY* and *YZ*. The former is tangential to the path and, thus, is *the* path-effect of the cause of the just-mentioned possible changes of speed of the body, while the latter is the stretching-effect of the cord under the impact of the attracting force which, under the idealization that the length of the cord does not change under the impact of forces, can be neglected because it does not change the speed of the body. And he puts *TY* equal to the arc *TR* in order to claim that since force represented by *TR* [75] (1999, 556–557):

is as the projection TR to be described, the body's accelerations or retardations in describing proportional parts of two oscillations (a greater and a lesser oscillation) will always be as those parts and will therefore cause those parts to be described simultaneously. And bodies that in the same time describe parts always proportional to the wholes will describe the wholes simultaneously.

B. From the Forces of Nature to the Same Phenomena of Motion

It is worth noting that even in section 2, which as a whole deals with the discovery of forces from their respective effects, there are several places where Newton reverses the order of his thought-movements. It first occurs in Newton's differentiation between proposition 1 and proposition 2 given in part 3.1.2 above. A The second time, it appears in corollaries 3 through 6 of proposition 4 in the *Principia*, where he adds at the very end the phrase "and conversely" (1999, 451), that is to say, he claims that *once we are able to determine that the ratio of forces is such and such, then we can determine as well the specific ratio of the characteristics of movement of the bodies upon which these forces act.* The third time, the reversal appears in the scholium to proposition 8, where he states (1999, 458):

A body will be found to move in an ellipse, or even in a hyperbola or a parabola, under the action of a centripetal force that is inversely as the cube of the ordinate tending toward an extremely distant center of forces.

This scholium is the converse of proposition 8, in which from the movement of a body in a semicircle and the "fact" that the center of force acting on it is very distant, he derives the distance-dependence of the force.

Finally, in corollary 1 of proposition 10 he states not only that "the force is as the distance of the body from the center of the ellipse," but also "and conversely, if the force is as the distance, the body will move in an ellipse having its center in the center of forces" (1999, 460).

A profound reversal, which I will discuss in all its details in section 3.1.4 below, is accomplished by Newton in Book I, starting from corollary 1 of proposition 13 and ending in proposition 17. While in propositions 11 through 13 he moved *from the conic character of the trajectory of the body to the force producing it*, corollary 1 states (1999, 467):

---

[75] Professor C. A. Wilson in written correspondence with me was very helpful in this reconstruction of Newton's ideas in proposition 35 of Book I.

From the last three propositions it follows that if any body P departs from the place P along any straight line PR with any velocity whatever and is at the same time acted upon by a centripetal force that is inversely proportional to the square of the distance of places from the center, this body will move in some one of the conics having a focus in the center of forces; and conversely.

A similar movement from forces acting on bodies to the trajectory of these bodies is accomplished by Newton also in corollary 2 of proposition 13. Newton claims that (1999, 467)[76]

If the velocity with which the body departs from its place P is such that the line element PR can be described by it in some minimally small particle of time, and if the centripetal force is able to move the same body through space QR in the same time, this body will move in some conic whose principal latus rectum is the quantity QT^2/QR.

In proposition 14, Newton presupposes the existence of several (mutually not interacting) bodies that orbit about the same center and, then, continues (1999, 467):

If ... the centripetal force is inversely as the square of the distance of places from the center, I say that the principal latera recta of the orbits are as the square of the areas which the bodies describe in the same time by radii drawn to the center.

Proposition 15, presupposing the same like proposition 14, states that *"the squares of the periodic times in ellipses are as the cubes of the major axis"* (1999, 468). So, while harmonic law was in proposition 4 *presupposed* (for circular trajectories) and enabled to derive the character of the force, in proposition 15 it becomes a *consequence* (for elliptical orbits) by presupposing that $F \propto 1/SP^2$.

Proposition 17, concluding section 3, states (1999, 470):

Supposing that the centripetal force is inversely proportional to the square of the distance from the center and that the absolute quantity of this force is known, it is required to find the line which a body describes when going forth from a given place with a given velocity along a given straight line.

Newton then proves that the orbit can be either an ellipse, or a parabola, or a hyperbola (1999, 471).

Finally, in corollary 1 of proposition 53 Newton, drawing upon proposition 53 (reconstructed in section 3.1.2.A above) proves that for a pendulum under the

---

[76] The line element *PR* and the space *QR* are displayed in Figure 3.2. The latus rectum *L* of a conic is the chord perpendicular to the principal axes that passes through a focus of the conic, so that if *AB* is the major diameter and *PD* its minor diameter, it holds that $L=PD^2/AB$.

action of a uniform force of gravity, its oscillations will have the same period. On the basis of this knowledge Newton explains in corollary 2 how to construct pendulum clocks with isochronous oscillations.

C. From the Forces of Nature to New (Different) Phenomena

The movement from the forces of nature to types of phenomena, where the latter never before appeared in the *Principia*, can be identified for the first time already in section 2, namely, in the scholium to proposition 10. It states (1999, 460):

If the center of the ellipse goes off to infinity, so that the ellipse turns into a parabola, the body will move in this parabola, and the force, now tending toward an infinitely distant center, will prove to be uniform. This is Galileo's theorem.

I view this scholium as a direct consequence of the "converse" section of corollary 1 of proposition 10, where Newton states that "if the force is as the distance, the body will move in an ellipse" (1999, 460). So, if the center of force is, with respect to the body on which the force acts, very distant, this body will be subject to a constant force and will move in a parabolic trajectory, for example, like a projectile.

From section 3, I view proposition 16 as standing for the force → other phenomena type of thought-movement, because it explains how to find a previously unknown characteristic of the moving body – its velocity – and where the proof explicitly draws upon proposition 14 which starts/proceeds from the inverse-square character of the centripetal force.

Section 7 considers also a previously unanalyzed (i.e., a new) phenomenon, namely, the *"rectilinear descent and ascent of bodies"* (1999, 518). Starting (implicitly) from proposition 17, where the conic orbit is understood as an effect of the action of an inverse-square centripetal force, Newton states the problem to be solved in proposition 32 as follows (1998, 518):

Given a centripetal force inversely proportional to the square of the distance of places from its center, to determine the spaces which a body in falling straight down describes in given times.

In the solution, Newton, first supposes that the body does not fall perpendicularly. By drawing, explicitly, on corollary 1 of proposition 13, he views its trajectory as a conic and, thus, considers three possible cases of the fall of a body along an orbit: ellipse (case 1); hyperbola (case 2); parabola (case 3).

Propositions 33 through 38, drawing upon the results from proposition 32, then deal with various aspects of the descent and ascent of bodies (their velocity, fall along a parabola, time of descent from a certain point, time of ascent/descent of a projected body). In proposition 39, concluding section 7, Newton presupposes the action of a centripetal force of any kind (not only an inverse-square one) and then determines the time to reach a place and the speed of a body in any place it reaches when ascending/descending straight up/down.

Section 8 with its propositions 40 through 42 has as its aim "*to find orbits in which bodies revolve when acted upon by any centripetal force,*" (1999, 528) that is to say, the force is not an inverse-square one, but in general some unspecified function of distance. For example proposition 41 states (1999, 529)[77]

Supposing a centripetal force of any kind and granting the quadratures of curvilinear figures, it is required to find the trajectories in which bodies will move and also the times of their motions in the trajectories so found.

Proposition 45 of section 9 is a continuation of the propositions 43 and 44 from this section which I (mentioned in part 3.1.2.A above). But while in propositions 43 and 44 Newton determined the character of a centripetal force from the fact that the orbit was a moving orbit, in proposition 45 Newton derives the angle that a body completes when it descends from the upper to the lower apsis.[78] It states that: "*it is required to find the motions of the apsides of orbits that differ very little from circles*" (1999, 539).[79] Based on the derived relations between forces and the characteristics of the orbits they produce, Newton states in corollary 1 that "if the centripetal force is as some power of the height, that power can be found from the motion of the apsides, and conversely" (1999, 543). Similarly, in corollary 2, he states that (1999, 544)

if a body, under the action of a centripetal force that is inversely as the square of the height, revolves in an ellipse having a focus in the center of forces, and any other extraneous force is added to or taken away from this centripetal force, the motion of the apsides that will arise from that extraneous force can be found out (by instances of ex. 3) and conversely.

This means that once Newton was able to derive in the beginning of proposition 9 the character of force from certain characteristics of specific types of trajectories, he can – as in the corollaries of proposition 45 – propose a certain type of the centripetal force, and then derive from it quantitative characteristics of its apsidal motion and, finally, compare them with the actual data for a certain celestial object, say, the Moon in corollary 2.

---

[77] I. B. Cohen provides in (1999) a detailed proof of this proposition.
[78] Apsis is the point on the orbit of a body, which is the most distant from or closest to the center of the orbit.
[79] The apside is the line that connects the apsises.

Section 10 considers, in addition to the effects of motion laying in a plane that either passes through the center of forces (proposition 54) or the axis of which passes through that center (propositions 55 and 56), a new type of effects on bodies that comes into play when their centers do not lay in the planes in which the bodies move. The thought-objects that Newton considers are a wheel moving upon the inner/outer surface of a globe and a bob oscillating in a given cycloid, and then he finds out different characteristics of their movement. In the case of the wheel it is, for example, the length of the curvilinear path traced by any point in the wheel, while in the case of the pendulum bob, Newton finds out, for example, the conditions under which it oscillates in a given cycloid.

Section 11 (propositions 57 through 69) brings in an important change in Newton's derivation of effects of centripetal forces. In sections 1 through 10, he presupposed, first, that the orbiting body was a point; second, that the only physical properties the orbiting body had were that of velocity and acceleration; third, that if several bodies were orbiting the same center, they did not interact mutually (as in proposition 14); and, fourth, that the center around which the body moves, and toward which the centripetal force tends is also a point devoid of any physical properties. In section 11, Newton, instead of talking about a body as a point (with the property of velocity and acceleration) orbiting another point, starts talking about the movement of a body orbiting another, each with the property of mass, because "attractions are always directed toward bodies" and for this reason he then goes on to "set forth the motion of bodies that attract one another, considering centripetal forces as attractions" (1999, 561).

In propositions 57 and 58 he is dealing with a two-body problem and proves that "*two bodies that attract each other describe similar figure about their common center of gravity and also about each other,*" (1999, 561) and that (1999, 562)

If two bodies attract each other with any forces whatever and at the same time revolve about their common center of gravity, I say that by the action of the same forces there can be described around either body if unmoved a figure similar and equal to the figures that the bodies so moving describe around each other.

Corollaries 1 and 2 of proposition 58 then show that once two bodies attract each other by a certain type of force (proportional to their distance; inversely proportional to the square of their distance), they will describe around each other a certain type of orbit (concentric ellipses; conic with a focus in that center).[80] Corollary 3 of proposition 58 then affirms Kepler's area law as holding/valid both for

---

80 It is worth noting here that these three corollaries contain the phrase "and conversely," i.e., Newton accomplishes here also the thought-move from a certain type of orbit described to the generating force.

the movement of each body around the other as well as for both of them orbiting their common center of gravity.

Newton then compares in proposition 59 the period of orbital motion of two bodies around their common center of gravity C to the period of body P revolving around body S which is at rest, so that P describes a figure similar and equal to that which S and P describe around each other. The conclusion he draws is that the ratio of these two periods is "as the square root of the ratio of the mass of the second body S to the sum of the masses of the bodies $S+P$" (1999, 564), which represents a modification of the harmonic law.

Propositions 60 through 63 explain further characteristics of the S-P system of bodies: the geometrical characteristics of their orbits and the character of their motion under the impact of forces that belong to them and that are of the inverse-square type. Proposition 64 considers the effects of a force, directly proportional to the distance, on a system composed first of two bodies, then of three, and, finally, of four. Proposition 65 considers again the action of an inverse-square force on a multiple-body system and shows that with respect to each other, Kepler's first and second laws hold.

Proposition 66 in its corollaries 2 through 11 provides a *phenomenological*,[81] *nonqualitative* account of how the force of a distant body causes, by acting on a second body that is orbiting a third body,[82] the second body's angular velocity, its period, the eccentricity of its orbit, the motion of the lines of apsides, and so forth. Worth mentioning here is that while Newton holds, up to corollary 14, to the concept of the accelerative measure of forces, in corollary 14 (1999, 579) he starts using the term "absolute force," (i.e., "absolute quantity of the centripetal force") and then states that "the magnitude of body S is proportional to its absolute force" (1999, 580). Newton also starts considering in corollaries 18 through 22, as a preparation for Book III, the composition (density) and shape of the perturbed body.

Propositions 67 and 68 deal with further force-effects of a distant body on a second body orbiting, in turn, a third body – these effects being the area law and the nearly elliptical shape of the orbit of the second body.

Proposition 69 states that "*the absolute forces of attracting bodies A and B will be to each other in the same ratio as the bodies A and B themselves to which these forces belong*" (1999, 587). Its proof runs as follows. Let us have two bodies A and B in a certain mutual distance so that body A, with mass m_A and force F_A belonging to it, accelerates body B toward A to an acceleration a_B; F_{aA} is the ac-

---

81 What I mean by this will be explained in part 3.1.4 below.
82 They are represented in the corresponding figure (Newton 1999, 571) by the letters S, L, and T.

celerative – with respect to B – force belonging to A, so that it holds that $F_{aA} \propto a_B$. Body B, with mass m_B and force F_B belonging to it, accelerates body A toward B to an acceleration a_A; F_{aB} is the accelerative – with respect to A – force belonging to B, so that $F_{aB} \propto a_A$ holds. We can thus state that the following ratio does hold: $F_{aA} : F_{aB} = a_B : a_A$. If F_{mA} denotes the *motive* force belonging to body A, which confers to body B with mass m_B the acceleration a_B, then it holds that $F_{mA} \propto m_B \times a_B$. F_{mB} is the motive force belonging to body B which accelerates body A with mass m_A to a_B; here $F_{mB} \propto m_A \times a_A$ holds. Newton states that because of definitions 2, 7, and 8, it holds that $F_{mA} : F_{mB} = (F_{aA} \times m_B) : (F_{aB} \times m_A)$ (1999, 587). By the third law it holds that $F_{mA} : F_{mB} = 1$, thus $(F_{aA} \times m_B) : (F_{aB} \times m_A) = 1$ holds as well. From this it follows that it holds that $F_{aA} : F_{aB} = m_A : m_B$, where F_{aA} is the force belonging to A body and accelerating body B, while F_{aB} is the force belonging to body B accelerating body A. They can therefore be viewed as the absolute force of the respective body, and thus Newton can claim that "the absolute attractive force of body A is to the absolute attractive force of body B as the mass of body A is to the mass of body B" (1999, 587).

Section 12, contrary to the previous sections in Book I, starts considering the dimensions of bodies, which are under the impact of a centripetal force, and where these bodies are supposed to be spherical. Newton proves in proposition 70 that the force acting on a body placed in a hollow sphere, the particles of which act on each other by inverse-square forces, is zero, while on its outside that particle would be (by proposition 7) under the impact of an inverse-square force tending to the center of the sphere.

Starting with section 12 Newton gives up the presupposition that bodies have no dimensions and shapes. He presupposes, instead, that bodies are spherical and act with inverse-square forces. He then considers their effects, for example, on a corpuscle placed inside or outside of a hollow sphere, their action on another hollow sphere, and so forth.

In section 13, Newton considers nonspherical bodies and in section 14, at the end of Book I, he targets the motion of minimally small bodies on which act centripetal forces tending toward each of the individual parts of some great body.

3.1.3 Book III of the Principia

In Book III of the *Principia* it is possible, as in Book I, to discern the phenomena of motion-forces of nature-phenomena of motion-type of thought-movement. In fact Newton himself in the introduction to the 1687 edition of the *Principia*, after mentioning this type of thought-movement, states that "in Book III our explanation of the system of the world illustrates these propositions. For in Book

III ... we derive from celestial phenomena the gravitational forces by which bodies tend toward the sun and toward the individual planets. Then the motions of the planets, the comets, the moon, and the sea are deduced from these forces" (1999, 382).

A. From the Phenomena of Motion to the Forces of Nature

Newton starts Book III by stating four rules of reasoning and six phenomena, where the latter pertain to the "Keplerian"[83] characteristics of the primary planets circling the Sun and of the satellites of Jupiter and Saturn. The first rule states that *"no more causes of natural things should be admitted than are both true and necessary to explain their phenomena"* (1999, 794), while the second rule states that *"the causes assigned to natural effects of the same kind must be, so far as possible, the same"* (1999, 795).

From these phenomena and rules, together with the knowledge obtained/derived in Book I, Newton then starts his thought-movement *to* the forces that cause those "Keplerian" characteristics. Proposition 1 states (1999, 802)

The forces by which the circumjovial planets are continuously drawn away from rectilinear motions and are maintained in their respective orbits are directed to the center of Jupiter and are inversely as the squares of the distances of their places from that center.

The first part of this proposition – that there is a force directed to the center of Jupiter – is proved on the basis of proposition 2 or 3 (Book I), because for the satellites of Jupiter the area law holds, by phenomenon 1. The second part of the proposition, that that force is an inverse-square one, is proved by corollary 6 of proposition 4 (Book I) because, by the second part of phenomenon 1, the harmonic law holds for the satellites of Jupiter. Newton then adds, on the basis of both parts of phenomenon 2, that "the same is to be understood for the planets that are Saturn's companions" (1999, 803).

Proposition 2 states the same for the planets, while the proof follows the same pattern and draws upon the phenomenon 5 (the area law for the planets) and phenomenon 4 (the harmonic law for the planets). Newton proves its second part also by considering corollary 1 of proposition 45 (Book I), namely, that "the slightest departure from the ratio of the square would ... necessary result in a noticeable motion of the apsides in a single revolution and an immense such motion in many revolution" (1999, 802). But because the aphelia of the planets are at rest, the centripetal force tending toward the Sun is really an inverse-square one.

---

83 Newton mentions in the phenomena 1 through 6 what is today labeled as "Kepler's second" and "Kepler's third law," but not what is today labeled as "Kepler's first law."

Proposition 3 states for the Moon what proposition 1 and 2 stated for the satellites of Jupiter and Saturn, and for the planets. Its first part is proved in the same way as are the proofs performed for the first parts of propositions 1 and 2 (by drawing upon the area law for the Moon stated in phenomenon 6). But the proof of its second part follows another path, simply because the Moon was in the time of Newton the only satellite of the Earth and, thus, Newton could not apply Kepler's area law to the system Earth-Moon. He draws, instead, again – as in proposition 2 – on corollary 1 of proposition 45 (Book I). Even if the apogee of the Moon is moving, it is still "... very slow ..." and "this motion of the apogee arises from the action of the sun ... and accordingly is to be ignored here," so that "the remaining force by which the moon is maintained in its orbit will be inversely as D^2" (1999, 802–803), where D expresses the distance of the Moon from the center of the Earth, while the semidiameter of the Earth is put equal to 1.

Proposition 4 states that "*the moon gravitates toward the earth and by the force of gravity is always drawn back from the rectilinear motion and kept in its orbit*" (1999, 803). Newton proves it by means of the well-known first[84] Moon-test[85] and, then, claims that "that force by which the moon is kept in its orbit, in descending from the moon's orbit to the surface of the earth, comes out equal to the force of gravity here on earth, and so (by rules 1 and 2) is that very force which we generally call gravity" (1999, 804).

Proposition 5 then states that the satellites of Jupiter and Saturn and the planets orbiting the Sun "gravitate" to the respective celestial object because the revolutions (1999, 806)

are phenomena of the same kind as the revolution of the moon about the earth, and therefore (by rule 2) depend on causes of the same kind, especially since it has been proved that the forces on which those revolutions depend are directed toward the centers of Jupiter, Saturn and the sun, and decrease according to the same ratio and law (in receding from Jupiter, Saturn, and the sun) as the force of gravity (in receding from earth).

In the scholium of this proposition Newton then performs an important shift from the *concept of centripetal force* to that of *force of gravity*, "for the cause of the centripetal force by which the moon is kept in its orbit ought to be extended to all the planets" (1999, 806).

Proposition 25 sets as its aim "*to find the forces of the sun that perturb the motions of the moon*" (1999, 839). This aim is achieved (by drawing upon proposition

---

84 The second Moon-test is performed in corollary 7 of proposition 37.
85 On the Moon-test, see S. Aoki's paper (1992). Newton performs in the scholium of proposition 4 a different proof by considering an imagined second satellite orbiting the Earth (1999, 805).

66 in Book I) via a decomposition of the effect of the force of gravity of the Sun, which is perturbing the Moon. He represents this force by lines and views them as representing "accelerative gravity" (1999, 840), that is to say, he is able to deal with the perturbing force of the Sun by representing it via its effect.

Proposition 36, then, as a continuation of proposition 25, sets as its aim "*to find the force of the sun to move the sea*" (1999, 874), while proposition 37's aim is "to find the force of the moon to move the sea" (1999, 875). It is, with respect to the latter, worth noting that Newton performs here the computation of the size/quantity of forces by means of the size/quantity of their effects. He states that "the force of the moon to move the sea is to be reckoned from its proportion to the force of the sun, and this proportion is to be determined from the proportion of the motions of the sea that arises from these forces" (1999, 875).

B. From the Forces of Nature to the Same Phenomena of Motion

The first and only proposition that is the result of the thought-movement forces of nature → same phenomena of motion in Book III is proposition 13. It states "*The planets move in ellipses that have a focus in the center of the sun, and by radii drawn to that center they describe areas proportional to the times*" (1999, 817). This proposition is viewed by Newton himself as a result of such a type of movement. He claims (1999, 817–818)

> We have already discussed these motions from the phenomena. Now that the principles of motions have been found, we deduce the celestial motions from these principles a priori. Since the weights of the planets toward the sun are inversely as the squares of the distances from the center of the sun, it follows (from Book I, prop. 1, and prop. 13, corol. 1) that if the sun were at rest and the remaining planets did not act upon one another, their orbits would be elliptical, having the sun in their common focus, and they would describe areas proportional to the times.

C. From the Forces of Nature to New (Different) Phenomena

The first proposition belonging to the forces of nature-new (other) phenomena type of thought-movement is proposition 6. It states that (1999, 806)

> *all bodies gravitate toward each of the planets, and at any given distance from the center of any one planet the weight of any body whatever toward the planet is proportional to the quantity of matter which the body contains.*

Newton's proof consists in fact of five different proofs. I will analyze only the first one to show that proposition 6 really belongs to that type of thought-movement but that it differs from those analyzed in part 3.1.2.C, as well as from the type of thought-movement in propositions 8 through 39 in Book III, which I will analyze later.

The first proof is *theoretically* based on proposition 24 of Book II which states that for a pair of pendulums "*whose centers of oscillation are equally distant from the center of suspension, the quantities of matter are in a ratio compounded of the ratio of the weights and the squared ratio of the times of oscillation in a vacuum*" (1999, 700), from which he then derives corollary 1, stating "and thus if the times are equal, the quantities of matter in the bodies will be as their weights," as well corollary 6, which runs as follows (1999, 701):

But in a nonresisting medium also, the quantity of matter in the bob of a simple pendulum is as the relative weight and the square of the time directly and the length of the pendulum inversely. For the relative weight is the motive force of a body in any heavy medium ... and thus fulfills the same function in such a nonresisting medium as absolute weight does in a vacuum.

Experimentally the proof of proposition 6 is based on Newton's experiments with a pair of pendulums of the same length with equal hollow bobs, filled with different materials (silver, gold, glass, etc.). These experiments show that their periods are the same, and thus the weights of the bobs are as their masses.[86]

If one looks into the structure of proposition 24 of Book II, it is readily seen that it is based on the following chain of thoughts:

i. velocity generated by a force in a given time (i.e., acceleration) in a given quantity of matter is proportional to the force and time and inversely as that quantity of matter; "this is manifest from the second law of motion" (1999, 700).
ii. For two pendulums of the same length, the motive forces are as the weights.
iii. If two oscillating bodies describe equal arcs and if the arcs are divided into equal parts, then the velocities in corresponding parts of oscillation will be to one another as the motive forces and the whole times of the oscillations directly and the quantities of matter inversely.
iv. Therefore, the quantities of matter will be as the forces and the times of the oscillations directly and the velocities inversely. But the velocities are inversely as the times, so, the quantities of matter are as the motive force and the square of the times, that is, "as the weights and the square of the times" (1999, 700).

From steps i through iv it is readily seen that Newton draws here upon the second law, which (as shown earlier) is just the converse of definition 7, and where the

---

[86] A reconstruction of these experiments is given in (Wilson 1999).

latter derives force from its effect. This is also apparent from the fact that Newton substantiates the concept of motive force by means of the concept of weight, which is the type of force theoretically assigned to acceleration. From this I thus draw the conclusion that after Newton has accomplished his thought-movement from space and time, velocity and acceleration, and accelerative measure of force-weight, he can theoretically deal with the motive quantity of force (definition 8), then move to the second law, and finally, via proposition 24 and its corollaries in Book II, arrive at proposition 6 of Book III.

What has to be emphasized here is, first, that proposition 6 belongs to that part of the *Principia*, in which Newton, *after he initially moved from phenomena of motion to the forces causing them, moves from the forces to other (new) phenomena*. But, second, the thought-movement from the concept of force as weight of a body to the discovery of its proportionality to the mass of this body is *not* the discovery of a new phenomenon as an *effect* of this force. In the conceptual hierarchy of the *Principia*, mass (quantity of matter) of a body on which the force of gravity acts is not an effect of this force. We only discover, via that thought-movement from the phenomenon of motion to the forces causing them by acting on that body, what that force (also) depends on from the "side" of that very body. Mass, as stated by Cohen, is in the *Principia* a *primary quantity* (1999, 92), but we are able to discover it as a magnitude in the framework of the *Principia*, only via the derived concept of force, so that the concept of mass is in that framework also a derived concept.

Proposition 7 makes the universal claim that "gravity exists in all bodies universally and is proportional to the quantity of matter in each" (1999, 810). While proposition 6 dealt with the force of gravity acting on a body, here the force of gravity is considered as having its source in a body and is related to the latter's mass. Newton proves this proposition by drawing upon the claim from proposition 69 of Book I, reconstructed above in 3.1.2.C, which states that "*the absolute forces of attracting bodies A and B will be to each other in the same ratio as the bodies A and B themselves to which these forces belong*" (1999, 587). Because the force labeled in Book I by Newton as the "force of attraction" is labeled in Book III as the "force of gravity," he can state proposition 7.

For proposition 7 it holds what was stated already for proposition 6. Newton, after he accomplished the thought-movement from phenomena-effects of motion in definitions 7 and 8 can, via the third law – which expresses the mutual ratio of the forces of two bodies acting on each other via the ratio of the effect each of them has on the other – derives the proposition stating the relation between the (absolute) force belonging to a body and the mass of this body.

Propositions 8 through 42 of Book III contain Newton's reflections on various phenomena not previously discussed or used in the derivation of the forces.

These phenomena are characteristics of planets (their masses, sizes, shapes, etc.), the motion of the Moon, the motion of comets, and the phenomenon of tides of the sea.

Proposition 8 states that (1999, 811)

If two globes gravitate toward each other, and their matter is homogeneous on all sides in regions that are equally distant from their centers, then the weight of either globe toward the other will be inversely as the square of the distance between the centers.

On its basis Newton then states in its corollaries the recipe for the determination of the weights and densities of planets. I will analyze Newton's method given in this proposition in section 3.3 when dealing with the structure of nomic measurement.

Proposition 9 considers the gravity inside the planets. While in proposition 73 (Book I) force inside a sphere was considered to be proportional to its diameter, Newton brings here in the impact of the factor of density of the planets. Only if "the matter of the planets were of uniform density, this proposition would hold true exactly ... Therefore the error is as great as can arise from the nonuniformity of the density" (1999, 815). This means that from now on attraction (gravity) toward a body as well as inside a body is, not only a function of the distance from this body (from its center), as initially claimed in Book I, but also a function of the mass of this body and, thus, of its density.

Propositions 10, 11, and 12 deal with the motion of planets about the common center of gravity of our solar system.[87] Proposition 10 draws from the fact that "the heavens ... are void of air and exhalations," (i.e., that there is vacuum) the conclusion that "the planets and comets, encountering no sensible resistance, will move through those spaces for a long time" (1999, 816).

Proposition 12 states that "*the sun is engaged in continual motion but never recedes far from the common center of gravity of all the planets*" (1999, 816). The proof of this proposition is based on taking into account the factor of ratio of the mass (and thus of the density) of the Sun to that of Jupiter, of the Sun to that of Saturn, and so forth. This means that Newton's derivation of *new* phenomena in Book III is based on two closely related steps. First, Newton starts considering a new feature which by itself is not an effect of the very force of gravity (e.g., the resistance of a medium, as in proposition 10), or of the mass of a body and in the second step, he tries to find out how this factor together with gravity enables one to derive a new phenomenon.

---

[87] Proposition 11 about the state of rest of the common center of gravity in our solar system is based on the hypothesis that the center of the system of the world is at rest (Newton 1999, 816).

Propositions 14 through 18, together with proposition 21, deal with planetary orbits (their aphelia, nodes, their diameters, eccentricities, etc.) or the shapes of planets. In propositions 14, 15, and 18 Newton again unifies the results obtained in Book I with the knowledge that the gravity of a body is proportional to its own mass and, then, derives the respective new characteristics of planetary motion or of the shape of the planets.

Proposition 19 sets as its aim to *"find the proportion of a planet's axis to the diameters perpendicular to that axis,"* (1999, 821) while proposition 20 has as its aim *"to find and compare with one another the weight of bodies in different regions of our earth"* (1999, 826). The former proposition is derived[88] by a unification of data available at the time of Newton (measures of the Earth and the distance a body falls in the first second) with the already derived theory of forces. He computes the accelerative forces causing the free fall of a body at the latitude of Paris and the centrifugal force at the equator. He then computes the ratio of the total force of gravity at the latitude of Paris to the centrifugal force at the equator and arrives at a ratio of 289 : 1. Newton then constructs a thought-model of the Earth so that there are two canals, each of them equal in length to the radius of the Earth, so that one is going vertically from the north pole to the center, while the other runs along Earth's horizontal axis. Because of the computed ratio of 289 : 1, the weight of the water in the horizontal channel will be greater than in the vertical one. Newton then computes the gravity at the pole and at the equator, and by taking into account the centrifugal force of 1/289, finds out, finally, that the ratio of the diameter of the Earth at the equator will be to the diameter through the poles as 230 : 229. Newton does not content himself with this ratio and brings in J. Picard's value of the mean semidiameter of the Earth and then concludes that "the earth will be ... $17^{1}/_{10}$ miles higher at the equator than at the poles" (1999, 824). Then in a similar manner he determines the measures of Jupiter.

In proposition 20,[89] Newton relates data on weight at various points on the surface of the Earth to the latitude of these points, where weight is understood as "the combination of force of the earth's gravity and the centrifugal effect of the earth's rotation" (Cohen 1999, 350). He uses the thought-model of the Earth from proposition 19, as well as the ratio 230 : 229 for the diameters of the Earth and, then, states that the "increase of weight in going from the equator to the poles is very nearly as the versed sine of twice the latitude or (what is the same) as the square of the sine of the latitude" (1999, 827).[90]

---

[88] A detailed reconstruction of the derivation of proposition 19 is given in (Cohen 1999, 233–235, 347–350).

[89] A reconstruction of Newton's derivation is given in (Cohen 1999, 350–355).

[90] On Newton's understanding of the terms sine and versed sine and their relation to angular functions used today see (Cohen 1999, 305–306).

Starting with proposition 22 through proposition 35 and its scholium (with the exception of propositions 24 and 25) Newton outlays his theory of Moon's motion.[91] It is based on his theory of gravity, about which he claims in proposition 22 that *"all the motions of the moon and all inequalities in its motions follow from the principles that have been set forth"* (1999, 832), by which he refers to his theory of gravity, because he states in the scholium to proposition 35, "I wished to show ... that the lunar motions can be computed from their causes by the theory of gravity" (1999, 869).

Once Newton was able to determine in proposition 25, with which I dealt in section 3.1.3.A, the force of the Sun perturbing the motion of the Moon, he derives in proposition 26 one of its effects not discussed previously, namely, the area the Moon describes by a radius drawn to the Earth once it is under the impact of the force from the Sun. On the bases of this derivation he, then, derives in proposition 27 the distance of the Moon from the Earth. and in proposition 28 the diameters of a noneccentric orbit of the Moon. Proposition 29 concludes the computation of the variation anomaly of the Moon (i.e., "a fluctuation in the rate at which a radius to the moon sweeps out areas" [Cohen 1999, 251]). Propositions 30 through 33, then, provide derivations of the motion of the nodes, while propositions 34 and 35 deal with the varying inclination of Moon's orbit to the plane of the ecliptic.

Proposition 24 represents a derivation of yet another effect of the force of gravity not discussed before, namely, the movement of the sea under the combined impact of the force of gravity from the Sun and the Moon.[92] From the force of gravity, having its source in the Earth, Newton derives its effect on the shape of the Moon in proposition 38 and in proposition 39 he derives the precession of Earth's equinoxes taking into account not only the gravity of the Earth but also of the Sun, while also taking into account the knowledge (derived already in proposition 19 of Book III) about the shape of the Earth.

Propositions 40 through 42 deal with the motion of comets. According to Newton "comets are a kind of planet and revolve in their orbits with a continual motion" (1999, 895), so that the whole theory of the inverse-square centripetal (gravitational) force can be applied to them. In proposition 40, drawing upon corollary 1 of proposition 13 (Book I) and upon propositions 8, 12, and 13 (Book III) he claims the validity of Kepler's first and second law for the comets and, in its first corollary states the validity of Kepler's third (or harmonic) law for them. In proposition 41 he sets as his aim to determine the trajectory of a comet, given three observations of its position. Proposition 42, finally, as a conclusion of Book

---

91 A detailed reconstruction of Newton's theory of the Moon is given in (Kollerstrom 2000), (Smith 1999a), (Smith 1999b), (Waff 1976), (Whiteside 1975), (Wilson 1989), and (Wilson 2001).
92 An analysis of Newton's theory of tides is given, e.g., in (Aiton 1955).

III, provides a sequence of operations to correct the trajectory that has been found by the method given in proposition 41.

The results of my analysis of the method by means of which Book I and Book III of the *Principia* are built from the point of view of the cyclical movement between the phenomena of motion and their corresponding forces can be expressed as shown in Table 3.1.

Type of thought-movement in the *Principia*	From the phenomena of motion to the forces of nature	From the forces of nature to the same phenomena of motion	From the forces of nature to new phenomena of motion
Book I	prop. 2 through 13, propos. 43, prop. 44, prop. 53	prop. 1, corollaries 3 through 6 of prop. 4, scholium to prop. 8, corollary 1 to prop. 10, corollary 1 to prop. 13, prop. 14, prop. 15, prop. 17, corollary 1 to prop. 53	scholium to prop. 10, prop. 16, prop. 32 through 42, prop. 54 through 69
Book III	prop. 1 through 5, prop. 25, prop. 36	13	prop. 6 through 24, prop. 26 through 35, prop. 37 through 42

Table 3.1 The structure of Books I and III of the **Principia** from the point of view of their method of construction

3.1.4 Metareflections

After such a lengthy and detailed analysis of the internal construction of the *Principia* I will now try to make some generalizations about this construction. I will, first, reconstruct three types of measure used by Newton in his thought-movement from the phenomena-effects to their cause. Second, I will reconstruct the specific characteristics of this movement. Third, I will show that Newton's measures of force are of a specific type, namely, that of an *external measure* and that, because of this, Newton's thought-movements display a certain type of incompleteness.

A. The Three Dynamic Measures of Force

If one looks at 6 and 7 in Book I, he or she can distinguish three ways of determining the force by means of its effects. One, analyzed earlier, arrives at the result that $F \propto PX/\Delta t^2$. This relationship is based on the idea that *uniform rectilinear accelerated motion* can provide us the quantitative data enabling us to find the quantity of the force causing it. According to I. B. Cohen (1999, 321)

> this is Newton's dynamical measure of a force ... It is a dynamical measure of force because it measures the force by its dynamical effect, the rate at which the action of the forces causes the moving object to deviate from a linear path,

while J. Brackenridge labels this measure as the *linear dynamic ratio* (Brackenridge 1995, 7, 171).

In proposition 6, Newton draws not only on uniform linear motion but also on uniform circular motion. Figure 3.5 shows this idea. In the figure, Q is any point on the curve; R is a point on the tangent so that QR is a parallel to SP and QT is a perpendicular to SP,

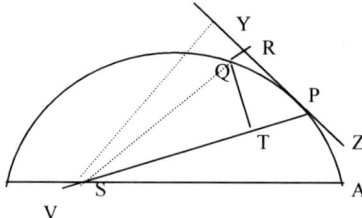

Figure 3.5 Newton's dynamical measure of force for uniform circular motion

where SY is a perpendicular to the tangent YZ and passes through the center S of force (1999, 454). In corollary 2, he states that $F \propto QR/(SY^2 \times QP^2)$. The proof is as follows. If Q approaches P, QP approaches RP and then, because the triangles TQP and YSP are similar, $SY : SP = QT : QP$. But if $SP \times QT = SY \times QP$, the result from proposition 6 that $F \propto QR/(SP^2 \times QT^2)$ is equivalent to $F \propto QR/(SY^2 \times QP^2)$.

As a next step one should notice the line PV in figure 3.5. It is the chord of a circle that approximates the curve APQ in point P. Combining of Figure 3.5 with Figure 3.2 we obtain figure 3.6 (X is here the intersection of QQ' and VP).[93]

---

93 I draw here partially upon the figure in (Brackenridge 1995, 195).

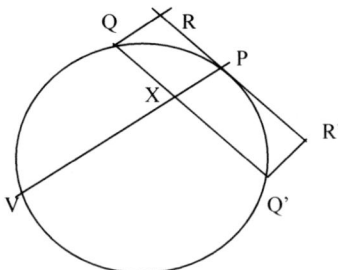

Figure 3.6 Determination of the circular dynamic ratio for the centripetal force

It holds[94] that $QX \times XQ' = VX \times XP$. But because $XP = QR$ we obtain $QX \times XQ' = VX \times QR$. If Q approaches P, then VX approaches PV while QX and XQ' approach QP, and we obtain $QP^2 = PV \times QR$. Because, by corollary 2 of proposition 6, $F \propto QR/(SY^2 \times QP^2)$, we obtain, finally, $F \propto 1/(SY^2 \times PV)$. Brackenridge labels $1/(SY^2 \times PV)$ as the "circular dynamic ratio" (1995, 37), which serves Newton as yet another *measure of the centripetal force.*

Finally, in corollary 2 and 3 of proposition 7, Newton employs, as an alternative solution, a third measure of force, namely, the so-called comparison theorem (Brackenridge 1995, 173).[95] It grows out of the linear dynamics ratio, but here Newton considers the ratio of a force by which a body P revolves in an orbit around the center S to a force by which this body revolves in the same orbit around any other center R.[96]

B. The Derivation of Forces from the Phenomena of Motion

The above given three measures enable Newton to derive forces from their effects. Let me now deal with these effects, with the place of the measures of force in these derivations, and, finally, with the cyclical "end" of these derivations.

An important characteristic of the effects-phenomena from which derivation starts is, with respect to the character of the thought-movements take their course from them, that they are idealized phenomena.[97] On the one hand, some of the idealizations are "short-lived" in Book I (idealization of the orbit, so that initially it has a circular character, then abolished in favor of an elliptical orbit; the initial supposition that the center of force is nonaccelerated, then abolished in favor of

---

94 It holds by proposition 35 of Book 3 of Euclid's *Elements*.
95 This third measure is applied in the alternative proof of proposition 11 of Book I.
96 The whole computation is given in (Brackenridge 1995, 172–174).
97 For a list of these idealizations see (Harper 1993, 148–153).

an accelerated one; etc.). On the other hand, Newton holds to the following, more "long-lived" idealizations which span several propositions in the *Principia*:

i. the center of force, orbited by a body, is devoid of any mass, therefore, it is not attracted by the orbiting body (up to proposition 57, Book I);[98]
ii. there is only one body orbiting the center of force, or, if there are several of them, then they do not mutually interact;[99]
iii. the orbiting body is devoid of any dimensions, that is, it is a mass-point (up to proposition 19, Book III);
iv. the central body is devoid of any dimensions, that is, it is a mass-point (up to Section 12, Book III).

Based on these idealizations Newton employs his strategy of deriving from the knowledge of quantity given in the idealized phenomena-effects the knowledge about the quantity of the forces causing them. Four cases of applying this strategy are worth mentioning:

First, the derivation (proposition 2 of Book I) of the centripetal character of a force acting on a body from the fact that the latter's orbital motion satisfies Kepler's area law, that is, in the same time it will describe the same areas, or, that its area rate is constant. In corollary 1 of proposition 2, Newton states that if the area rate decreases, the force is directed off the center in the direction of motion, and if the area rate increases, the force is directed off the center against the direction of motion. Proposition 2 together with corollary 1 thus give us "systematic dependencies, which make a constant area rate measure the centripetal direction of force maintaining a satellite in its orbit" (Harper 1999, 77).

Second, the derivation (proposition 4 of Book I) of the inverse-square character of the centripetal force from Kepler's harmonic law, that is, from the fact that the periodic time of a group of bodies orbiting the same center are as the 3/2 power of their distances from that center. In corollary 7 of proposition 4 Newton broadens the relation between the orbital characteristics of the bodies and the centripetal force to which they are subjected, so that if the periodic times are as the n-th power of the distances, the centripetal force will be inversely as the $2(n-1)$ th power of the distances. Here again it is readily seen that the area law measures the power-of-the-distance-dependence of the centripetal force.

Third, the derivation (proposition 45 of Book I) of the power-of-the-distance-dependence of the centripetal force from the precession characteristics of the or-

---

[98] From proposition 57 to proposition 63, Newton considers the case when the center of force has mass and thus both the central body and the other body are orbiting a common center of force.

[99] For this see, e.g., propositions 14 and 15, Book I.

bit under the impact of this force. Newton proves here that "zero orbital precession measures inverse square law for distances explored by orbit" (Harper 1999, 87), and generally that "if the centripetal force is as any power of the radius, that can be found from the motion of the apsides" (1999, 87).

Fourth, a specific form of inference of the quantitative characteristic of the centripetal force, namely, not from one but simultaneously from two phenomena-effects. Such a type of derivation was accomplished by Newton in proposition 2 of Book III, where he derived the inverse-square character of the centripetal force for the primary planets from Kepler's harmonic law holding for these planets and simultaneously from the zero-orbital precession of their orbits. In a similar manner Newton accomplished also his famous first Moon-test in Book III. Here he used two phenomena: the fall of a body on the Earth and the acceleration of the Moon toward the Earth. By the computation for the Moon, Newton arrives at "data" which "measure a force producing accelerations at the surface of the earth. These accelerations are equal and equally directed toward the center of the earth" (Harper 1993, 161). For the fourth case of derivation thus holds in general that (Harper 1993, 159)

there is a *special advantage to inferences to a proposition from alternative phenomena*. Each such inference is a measurement of the value of the relevant magnitude specified in the proposition. An inference to this same proposition from another phenomenon *is an independently agreeing measurement of this same magnitude*.

From the above reconstructed four cases of thought-derivation of forces from idealized phenomena it follows that "the phenomenal parameters *measure* corresponding values of the theoretical parameters that are inferred" (Harper 1999, 74), and that "values of the phenomenal magnitude carry the information that corresponding values of the theoretical magnitude obtain" (Harper 1993, 147). As I will show below, by bringing in categories from Hegel's *Science of Logic*, that type of measuring information can be labeled as *the phenomenal* or *external measure of the respective cause*. That Newton consciously uses this method of measuring the respective cause is readily seen from his claim: "The representatives of times, spaces, motions, speeds and forces are any quantities whatsoever proportional to things represented" (Newton 1965c, 312).

C. The Cyclical Method of Theory Construction and the Change of Meaning of the Harmonic Law in the Principia

Newton, as shown in 3.1.4.B, derives the centripetal nature of the force acting on the orbiting body as well as the dependence of the size/quantity of this force on the distance from the center of force, by drawing on certain idealized phenomena. This has, as I will show now, a very interesting "feedback" consequence on what we today label as "Kepler's third (or harmonic) law" of planetary motion, which is one of the starting points of Newton's movement *to* the characterization of centripetal force (in Book I) and of the force of gravity (in Book III).

Let me start with Book I. Here Newton in proposition 4, corollary 6, as shown above in 3.1.2.A, derives from the claim that "the periodic times are as the 3/2 powers of radii" (1999, 451) the claim that the centripetal force will be inversely as the square of the radii, while the term "radius" is here understood as the *line from the orbiting body, viewed as mass-point, to the center of force, viewed as a point without a mass and any spatial dimensions.* So I can give the following concise representation of the harmonic law (T_1, T_2 stand for the periodic times of two bodies orbiting the same center of force; r_1 and r_2 are their respective distance from this center):

$$T_1^2/T_2^2 = r_1^3/r_2^3. \qquad /1/$$

If we consider the case of just one orbiting body we can state the harmonic law as follows ("$\propto$" stands for "is proportional to"):

$$T^2 \propto r^3. \qquad /2/$$

In a next step Newton at the very beginning of Section 11, Book I, states the following (1999, 561):

Up to this point, I have been setting forth the motions of bodies attracted toward an immovable center, such as, however, hardly exists in the natural world. For attractions are always directed toward bodies, and – by the third law – the actions of attracting and attracted bodies are always mutual and equal; so that if there are two bodies, neither the attracting nor the attracted body can be at rest, but both (by corol. 4 of the laws) revolve about a common center of gravity as if by a mutual attraction. ... For this reason I now go on to set forth the motion of bodies that attract one another, considering centripetal forces as attractions,

and where, according to the third law, by mutual actions of these bodies "equal changes occur ... [in their] motions" (1999, 417) and where motion is understood as arising "*from the velocity and quantity of matter jointly*" (1999, 404).

By bringing in the masses (quantities of matter) of the orbiting body and of the other body which I label here, tentatively, as "Sun," he then proceeds in proposition 60 (Book I) to the following statement of the harmonic law (1999, 564):

If two bodies S and P, attracting each other with forces inversely proportional to the square of the distance, revolve about a common center of gravity, I say that the principal axis of the ellipse which one of the bodies P describes by this motion about the other body S will be to the principal axis of the ellipse which the same body P would be able to describe in the same periodic time about the other body S at rest as the sum of the masses of the two bodies S + P is to the first of two mean proportionals between this sum and the mass of the other body S.

So, we have here a change in the initial statement of the harmonic law so that it now holds that[100]

$$T_1^2(S + P_1)/T_2^2(S + P_2) = r_1^3/r_2^3, \qquad /3/$$

where P_1, P_2 are the masses of the orbiting bodies and S is the mass of their common "Sun." In the case of just one body with mass P orbiting its "Sun" the following harmonic law can be derived:[101]

$$T^2(1 + P/S) \propto r^3. \qquad /4/$$

What has to be emphasized here is, first, that the radii are here still understood as the lines connecting the "dimensionless" bodies with their dimensionless "Sun." Second, Newton does not draw here on the harmonic law as one of the presuppositions of his thought-movement to the force but already on proposition 15 (Book I), where the harmonic law is already the "end"-point of the derivation starting *from* the inverse-square-distance dependence of the centripetal force.

Let me now move to Book III. Here the harmonic law appears for the first time already in its very beginning in the statements for the Phenomenon 1 (for the satellites of Jupiter), Phenomenon 2 (for the satellites of Saturn), and Phenomenon 4 (for the primary planets) (1999, 797–800):

The circumjovial planets ... their periodic times ... are as the 3/2 powers of their distances from that center [the center of Jupiter]. *... The circumsaturnian planets ... their periodic times ... are as the 3/2 powers of their distances from that center* [the center of Saturn]. *... The periodic times of the five primary planets ... are as the 3/2 powers of their mean distances from the sun.*

Unlike the understanding of the term "distance" in the harmonic law in propositions 6 and 60 in Book I, in Phenomena 1 and 2 "distance" is understood already as the line connecting the satellites with the center of the respective planet (i.e., the latter are now viewed as having already spatial dimensions), and in Phenom-

---

100   Here I draw on (Wilson 1989, 260).
101   Here I draw on (Cohen 1980, 224).

enon 4 as the "mean distance" from the Sun, that is, as the semimajor axis of the elliptical path of the planet. A concise representation of the harmonic law in Phenomena 1, 2, and 4, thus, is as follows (R_1, R_2, and R stand here for the semimajor axes):

$$T_1^2/T_2^2 = R_1^3/R_2^3 \qquad /5/$$

$$T^2 \propto R^3. \qquad /6/$$

Newton then modifies the harmonic law in proposition 15, where his aim is "*to find the principal diameters of the* [planetary] *orbits*" (1999, 819), and then states (1999, 819–820):

These diameters are to be taken as the 2/3 powers of the periodic times by book 1, prop. 15; and then each one is to be increased in the ratio of the sum of the masses of the sun and each revolving planet to the first of two mean proportionals between that sum and the sun, by book 1, prop. 60.

This means that similar to the status of the harmonic law in proposition 60, Book I, where it was already the "end"-point of the derivation starting *from* the centripetal force, here the harmonic law is viewed already as the "end"-point of the derivation proceeding *from* the force of gravity. The following expressions can then be derived:

$$T_1^2(S + P_1)/T_2^2(S + P_2) = R_1^3/R_2^3, \qquad /7/$$

$$T^2(1 + P/S) \propto R^3. \qquad /8/$$

The transformation of the harmonic law in Book I (from proposition 6 to proposition 60) and Book III (from Phenomena 1, 2, and 4 to proposition 15) of the *Principia* can thus be represented as follows (the content of the frames stands for the context via which the harmonic law has to be transformed):

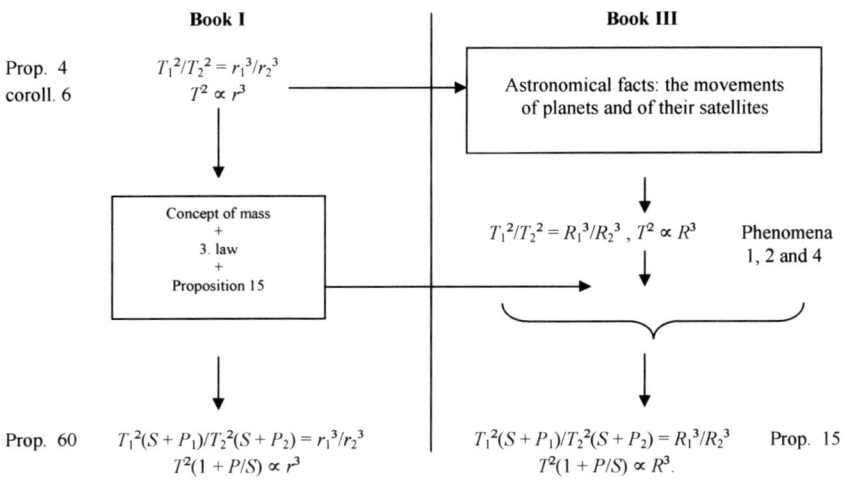

*Figure 3.7 Transformations of the harmonic law in Book I and Book III of the **Principia***

From this it is readily seen that once Newton starting from idealized phenomena derived in Book I the characteristics of the centripetal force and, in Book III, the characteristics of force of gravity, he had to take into account these forces as a perturbing factor, for example, of the orbiting body (bodies) acting on the central body, and thus had to *return* to those phenomena and correct and modify them. This feature of the *Principia*, together with my previous reconstruction of its internal construction, shows that it is built by a *cyclical method of theory construction* and thus contradicts the "Standard Conception." In fact, *Principia* contradict that conception also in yet another important aspect. While according to the "Standard Conception," one should end up at the level of statements referring to the observable state of affairs, in the cases of Book I and Book III of *Principia*, it is readily seen that the statements pertaining to the harmonic law (proposition 60 from Book I and proposition 15 from Book III) contain in an irreducible manner the term "mass," thus a term referring, from the point of the *Principia*, to something *unobservable*.

What has to be emphasized is that such a cyclical return to the phenomena from which the theory-construction initially started and their successive correction does invalidate neither Newton's thought-movement from phenomena of motion to the forces causing them, nor this method of theory construction as such. "So long as the corrections are perturbations attributable to other forces – whether other components of gravitational force or even foreign forces – the inferences to the original phenomena can be construed as components of the perturbed motion" (Harper and Smith 1995, 144). With respect to Kepler's laws, which initially served as a basis of derivation of the inverse-square character of the centripetal force, this means that (Harper 1993, 156)

the formula ... for ... a perturbed orbit is properly conceived as a formula for a composition of motions one of which is the Keplerian orbit that fits the [initial] law and the other is the perturbation produced by the interaction. According to such a conception the Keplerian phenomenon [from which the inference initially started] is there to be found ... It is, however, transformed from a claim about the total motion to a claim about that component of the total motion caused by the inverse square centripetal force.

Figure 3.7 above leads now to the following question. Does the term "harmonic law" transform its meaning in the course of theory-construction as given in Book I and Book III of the *Principia*? In order to exclude beforehand any possible misunderstanding with respect to this question I emphasize that I am investigating here only into the possible meaning-changes of this term as given in the framework of Book I and Book III of the *Principia* and, thus, not into the changes of the meaning of this term as it was initially stated by Kepler.[102]

In order to find out the meaning of the term "harmonic law" I apply Transparent Intensional logic (hereafter, TIL) to the above given reconstructions /2/, /4/, /6/ and /8/ of this law.

I reconstruct /2/ as follows:

$$\lambda w \lambda t [^0\forall \ [\lambda x \ [^0\propto \ [[^02^0T_{wt} x][^03 \ ^0r_{wt} x]] \ ]] \qquad /2'/$$

The harmonic law in the form of /4/ I reconstruct as follows:

$$\lambda w \lambda t [^0\forall \ [\lambda x \ [^0\propto \ [^0\times \ [^02^0T_{wt} x][^0+ \ ^01[^0: \ [^0P_{wt} x^0S_{wt} x]]]][^03^0r_{wt} x]]] \ ]] \qquad /4'/$$

In analogy with /2'/ I reconstruct /6/ as follows:

$$\lambda w \lambda t [^0\forall \ [\lambda x \ [^0\propto \ [[^02 \ ^0T_{wt} x][^03 \ ^0R_{wt} x]]]] \qquad /6'/$$

In analogy with /4'/ I reconstruct /8/ as follows:

$$\lambda w \lambda t [^0\forall \ [\lambda x \ [^0\propto \ [^0\times \ [^02 \ ^0T_{wt} x][^0+ \ ^01[^0: \ [^0P_{wt} x^0S_{wt} x]]]][^03 \ ^0R_{wt} x]]] \ ]] \qquad /8'/$$

---

102   On Kepler's approach to his three rules or "laws" see, e.g., (Russell 1964), (Aiton 1969), (Aiton 1969), (Aiton 1973), (Aiton 1975a), (Aiton 1975b), (Wilson 1970, 92–106), (Wilson 1972a), (Wilson 1972b), (Wilson 1974) and (Whiteside 1974).

From this reconstruction I draw the conclusion that Book I and Book III of the *Principia* can be viewed as a cyclically organized sequence of gradually shifting constructions. And so, as TIL identifies constructions represented by language expressions as the latter's meanings, I arrive at the conclusion that the term "harmonic law" inside the *Principia* gradually changes its meaning. *Thus, cyclically built scientific theories do not fulfill one of the central requirements of the "Standard Conception" of scientific theories, namely, that inside scientific theories their terms do not change their meanings.*

My reconstruction of the cyclical nature of Book I and Book III of the *Principia* together with the TIL-reconstruction of shifts of meaning of the term "harmonic law" provides already here a first answer to the question stated by Fay and Moon and quoted in the Introduction of this book. It shows that *even a natural science theory, once it is built in a cyclical manner, can fulfill a critical function.* By transforming appearances into manifestations via their *cause/essence/ground*, which involves a shift of meanings of terms initially used in the description of the appearances, one is able to explain, even if only partially at the level of *formal* ground, the origin of those appearances, that is, to provide answers to the questions about the *how and why* of their origin.

D. External Measure and Newton's Negatively Closed Cycle of Cognition

According to Harper, and here I am in complete agreement with him, the original phenomena of motion, which initially served Newton in the derivation of the forces, *reappear* again at the "end" of the cycle of theory construction, where they become just one component of the total motion of the orbiting bodies. While the *other* components of this total motion are understood via derivation as the *effects of forces*, the question I will deal with now and here is *if, and if yes, then how the initial (original) phenomena are explained by forces*. Answering this question is of crucial importance for the *positively cyclical* character of the *Principia*, that is to say, that it is not viewed as flawed because of this way of its construction. Flawed, first, because the whole cycle *starts* from them, so that once they are not explained by derivation from their causes at the end of the cycle, the whole cycle *is based* on pre-*Principia*[103] (Keplerian and Galilean) forms of knowledge that it cannot get rid of. Flawed, second, because the cycle *ends* with such a type of knowledge, where the original phenomena are included as a component of the total motion of the orbiting body, so, once the former are left unexplained by their causes, the total movement of the bodies is as well left *partially unaccounted* for by their causes. Thus, again, pre-*Principia* forms of knowledge would be given at the *end* of these cycles of theory construction in a noneliminable manner.

---

103 An even better term would be "pre-De-Motian."

While Newton speaks in his 1687 introduction to the *Principia* about the cycle of cognition of the type "phenomena of motion → ... → *other* phenomena of motion" (Newton 1999, 382), my problem pertains here to the question of how the cycle of cognition "phenomena of motion → ... → *same* phenomena of motion" is dealt with in the *Principia*.

In order to answer this question as well as the question *of how the intial (original) phenomena are explained by forces*, one has to compare the phenomena that served as the basis from which the inverse-square character of the centripetal forces was derived with those phenomena that were derived by Varignon, Hermann and Bernoulli[104] by methods partially outlined by Newton in propositions 39, 40 (section 7), and 41 (section 8) of Book I.[105]

For the sake of brevity I reproduce here, drawing upon E. J. Aiton's (1964), only the approach of P. Varignon to propositions 39 and 40 as well as of J. Hermann to proposition 41.

Varignon defined velocity v via the relation $v = \frac{dx}{dt}$ and force f via the relation $f = \frac{dx}{dt}$, the second being the application of definition 7 from the *Principia*. From the latter relation it is possible to derive $fdx=vdv$ or $\int fdx = \int vdv$ and thus obtain $\int fdx = \frac{1}{2}v^2$, which corresponds to proposition 39 of Book I. From this result, because $v^2 = (\frac{dx}{dt})^2$, we obtain $(2\int fdx)^{1/2} = \frac{dx}{dt}$, and thus the time for any position of the moving body is $t = \frac{\int dx}{\sqrt{2\int fdx}}$, which also corresponds to proposition 39.

The result of proposition 40 was derived by Varignon drawing upon the diagram (in modern notation) of Figure 3.8 (Aiton 1964, 82).

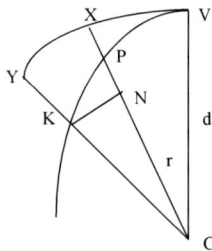

Figure 3.8 Result of proposition 40 of Book I as derived by P. Varignon

Here a body is falling from the point V under the impact of force f with its center at C, so that f_1 is the component of this force in the PK-direction while $PK = ds$ and $PN = dr$. It holds $f : f_1 = \frac{ds}{-dr}$, thus $f = -f_1 \frac{ds}{dr}$. From the previous proof we know

---

104 On their contributions see Aiton's papers (1964; 1988; 1989).
105 On Newton's method applied in proposition 41 see (Brackenridge 2003).

already that $f_1 = \frac{dv}{dt}$, thus $f = -(\frac{dv}{dt})/(\frac{ds}{dr})$, so $f = -\frac{dv}{dr}\frac{ds}{dt}$, or, because $v = \frac{ds}{dt}$, $f = -v\frac{dv}{dr}$. By multiplying the last equation by dr we end up with $\int f dr = -\int v dv$.

Proposition 41 was derived by Hermann for the case of a single centripetal force varying inversely as the square of the distance (see Figure 3.9) (Aiton 1964, 94).

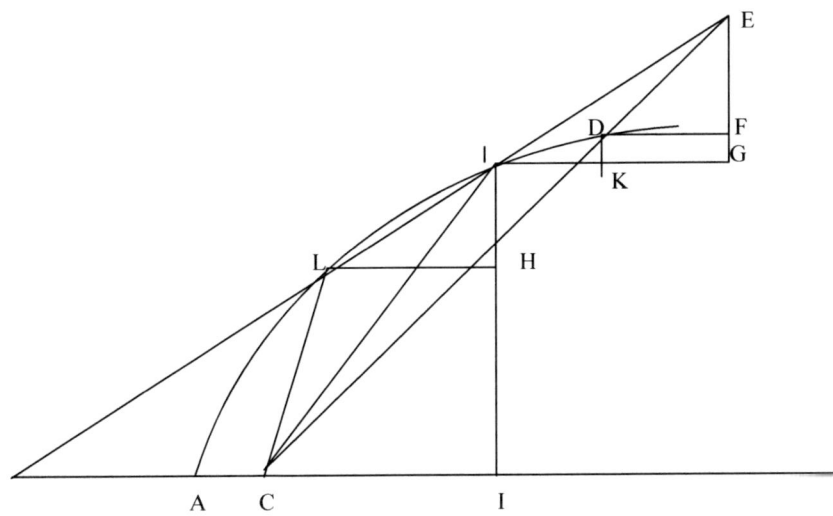

Figure 3.9 J. Hermann's derivation of proposition 41 of Book I

Here C is the center of force f, so that $f \propto ED$, $Ll = lE = ds$, $CI=x$, $Il = y$, $LH = IG = dy$, ED is parallel to IC, and it holds as well that $KG = DF = -ddx$ and $EF = -ddy$. It can be proved that for the area of the triangle KCl it holds that $\frac{1}{2}(ydx - xdy)$. From this one can derive, via a repeated differentiation, that $ED = ddx(x^2 + y^2)/x$, thus $f \propto ddx(x^2 + y^2)/x$. Taking now into consideration an inverse-square centripetal force, that is, $f \propto 1/(x^2 + y^2)$, one can put these two relations together and thus obtain a differential equation. Its solution yields the result $a \pm cx/b = (x^2 + y^2)^{1/2}$, where a, b, and c are constants. This result stands for the equation of a conic section, so that if $b = c$, the conic is a parabola; if $b > c$, it is an ellipse; and a hyperbola if $b < c$.

From the result obtained by Hermann, when combined with proposition 1 of Book I and proposition 15 of Book I, it is readily seen that Keplerian phenomena (Kepler's first, second, and third laws) are *not only the presupposition of the thought-movement to the centripetal forces, but also the result* of the thought-movement from the forces to their effects, *where the concept of centripetal force* and also the process as well as the results of derivation of effects from forces, belong – with respect to that presuppositions – to contraintuitive forms of knowledge. I label this aspect of the *Principia* as *the positively closed cycle of cognition* in the *Principia*.

But if we turn to the results of Varignon's derivations $\int f dr = -\int v dv$ and $\int f dx = v^2$, we discover that Newton's *Principia* display also a feature that can be labeled *as the negatively closed cycle of cognition*. The "problematic" part in those two formulas is force f. What can one substitute for it, given the attained level of knowledge in the *Principia*? Because Newton deals up to section 11 of Book I only with the accelerative measure of force, and taking into account the fact that mass in the *Principia*, in the long run, is derived from considerations of the force of gravity, then f can be expressed only by its proportion to acceleration, that, as $f \propto a$. But this means that while the Keplerian phenomena are both the initially unexplained/nonderived *and* the contraintuitive, explained/derived results of the thought-movements in the *Principia*, the phenomenon of acceleration is *only* the unexplained/nonderived presupposition and *never* the contraintuitive, explained/derived result of thought-movements in the *Principia*.

To understand the phenomenon of acceleration as the presupposition of Newton's thought-movement to the centripetal forces, let me deal again with proposition 4 of the *Principia* (theorem 2 of *De Motu*). Here the thought-movement is based on the following chain of thoughts. Force f can, according to definition 7, be measured by its acceleration of the body on which it acts. By this action, the body is deflected from its rectilinear inertial motion.[106] This acceleration of the body can thus be represented, following Galileo, by the square of the time in which the body falls, supposing that the acceleration of the body is constant, that is to say, in the case of a planetary centripetal force acting on very small, nascent "distances," or as Newton puts it, "at the very commencement of motion."

Because the phenomenon-effect of acceleration has in the *Principia* only the status of an unexplained/nonderived presupposed phenomenon, it is possible to view Newton's cyclical movement from definition 8 to the second law, reconstructed above in section 3.1.1, also as *negatively* closed. This is readily seen in the mutual relation between that definition and that law. Because *nowhere* between this definition and this law does Newton give a measure of force other than the change of motion, what one can substitute for the phrase "motive force impressed" in this law is just the phrase "the motion that it generates in a given time" from this definition, and so the whole circular movement ends up in the claim, "A change of motion is proportional to the motion generated in a given time."

---

106 In the concept of rectilinear motion, J. Brackenridge sees the debt of Newton to Descartes (1995, 17–18).

The reason why *Principia* display, not only a positively circular character, but also a negatively circular character can be understood also through an understanding of the type of knowledge which *Principia* provides with respect to forces and to the force of gravity. Regarding the former, Newton speaks about "considering in this treatise not the species of forces and their physical qualities but their quantities and mathematical proportions, as I have explained in the definitions" (1999, 588), and says in his *System of the World*, that (1946, 550):

our purpose is only to trace out the quantity and properties of this forces from phenomena, and to apply what we discover in some simple cases as principles, by which, in a mathematical way, we may estimate the effects thereof in more involved cases ... We said, in a *mathematical way*, to avoid all questions about the nature or quality of this force.

In a similar manner he states, regarding the force of gravity that, "thus far I have explained the phenomena of the heavens and of our sea by the force of gravity, but I have not yet assigned a cause to gravity" (1999, 943).

This means that in the *Principia* we will not find any explanation of the forces and of the force of gravity by means of an understanding of their *quality*, and thus not the explanation *why it manifests* itself as acceleration, and, therefore, also not the explanation why for acceleration, as an effect of force, holds $a \propto 1/r^2$. Then, the only option Newton has at his disposal is to determine, as reconstructed earlier, the quantity of the force by means of the quantity of its effects, that is, by its *external measures*. The fact that the external measure clearly dominates in the derivation of the quantity of forces and of the quantity of the force of gravity is readily seen also in Newton's *linguistic practice* – in propositions 4 and 5 of Book III – to assign to the very planetary force the name of its effect on celestial and terrestrial objects. So as the primary planets gravitate (are attracted) to the Sun, and the Moon and the terrestrial objects gravitate (fall) to the Earth, he assigns to this force the name *gravity*.

This, in turn, leads in the *Principia*, at least in my view, to a distorted understanding of the concept of the force of gravity and, at the level of categories, to a distorted understanding of what properties of objects are their *essential* properties. In the commentary to Rule 3 (Book III) Newton states, "I am by no means affirming that gravity is essential to bodies. By inherent force I mean only the force of inertia. Gravity is diminished as bodies recede from earth" (1999, 796). What unifies, in the framework of the *Principia*, the force of gravity and the force of inertia is that both are, with respect to a body, proportional to its mass; according to Newton: "Inherent force of matter: this force is always proportional to the body ... inherent force may also be called by the very significant name of force of inertia" (1999, 405) and "*gravity exists in all bodies and is proportional to the quantity of matter in each*" (1999, 810). But there is, *according to Newton*, a principal difference between these two forces. While for a given body with a mass of a certain size/quantity its inertia is of a constant size/quantity which does not change, the size/quantity of gravity even for a body of a certain size/quantity is inversely proportionally as the square of distance from the center of this body. So, the concepts of these forces can be compared as indicated in Table 3.2.

Type of force existing in a body as function a of	Inertia	Gravity
the mass of the body	Yes	Yes
the distance from the center of the body	No	Yes

Table 3.2 A comparison of the concepts of inertia and gravity as given in the **Principia**

But when one mutually compares the two characterizations of the force of gravity as given in the right column, it can be readily seen that Newton *confounds two different characterizations of the force gravity*. In one characterization, its proportionality to the mass of the central body pertains to its characterization as an absolute force described in Definition 8 as "*as absolute quantity of centripetal force*" (1999, 407) and with regard to which he states, in the commentary to this definition, that it has "some cause without which the motive forces are not propagated through the surrounding regions whether this cause is some centripetal force … or whether it is some other cause which is not apparent" (1999, 407).

The other characterization of the force of gravity, namely, that it changes, according to Newton, inversely as the square of the distance from the center of the body, has not its origin in the concept of absolute force but rather in the chain of reasoning starting from the concept of accelerative force. It runs as follows: $F_A \propto a$ (Definition 7), via the concept of centripetal force for which (by proposition 6, Book I) it holds that $F_C \propto 1/r^2$, up to the conclusion (in the scholium to proposition 5, Book III) that, in the case of the movement of the celestial objects, F_C is the force of gravity, that is, $F_C = F_G$, and thus $F_G \propto 1/r^2$. But so as the concept of gravity is based here on the concept of centripetal force that, in turn, is based on the concept of accelerative force, it holds in Book III of the *Principia* that what is inversely proportional as the square of the distance is *not gravity as an absolute force* but only one of gravity's *effects*, namely, *acceleration*. This confusion between, on the one hand, the characterization of the force of gravity as an *absolute* force, the size/quantity given of which is in a body proportional to the size/quantity of the mass of the body and, on the other hand, the characterization of the *effect* of the absolute force by means of its acceleration for which holds that $a \propto 1/r^2$ is the result of two facts. First, by the indicated conceptual path, Newton initially finds out that the primary planets are attracted by the Sun, while the Moon and terrestrial objects fall with acceleration to the Earth, and this effect – acceleration – is inversely proportional to the square of the distance from the center of the Sun and the center of the Earth. And then he *moves in thought from this effect to the force of gravity*. Second, and in distinction to this thought-movement, Newton finds out that gravity as an *absolute force* tied to a body depends on the size/quantity of its mass, independent of the effects that that body's absolute force of gravity generates in the surrounding bodies. But Newton is not able, even if he moves from the phenomenon-effect (i.e., acceleration) to its ground (i.e., the force of gravity), to mediate conceptually between these two characterizations of the force of gravity in the sense that he *cannot accomplish the thought-movement going in the opposite direction*: to derive the first of these two characterization of the force of gravity, namely, that the acceleration a body causes is as $1/r^2$, from the its second characterization, namely, that the absolute force of gravity of this body is as its mass, that is, that $F_G \propto m$.

In order to remove, at least partially, Newton's confusion between these two, at least in the framework of the *Principia*, nonmediated, approaches to the concept of the force of gravity I reorganize the previous table; Table 3.3 is the result.

Magnitude as a function of	(Absolute force) Inertia of a (central) body	(Absolute force) Gravity of a body	Accelerative effect of the absolute force of gravity of a central body on other bodies
mass of the central body	YES	YES	YES
distance from the center of the central body	NO	NO	YES

Table 3.3 *A revised comparison of the concepts of inertia and gravity*

The previously quoted denial that gravity is essential to bodies is framed by the general claim of the third rule of natural philosophy that "those qualities of bodies that cannot be intended and remitted ... should be taken as qualities of all bodies universally" (1999, 795). What Newton means here by the expressions "intended" and "remitted" is explicated by Cohen as "increased" and "diminished" while Andrew Motte's 1729 translation[107] of Newton's Latin terms "intendi" and "remitti" is "intensification of degrees" and "remission of degrees" (1946, 400). Thus we have here a *quantitative* approach to the differentiation between essential and nonessential properties. According to Newton, only those properties of objects that do not allow a *difference of degrees can be viewed as their essential properties.*

What is at the basis of this view of Newton is that he is not able to find out the quality of the absolute force of gravity tied to a body and on the basis of the knowledge of this quality, then, to explain why it manifests itself in interaction between this body and another body as the latter's acceleration and, thus, to explain why $a \propto 1/r^2$. But Newton still wants to differentiate between phenomena-effects of forces from the very forces, and because the former vary quantitatively (i.e., by degrees), he wrongly concludes that the specificity of the latter and their difference from the former lie in their quantitative invariability. Stated otherwise: Newton's understanding of the search for essential properties of objects as a search for their quantitatively invariable properties is simply the (wrong) substitute for the lack of knowing the qualitative characteristics of the cause underlying the phenomena-effects and, thus, also for the inability, in the framework of the *Principia*, to derive the phenomena-effects from their cause as its *manifestations*, even if these phenomena-effects as *appearances* served him as the point of departure from which to arrive at the concepts of centripetal force, accelerative force, motive force, and force of gravity.

A paradigmatic example of that inability is Newton's treatment of the relation of two bodies, here designated as A and B, *after* he introduced the concepts of accelerative, motive, and absolute force. By choosing the mass m_A of body A as a unit it is possible to express the mass of B as a multiple or fraction of the mass m_A. Here, A is in such a position that B's mass can be expressed as *equivalent*, or as a *degree of equivalence* to the mass of A. Body B is here in another position; what is the size/quantity of its mass can be expressed only *relative to* the size/quantity of the mass m_A of body A. This relation can, of course, be *reversed*. One can take the mass m_B of body B and set it as a unit. So not only does A provide its body to express something belonging to B, but also B's body can be used to express something belonging to A. This something is, according to Newton's *Principia*, the absolute force f. In the former case, force f_B tied to body B "uses" the body A to manifest itself as acceleration a_A. In the latter case, force f_A tied to body A

---

[107] J. McGuire states about Motte's translation that it "admirably catches Newton's meaning" (McGuire 1968, 244).

"uses" the body B to manifest itself as acceleration a_B. The basis, or *ground* (to use a Hegelian term, I will analyze later), of the fact that in their mutual relations bodies A and B stand to each other as the *relative form* to a *form of equivalence* or[108] *vice versa* is the existence of something different from both mass and acceleration – it is the absolute force that exists as a common ground in both bodies. Stated otherwise: what is common to both of them is force as *quality*, or $[f]$, for short, given in body A in *quantity* f_A, or $\{f_A\}$ for short, and in body B in quantity f_B, or $\{f_B\}$, for short.

Newton's third law is based on this presupposition of the existence of a common ground of mutually interacting bodies; it states: "*the actions of two bodies upon each other are always equal*" (1999, 417), and can thus be written as the *equation* $f_A = f_B$ or as $\{f_A\}[f] = \{f_B\}[f]$, that is, on both sides we have one and the same quality and, because it is an equation, also one and the same size/quantity. That Newton's third law is really based on what he labels in definition VI as "absolute force" can be readily seen from his commentary to the third law, which is as follows: "If some body impinging upon another body changes the motion of that body in any way by its own force, then, by the force of the other body ... it will in turn undergo the same change in its own motion" (1999, 417). Once Newton in the third law and its commentaries starts from the equation $f_A = f_B$, he can, by bringing in the concepts of accelerative and motive force, claim that $dp_A/dt = dp_B/dt$ holds, that is, that $m_A a_A = m_B a_B$ holds and then, finally, state that "changes in velocities ... are inversely proportional to the bodies because the motions are changed equally" (1999, 417).

But even if Newton's thought-movement from the concept of absolute force to the third law accomplished by means of the categories of equivalent form and a form of equivalence is already a type of movement *from the cause to its effects as manifestations*, still it displays the same *fundamental incompleteness* we noticed in his movement from Definition 8 to the second law. For the same reason, namely, the absence of knowledge what is $[f]$, that is, the quality of force, prevents him to derive the equation $\{f_A\}[f] = \{f_B\}[f]$ as it is stated in the third law.

The fact that acceleration, and thus accelerated fall, has in the *Principia* only the status of *initially presupposed and then left unexplained external measure*, enables one then to understand J. Herivel's observation that nowhere in "the *Principia* is any justification given for the proportionality between the deviation and the centripetal force" (Herivel 1965, 290).

By reconstructing the negative feature of the cyclical construction of the *Principia* it is also possible to show that W. L. Harper's reconstruction of the feedback

---

108 The sentential connective "or" is here understood as "either ... or ..., but only one of them."

character of the *Principia*, while correct in its analysis, *errs in its epistemological generalization*. He claims that (Harper 1993, 163)

> Newton established a new mathematical ideal to guide our attempts to give causal explanations. According to this ideal our theory should deliver equivalencies that make the phenomenon to be explained into measurement of the relevant features of the magnitudes which figure in its explanation,

and (Harper and Smith 1995, 147)

> according to this ideal a theory explains a phenomenon when it delivers equivalencies that make the phenomenon measure a parameter of the theory which specified its cause. On this view, what counts as empirical success in a theory is to have its parameters to be accurately measured by the phenomenon which they purport to explain.

But from my reconstruction of the process of thought-movement, from the effects to their cause in the *Principia*, based on the external measure, it is readily seen that such a process is as yet *not a process of causal explanation*. Only once, and when the quantity of the cause is found out, can the "converse" process of thought-derivation from the quantity of the cause to the quantity of its effects begin; *and only this second type of thought-movement classifies as a type of causal explanation.*

Harper makes a similar mistake in his analysis of the case when Newton uses the identity of quantities of two different phenomena as a sign that both can serve us as external measures of one and the same cause. He correctly claims that "identifying the forces explains the agreement by the claim that each phenomenon measures the same force" (1993, 161), but from this he draws the wrong conclusion, namely, that "this makes these phenomena into effects of a single common cause" (1993, 161). For two or more phenomena-effects to be understood as the effects of a single common cause, they have to be *derived from the latter by causal explanation*. Here "explanation" stands for "substantiated," while in the previous quote from Harper "explain" stands for "reasoned," that is to say, we give the *reason for our claim* about the origin of that agreement by claiming "both phenomena measure one and the same (their common) cause." This giving of reasons and these claims by their character are as yet part of the process of thought-movement *from* effects *to* their common cause, and, in the framework of this type of thought-movement, they have as yet only the status of *unsubstantiated* reasons and of *hypothetical* claims; they still have to be *substantiated* via causal explanations.

My differentiation between the process that gives reasons for the thought-movement *from* phenomena-effects *to* their common cause, and the process of causal explanation that substantiates the phenomena-effects by their thought-derivation *from* a common cause, draws on Hegel, who in the *Science of Logic* deals with the "retreat into the ground and the coming out from it to the posited" (Hegel 1923, T. 2, 83; Hegel 1969, 462). The importance of the *Science of Logic* for the reconstruction of the structure of Book I and Book III of the *Principia* lies in the fact that Hegel was, to best of my knowledge, the first philosopher who dealt in detail at the level of *categories* with the issue of how knowledge (*Erkenntnis*) moves via thought-operations[109] beyond the level of the immediately given phenomena to their underlying common ground and, in fact, models, as I will show now, this movement on Newtonian mechanics.

Hegel's approach can be summarized and briefly explained by stating those categories which he views as crucial and central for those thought-operations. They are as follows: *phenomena as appearances (Schein)*,[110] *external measure, formal ground, real ground, the reason and the reasoned,* and *phenomena as manifestations (Erscheinung)*.[111] According to Hegel one can discern a cycle of knowledge starting from certain phenomena that, as the *point of departure* for the thought-movement to their common ground, have the epistemic/cognitive status of *appearances*, while the phenomena which are the *"end"-point* of the thought-movement from their common ground have the cognitive status of *manifestations*. The size/quantity of appearances can, according to Hegel, serve us the way to measure their common ground, that is, this size/quantity has the status of an *external* measure (Hegel 1969, 331, 346; Hegel 1923, T. 1, 341, 356–357). Once we are able to discover via the (size/quantity of the) phenomena the (size/quantity of their) ground, but at the same time are as yet not able to find out what is the very ground as something having its own being, we are – according to Hegel – still operating at the level of thought-operations characterized by the category of *formal ground.* As an exemplification of the latter Hegel regards the concept of centripetal force (and the concept of force of gravity) in Newtonian mechanics about which he states that "it has no other content than the phenomenon itself" (Hegel 1923, 79; Hegel 1969, 459) because (Hegel 1923, 79; Hegel 1969, 458)

the ground of the movements of the planets round the Sun is said to be the attractive force of the Earth and the Sun on one another. ... If one asks what kind of force the attractive force is, the answer is that it is the force that makes the Earth move round the Sun; that is, it has precisely the same content as the phenomena of which it is presupposed to be

---

109   Hegel uses for this the term "activity of knowing" (*Tätigkeit des Erkennens*) (Hegel 1923, T. 2, 3; Hegel 1969, 389).
110   A. V. Miller translated Hegel's term "Schein" as "illusionary being" (Hegel 1969, 394), while I translate it as "appearance."
111   A. V. Miller translated Hegel's term "Ercheinung" as "appearance" (Hegel 1969, 479, 499), while I translate it as "manifestation."

the ground; the relation of the Earth and Sun in respect to motion is the identical basis [*Grundlage*] of the ground and the grounded.

For the concept of force in Newtonian mechanics there obtains, in addition, a weird relation between the thought-movement phenomena → their ground, on the one hand, and the thought-movement ground → its phenomena, on the other hand; (Hegel 1923, 80; Hegel 1969, 459–460):

> in this way of explanation the two *opposite directions* of the ground-relation are present without being apprehended in their determinate relation. The ground is, on the one hand, ground ... of the phenomena which it grounds; on the other hand, it is the posited. It is that from which the content of the phenomena is to be understood; but *conversely* it is the ground that is *inferred* from the phenomenon and the former is understood from the latter ... consequently the ground, instead of being ... independent, is, on the contrary, the posited and the derived.

Finally, once we would arrive at the situation where "ground and grounded have a different content" (Hegel 1923, T. 2, 83; 1969, 462), for example, if we could substitute for accelerative force f – initially arrived at via $f \propto a$ – something other than a, then our thought-operations (for example explanations) would be already operating at the level of a *real ground*. In such a situation we could fulfill the request that "when we ask for a ground, we really demand a determination of the content different from that whose ground we are asking for" (Hegel 1923, T. 2, 83; Hegel 1969, 462).

Worth mentioning here is also the fact that Hegel, in addition to the differentiation between the formal and the real grounds differentiates, at least terminologically (Hegel 1923, T. 1, 341, 343; Hegel 1969, 331, 334), between the *external* and the *immanent* measure, that is, that by grasping the ground not as a *formal* but already as a *real* ground, it should be possible to measure the ground not by its appearances, but – somehow – by itself, that is, *inherently* or *immanently*. In this, Hegel clearly contradicts, at the level of the categories of thinking, William Whewell who claimed not only, in Axiom II of his article on the nature of laws of motion that "*causes are measured by their effects*" (Whewell 1834, 81), but also something much stronger, namely (italics are mine): "the existence of the cause is known *only* by the effects it produces. Hence the intensity or magnitude of the cause *cannot be known in any other manner* than by these effects: and, therefore, when we have to assign a measure of the cause, we *must take it from the effects produced*" (Whewell 1834, 81). It will be the task of chapter 4 to find out if this claim of Whewell really holds.

3.2 Scientific laws and scientific explanations based on formal ground

3.2.1 An alternative to Hempel's and Carnap's approach to scientific laws

In order to provide a more sophisticated reconstruction of the structure of scientific laws, as compared to Hempel's reconstruction in chapter 1 that would at them same time correspond also to the laws stated in Newton's *Principia*, allow me to reconstruct the structure of the law of the simple pendulum. The pendulum for which the equation $T = 2\pi\sqrt{\frac{l}{g}}$ holds meets the following eight idealizations:

1. The force of friction at the fulcrum equals zero, that is, the decrease of acceleration due to this force equals zero as well.
2. The whole mass of the pendulum is contained in the suspended body, that is, the acceleration of the suspension cord of the pendulum equals zero and the acceleration of the whole pendulum is due only to the acceleration of the suspended body.
3. Acceleration due to the action of non-gravitational forces is equal to zero because the non-gravitational forces are not acting.
4. The angle of deviation a of the pendulum is so small that it holds that $|a| \ll 1$.
5. Changes of acceleration of the pendulum due to changes of the length l of the suspension cord are equal to zero because l does not change under the influence of the force of gravity.
6. The volume of the suspended body equals zero; it is a mass-point.
7. The movement of the pendulum takes place in a vacuum, that is, the force of friction of the environment is equal to zero and so is the decrease of acceleration of the pendulum due to this force.
8. The acceleration of the pendulum due to the action of forces acting on the physical system within which the pendulum is situated is equal to zero because these forces are equal to zero.

Let me represent each of these idealizations as Id_i. The antecedent of the law of the simple pendulum involves a conjunction of eight idealizations Id_1 & Id_2 & ... & Id_8. If I express this conjunction as Id_{1-8}, while $T^{(8)}$ stands for the period if these eight idealizations hold, "P" for a pendulum located in a physical system, $l^{(2)}$ expresses that idealizations Id_2 and Id_5 pertain directly to the length l of the pendulum, and $g^{(6)}$ expresses that idealizations Id_{1-3}, Id_5, Id_7, and Id_8 pertain directly to acceleration g, then the structure of the law of the simple pendulum is as follows (here (x) stands for universal quantification over the individual variable x):

$$(x)(Px \; \& \; Id_{1-8}x \rightarrow T^{(8)}x = 2\pi\sqrt{l^{(2)}x/g^{(6)}x}) \qquad /9/$$

Since the law contains eight idealizations, we can label it as "the law in the eighth degree of idealizations," or $L^{(8)}$ for short. This law has been stated in the context of physical theories of gravitational as well as nongravitational forces. $g^{(6)}$ stands here for the acceleration due to the action of the force of gravitation, that, it is *the effect of a cause*. The period $T^{(8)}$ in /9/ stands for an effect of the force of gravitation acting on a pendulum with a length $l^{(2)}$. So, the law /9/, I would argue, is a *causal* type of scientific law. The causal nexus it states is represented symbolically in Figure 3.10 (where "F_{GR}" stands for the force of gravitation; "→" for the nexus):

$$F_{GR} \longrightarrow T^{(8)}$$

Figure 3.10 The causal nexus of the pendulum fulfilling eight idealizations

At the same time the law of the simple pendulum with the structure of /9/ can serve as the basis of explanation. We can, for example, suppose that the length of the suspension cord *does* change under the impact of the force of gravity from an initial length $l_0^{(1)}$ to the length $l^{(1)}$. This means that we have to abolish idealization Id_6, that is, it now holds $\neg Id_6$ (non-Id_6), and we have to determine anew the relation between period $T^{(7)}$, length $l^{(1)}$ and acceleration $g^{(6)}$. The law $L^{(7)}$ would give us a nexus, which is symbolically represented in Figure 3.11.

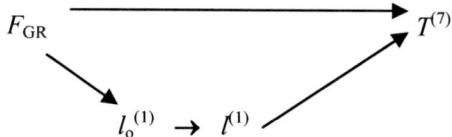

Figure 3.11 The causal nexus for a simple pendulum fulfilling seven idealizations

Another option is to suppose that the motion of the pendulum takes place under the influence of the resistance of the medium in which the pendulum moves. In this case we have to abolish, not only the idealization Id_7, but also we have to account for the force F_c acting against the force of gravity, so that $F_c = V \times d \times g$, where V is the volume of the suspended body and d is the density of the elements of the environment per unit of volume. Thus, we also have to abolish the sixth idealizations, that is, it holds $\neg Id_6$. We then obtain from the law with the structure of /9/ the following law (where "m" stands here for the mass of the suspended body, "m'" stands here for "$V \times d$"):

$$L^{(6)}: (x)[Px \text{ \& } Id_{1\text{-}5,8}x \text{ \& } \neg Id_{6,7}x \rightarrow T^{(8)}x = 2\pi\sqrt{mxl^{(2)}x/g^{(5)}x(mx - m'x)}] \qquad /10/.$$

Once we interpret /9/ and, therefore, also /10/, as a *causal* type of scientific law, it can be readily seen why Hempel, in his analysis of explanation based on the law for the simple pendulum, ends up with the problem reconstructed here. Even if he views g appearing in the equation $T = 2\pi\sqrt{\frac{l}{g}}$ as the acceleration due to the force of gravitation, (i.e., he views the law for the simple pendulum as being part of Newtonian mechanics), he does not understand it as a causal type of scientific law but conceives it as a law of coexistence. What is behind this approach is the fact that Hempel draws on a Humean regularity approach to causation and thus to causal laws, as well. So, for example, he characterizes the causal type of scientific law as being (1966, 53):

always presupposed by a an explanatory statement to the effect that a particular even of a certain kind G (e.g., expansion of gas under constant pressure; flow of current in a wire loop) was caused by an event of another kind F (e.g., heating of the gas; motion of the loop across the magnetic field). To see this, we need not enter into the complex ramifications of the notion of cause; it is sufficient to note that the general maxim, "Same cause, same effect," when applied to such explanatory statements, yields the implied claim that whenever an event of kind F occurs, it is accompanied by an event of kind G.

The problem Hempel encounters with respect to the law of the simple pendulum emerges from the impossibility of distinguishing, within the (Humean) regularity approach to scientific causal laws between the equation $T = 2\pi\sqrt{\frac{l}{g}}$ holds and the equation $l = T^2 g/4\pi^2$, obtained from the former by a simple mathematical manipulation. Newtonian mechanics, however, *respects the differentiation*. The periods $T^{(8)}$, $T^{(7)}$, $T^{(6)}$, and so forth are viewed in it as effects, with the force of gravity acting on the pendulum, as their principal cause. Because Hempel's approach to the law of the simple pendulum rests on a Humean approach to causation and causal types of scientific laws, there is no way for him to distinguish between an equation that expresses a bond of the genesis of an effect from its cause (e.g., the genesis of the period of the swing of the simple pendulum) and the original equation that has been amended and that expresses merely an approach allowing to trace back and compute the size/quantity of the cause issuing from the size/quantity of the effect.

To what scientific laws can the (Humean) regularity approach be applied? In my view it can be applied to those laws in which causation is understood as a recurring succession of events of a certain type, so that the cause is understood as an antecedent event and the effect as a consequent event. Let me return to Hempel's examples of scientific laws given earlier, for example,

Wherever the temperature of gas increases while its pressure remains constant, its volume increases.

Its structure can be represented as follows:

$$(x)(t)[G(x) \ \& \ t' \in <t, t + \Delta t> \ \& \ p(x, t') = \text{const} \ \& \ H(x, t) \rightarrow E(x, t + \Delta t)] \qquad /11/$$

Here "Gx" stands for "x is a gas", "$p(x, t') = \text{const}$" stands for "x's pressure is hold constant in time t'", "$H(x, t)$" stands for "x is heated in time t", and "$E(x, t + \Delta t)$" stands for "x expands in time $t + \Delta t$".

Thus, in my reconstruction, unlike in Hempel's superficial reconstruction as given in chapter 1, we take into account (i) *the universe of discourse*, that is, the kind of entities for which the law is stated[112] and (ii) *the conditions* under which the entities of this kind undergo (iii) certain changes – as *causes*, due to which they undergo (iv) other changes as *effects*. The general structure corresponding to /11/ can be represented as follows:

$$(x)(Nx \ \& \ Cn_{1-k}x \ \& \ C^{\circ}x \rightarrow E^{\circ}x) \qquad /12/,$$

that is to say, "Whenever and wherever there occur objects of a certain kind N under 1 through k conditions Cn as well as phenomena-events of type C°, then always and without exception phenomena-events of type E° occur." Here "C°" and "E°" stand for successive types of events, so that the former is a type of a cause-event and the latter a type of an effect-event. I label scientific laws with the structure corresponding to that of /3/ as *"Humean" type causal law*, or L_{cs} for short. What Hempel labels as the "law of coexistence," or L_{cx}, for short, can be reconstructed as follows:

$$(x)(Nx \ \& \ Cn_{1-k}x \ \& \ E_1x \rightarrow E_2x) \qquad /13/,$$

where "E_1" and "E_2" stand for *coexisting* types of phenomena.

If we compare my reconstruction of the law of the simple pendulum with its reconstruction by Hempel, we immediately notice the following difference: while Hempel views the law as expressing a *noncausal* description of coexisting phenomena-events, I view it as a *(non-Humean) type of causal law*. How is such a difference between two reconstructions of seemingly the same scientific law possible? Based on what Hempel states, for example, in his *Theoretician's Dilemma* (1958, 178), it is because the law of the simple pendulum he draws on is in fact the law stated before the advent of Newton's laws of motion, of the law of gravitation as well as of other derived laws of Newtonian mechanics. Given the structure of /9/, this law *could have at most*[113] the following structure:

---

112 That scientific laws are always explicitly stated as pertaining to a certain kind of entities was for the time explicitly taken into account, to best of my knowledge, in (Fisk 1970), (Nowak 1972), and (Stinchcombe 1973).

113 For an analysis of Galileo's formulation of the law for simple pendulum see, e.g., (Naylor 1974a) and for Huygens' approach to the simple pendulum see (Huygens 1986) and chapter 4 in (Yoder 1988).

$$(x)(P^*x \ \& \ Id_{1,4,6,7}x \rightarrow T^{(4)}x = 2\pi\sqrt{l^{(1)}x/k}) \qquad /14/.$$

Here "P^*x" stands for "x is a pendulum," while the four idealizations here are obtained by the following reductions: 1) idealizations Id_2 and Id_5 from /9/ drop out because the concept of mass appears for the first time, as shown earlier, in its modern form only in Newton's *Principia*. 2) idealization Id_3 drops out of /9/ because in the time of Galileo there were no physical theories about nongravitational forces; and 3) k is here just a constant of proportion and it does not stand here for acceleration like g in /9/, where acceleration is viewed as the effect of the force of gravitation.

Even if the reconstruction of the structure of Galileo's law for the pendulum before the advent of Newtonian mechanics cannot restrict itself to /11/ and requires a detailed analysis of the works of G. Galileo and C. Huygens,[114] nevertheless the comparison of /9/ and /11/ shows that the whole structure of the law for the simple pendulum underwent a complete restating once it was transformed from a *predecessor to* into an *inherent part of* Newtonian mechanics. Newton's second law of motion serves as one of the mediating links between these two forms of the law for the simple pendulum. Its equation $F = ma$ holds rigorously, if the following two idealizations hold: the accelerated body has a negligible volume, that is, it is a mass-point, and no forces are acting on the physical system where the body is located. Thus, the second dynamic law of Newtonian mechanics has the following form:

$$(x)(Ox \ \& \ Id_{1,2}x \rightarrow Fx = mxa^{(2)}x) \qquad /15/,$$

where "O" stands for an object with a nonzero mass occurring in a physical system, "$Id_{1,2}$" stands for the two already stated idealizations and "$a^{(2)}$" stands for acceleration given these two idealizations. Here it becomes readily apparent how the law of the simple pendulum, initially a *noncausal law of coexistence* (in Hempel's terminology) is transformed via /15/ into a (non-Humean) type of *causal* scientific law. Once we know the second law of motion then, via the law of gravitation, we can change the equation in /15/ into the form of $F = mg^{(2)}$, where $g^{(2)}$ stands for acceleration due to the force of gravitation given those two idealizations, that is, for *the effect of a cause*. It should now be clear why in /9/ I presupposed that the pendulum is in a physical system and why I introduced the idealizations Id_6 and Id_8; they are the result of a conceptual *transposition* from /15/ to /9/ occurring in

---

114 Ch. Huygens characterizes a simple pendulum as follows: "A simple pendulum is one which is understood to consist of a string, or an inflexible line devoid of gravity, and of a weight which is attached to the lowest part of the string. The gravity of the weight is understood to be located at one point" (Huygens 1986, 106).

the process of derivation of the law for the simple pendulum in the framework of Newtonian mechanics.

A similar conceptual transposition takes place in the derivation of the law of free fall in the framework of Newtonian mechanics. With respect to current physical knowledge it contains the following eight idealizations:[115]

1. The initial velocity of the falling body equals zero.
2. The body falls in a vacuum, that is, the decrease of acceleration due to the forces of friction equals zero.
3. Non-gravitational forces are not at work; that is, the acceleration of the falling body due to these forces equals zero.
4. Gravitational forces other than of the central body to which the body falls are not at work, that is, the acceleration of the falling body due to the action of these other gravitational forces equals zero.
5. The acceleration of the central body due to the action of the force of gravity of the falling body equals zero because the mass of the falling body is much smaller than that of the central body.
6. The physical system in which the falling body is placed is free from any acceleration due to the action of some external forces.
7. Acceleration of the falling body is constant at the same distance from the surface of the central body because the force of gravitation of the central body is constant at the same distance from the latter.
8. The volume of the falling body is zero; it is a mass-point.

The law of free fall can thus be stated as follows:

$(x)(Ox \ \& \ Id_{1\text{-}8}x \rightarrow s^{(8)}x = g^{(6)}xt^2x/2)$ /16/,

Here "O" stands for the falling object located in a physical system; g appears here as $g^{(6)}$ because idealizations Id_2 through Id_7 pertain directly to it.

The law of free fall with the structure of /16/ can serve as the basis of various explanations. For example, we can derive the law for cases when idealization Id_7 does not hold any more. The presupposition that acceleration due to the force of the central body is constant at the same distance from the surface of the central body is in fact equivalent to the conjunction of the following three idealizations:[116]

i. The angular velocity of the central body is equal to zero, that it does not revolve around its own axis.
ii. The deformation of the central body is zero, that is it is a perfect sphere with a volume of $V = \frac{4}{3}\pi R^3$.
iii. density of the central body is constant in space, that is, the distribution of mass in volume V is constant.

---

115 I draw here partially on (Such 1978).
116 On this see (Such 1978).

If we gradually abolish idealizations i. and ii. then we have to take into account changes in the right side of the equation of /16/ by introducing the following additional terms:

a. Once we abolish the idealization that the central body does not revolve we have to take into account the angle α determining the latitude at which the body falls via the additional term $-\cos^2\alpha\,/289$.
b. When abolishing the idealization that the central body is a perfect sphere we have to introduce the additional term - $98\cos^2\alpha\,/55199$. We then end up with the following law for the law of free fall

$$(x)[Ox\ \&\ Id_{1\text{-}6,8}x\ \&\ \neg Id_7x \rightarrow s^{(7)}x = g^{(5)}x(1 - \cos^2\alpha/289 - 98\cos^2\alpha/55199)t^2x/2]\quad /17/$$

Let me now try to propose a general scheme for scientific laws, taking into account the structure of scientific laws in Newtonian mechanics as given here previously as well as a model of explanation corresponding to explanations based on the law of the simple pendulum and on the law of free fall.

The second dynamic law of Newtonian mechanics can serve us as a paradigmatic example of a law that is used as a basis for the construction of a theory. Using it as a basis, it is thus possible to derive various laws of Newtonian mechanics; let us mention at least the following three laws. The *effect* of a force acting on a body at rest or already moving can be derived by means of the following deliberations:

a) the effect of a force acting on a body along a certain trajectory; we then obtain the magnitude of *work*;
b) the effect of a force acting on a body in a certain interval of time; we then obtain the magnitude of *momentum of force*.
c) Based on the second law of motion, initially stated for a mass-point, we can proceed to an object composed of several mass-points and then derive magnitudes that are viewed as the effect of force acting on a system of mass-points. We can, for example, determine the *turning-effect* of force acting on a system of mass-points via the relation (in vector-notation) **M** = **r** × **F**, where **M** is the moment of force and **r** is the perpendicular distance from the axis of turning to the line of action of the force.

As seen from these examples, as well as from my analysis of Newton's *Principia*, classical mechanics is a science that makes it possible to compute the *effects* of a force the size/quantity of which is, in turn, determined in the second law of motion by means of yet *another effect*, namely, acceleration. The general scheme that would account for the structure of the second dynamic law of classical mechanics is thus as follows:

$$(x)\ (Nx\ \&\ Cmod_{1\text{-}k}x = d_{1\text{-}k} \to f_1(Cx) = E^{(k)}x) \qquad /18/.$$

I will label it *the idealized type of law, in the kth degree of idealization, of the cause/ground underlying the phenomenon-effect $E^{(k)}$*, or $L^{(k)}$, for short. Here "N" stands for the type of entities for which the law is stated; "$Cmod_{1\text{-}k} = d_{1\text{-}k}$" stands for a conjunction of 1 through k idealization, namely, that 1 through k *modification conditions are put equal to zero*; "$E^{(k)}$" is the phenomenon-effect in the kth degree of idealization; and "$f_1(C)$" stands for a function of a cause defined by means of its effect. I view scientific laws with a structure corresponding to that of /18/ as *scientific laws of the causal type*, but that conceive causation in a manner different from that given in /12/. In /18/, cause is viewed as (i) that *what underlies* the phenomenon-effect chosen as the point of departure by means of which we can discover the cause and compute its size/quantity; and (ii) as the *ground* of all those phenomena-effects that can be derived by explanation from /18/. As shown earlier, we can – once we define and identify the force as the cause underlying acceleration – determine various *other* effects of the acting force: the period of a swinging pendulum, the trajectory of a falling body, the time-effect of a force acting on a body (the moment of force), the path-effect of a force acting on a body (work), and so forth. Once we derive these laws, in a manner I will reconstruct later, we can use these laws again for explanations obeying the following general structure. The point of departure is a type of law with the following structure:

$$L^{(l)}:\ (x)\ [N^*x\ \&\ Cmod_{1\text{-}l}x = d_{1\text{-}l} \to E_2^{(l)}x = f_1(E^{(k)})x] \qquad /19/$$

Here "N^*" stands for the type of entity for which the law is stated (e.g., falling body, pendulum, etc.); "$Cmod_{1\text{-}l} = d_{1\text{-}l}$" stands for the conjunction of 1 through l idealizations; "$E_2^{(l)}$" stands for a phenomenon-effect in the lth degree of idealization (e.g., the covered distance, the period of the swing, etc.); and "$E^{(k)}$" stands for the phenomenon-effect by means of which the underlying cause-ground was initially identified and defined in /18/. The type of explanation based on /19/, as shown already in the examples of explanations based on the law of the simple pendulum and the law of free fall, has the following structure: (i) one has to abolish the respective idealizations, thus to suppose that now the respective modifications condition are already at work, that is, that $Cmod_i \neq d_i$ holds, and (ii) one has to, at the same time, take into account the impact of the modification in order to derive the respective phenomena-effects one wants to explain. If all idealizations would be gradually abolished, one would obtain the following sequence of types of scientific laws $L^{(l\text{-}1)}, \ldots, L^{(0)}$.[117]

---

[117] For the sake of simplicity I suppose here that all idealizations are abolished. Usually, the law of the type $L^{(0)}$ is too complex to be derived and the concretization-sequence starting in $L^{(l)}$ ends in a law of the type $L^{(j)}$, so that $l > j > 0$.

$L^{(l-1)}$: (x) $[N^*x \ \& \ Cmod_{1-(l-1)}x=d_{1-(l-1)} \ \& \ Cmod_l x \neq d_l \rightarrow E_2^{(l-1)}x = f_{l-1}(E^{(k)}x, Cmod_l x)]$

× × × × × × ×
× × × × × × ×

$L^{(0)}$: (x) $[N^*x \ \& \ Cmod_{1-l}x \neq d_{1-l} \rightarrow E_2^{(0)}x = f_0(E^{(k)}x, Cmod_l x, ..., Cmod_l x)]$.

In the antecedent of $L^{(l-1)}$, the expression "$Cmod_{1-(l-1)}=d_{1-(l-1)}$" expresses that 1 through $(l-1)$ idealizations are still valid, while the inclusion of "$Cmod_l$" on the right side of the equation expresses that the impact of the lth modification conditions is already being taken into account. The type of explanation by means of which we derive $L^{(l-1)}$, ..., $L^{(0)}$ from $L^{(l)}$, I label – following (Nowak 1972) and (Such 1978) – as *explanation by gradual concretization*.

Let us now compare the reconstruction of the method of explanation by gradual concretization with Hempel's D-N model as well as by my reconstruction of the law of the cause/ground $L^{(k)}$ with Hempel's reconstruction of scientific causal laws.

First, the method of explanation by gradual concretization reconstructs, contrary to the D-N model, the case of explanation of scientific laws from other scientific laws. This case can be represented symbolically as follows ("$-_c-$|" stands here for gradual concretization; "$Cmod_i$" for the introduction of a modification condition, that is, for "$Cmod_i \neq d_i$"):

$L^{(l)} \ \& \ Cmod_l \ -_c-| \ L^{(l-1)} \ \& \ Cmod_{l-1} \ -_c-| \ L^{(l-2)} \ \& ... \ -_c-| \ L^{(1)} \ \& \ Cmod_1 \ -_c-| \ L^{(0)}$ /20/

The law of the type $L^{(0)}$, that is, where all idealizations have already been abolished can then serve as the basis of explanation which involves the introduction of 1 through s singular conditions $Csin_{1-s}$. In such an explanatory step we can explain a singular phenomenon $E^{(0)}a$. The whole sequence of explanations then is as follows ("⊃" stands here for logical consequence)

$L^{(l)} \ \& \ Cmod_l \ -_c-| \ L^{(l-1)} \ \& ... \ -_c-| \ L^{(0)} \ \& \ Csin_{1-s} \supset E^{(0)}a$ /21/

The fact that by means of the model of explanation by gradual concretization one is capable to reconstruct the case of explanation of a law from other laws is based on an enlargement of the conceptual framework upon which the concept of scientific law is built. In addition to the concept of singular conditions, it introduces the concept of *modification conditions*. The latter are causally relevant for whole classes of phenomena-effects and are stated *explicitly*, initially in the form of idealizations, *inside the structure of scientific laws*. Singular conditions,[118] on

---

118 On singular initial and boundary conditions, see, e.g., (Sklar 1991), (Wilson 1991), and (Frisch 2004).

the contrary, are relevant for a respective singular phenomenon and are not stated explicitly inside the structure of scientific laws. At the same time, one has to bear in mind that in my reconstruction the idealizations appearing in scientific laws are *not* identical with the so-called *ceteris paribus* (cp) qualifications[119] because it holds that (Earman and Roberts 1999, 457)

> in general the problem of *ceteris paribus* qualifications is distinct from the problem of idealizations. Often the idealization can be stated in a precise closed form ... Here the problem is not in saying precisely what is involved in the idealization but in relating it to the real world which is not ideal. By contrast, many cp laws claim to be about unidealized real world situations but make indefinite claims about these situations.

Second, the model of gradual concretization fulfills J. Woodward's requirement of functional interdependence (f). To show this let us take the law of the type $L^{(l)}$ given in /19/. Till now, for the sake of simplicity, I took into account only *one sequence* of gradual abolishment of idealizations, namely, $Cmod_l \neq d_l$, $Cmod_{l-1} \neq d_{l-1}$, ..., $Cmod_1 \neq d_1$, and therefore also only *one sequence* of less and less idealized scientific laws. But in fact we can by means of the method of gradual concretization derive several different sequences of less and less idealized laws depending on which idealization will be abolished as the first one, as the second one, as the third one, and so on from the set of all idealizations. We can thus obtain the following network of scientific laws as can be seen from Figure 15 (the lower index indicates which of the l idealizations has already been abolished; "⊥" stands here for gradual concretization).[120]

It is now readily seen that explanation by gradual concretization based on idealized laws of the type $L^{(l)}$ fulfills the (f)-requirement because it holds *that the more a scientific law of this type contains idealizations, the more different explananda (scientific laws and/or explanations of singular phenomena) can potentially be derived from that scientific law once it is given in an explanans.* I emphasize *potentially* – because each concretization step involves *an irreducibly heuristic moment, namely, the discovery of the type of causal impact of the respective modification condition.*

---

119 On the issue of cp-clauses see, for example, (Pietroski and Ray 1995), (Schurz 2001), (Rosenberg 1995), (Earman and Roberts and Smith 2002), (Lange 1993), (Lange 2002), (Drewery 2001), and (Smith 2002). On the issue of cp-laws see, for example, (Rupert 2007).

120 I suppose here, for the sake of simplicity, that one can abolish one idealization after the other and is thus not forced to abolish several idealizations at once as, e.g., in the case, given above, of the simultaneous abolishment of the idealizations that there is no resistance of the environment and that the suspended body is a mass-point.

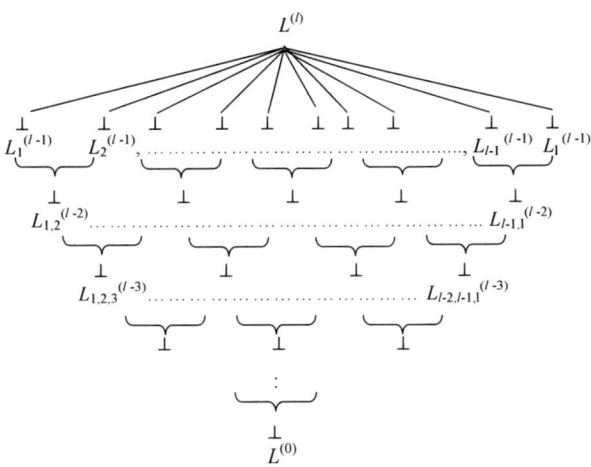

Figure 15 Network of scientific law obtained by gradual concretization

Third, it can be shown that explanation by gradual concretization is not a deductive argument by its very nature. While the *extra-logical* space of the D-N model is limited just to the *discovery* of a true scientific law (scientific laws) and true initial conditions, in the case of explanation by gradual concretization the extra-logical logical space is much larger. It involves, not only the introduction of the abolishment of the respective modification $Cmod_i$ but also the derivation of the *impact* of the modification condition which, once the idealization $Cmod_i = d_i$ is abolished, is supposed to be at work. This derivation involves, however, an irreducible *heuristic* moment; *in each concretization step one has to find out the impact of the respective modification condition.*

In addition, in explanation by gradual concretization the information given in the conclusions (*explananda*) goes beyond the given in the premises (*explanandum*). That this is so is readily seen when one compares the information given in the equation for the period of the pendulum in /9/ with the equation for the period of the pendulum in /10/. The latter contains, as compared to the former, new information about the inference of the force of gravity with the force of buoyancy.

Finally, for deduction it holds that the terms given both in its premises and conclusion have to have the same meaning. Let me now apply Transparent Intensional Logic to the equations for the period of the pendulum as given in /9/ and /10/ to find out if the term "period" changes its meaning in the movement from /9/ to /10/. So as T, l, and g are empirical expressions for the equation $T = 2\pi\sqrt{\frac{l}{g}}$ given in /9/ its TIL-reconstruction is as follows:

$\lambda w \lambda t \ [^0\forall \ [\lambda x \ [^0= \ ^0T_{wt} x \ [^0\times \ [^0\times \ ^02 \ ^0\pi][^0\surd \ [^0\div \ ^0l_{wt} x \ ^0g_{wt} x]]]] \ ]]$ /9*/

Let me now consider the equation $T = 2\pi \sqrt{\frac{ml}{g(m-m')}}$ given in /10/. Here m and m' are empirical expressions; then, its transcript by means of TIL is then as follows:

$\lambda w \lambda t \ [^0\forall \ [\lambda x \ [\ [^0T_{wt} x] \ ^0= \ [[^02 \ ^0\times \ ^0\pi] \ ^0\times \ [\ ^0\surd \ [\ [[^0m_{wt} x] \ ^0\times \ [^0l_{wt} x]] \ ^0\div \ [\ [[^0g_{wt} x] \ ^0\times \ [[^0m_{wt} x] \ ^0- \ [^0m'_{wt} x]]]]]] \ ] \ ]]$

$\lambda w \lambda t \ [^0\forall \ [\lambda x \ [^0= \ ^0T_{wt} \ [^0\times \ [^0\times \ ^02 \ ^0\pi] \ [^0\surd \ [^0\div \ [^0\times \ ^0m_{wt} \ ^0l_{wt}][^0\times \ g_{wt} \ [^0- \ ^0m_{wt} \ ^0m'_{wt}]]]]] \ ]]$
/10*/

The claims /9*/ and /10*/, as claims about identity, are based on two definitions. In the first one the construction ^0T is introduced by means of the definition

$[^0T_{wt} x] \equiv [^0\times \ [^0\times \ ^02 \ ^0\pi][\ ^0\surd \ [^0\div \ ^0l_{wt} x \ ^0g_{wt} x]]]$ /9**/;

in the second it is introduced as follows:

$[^0T_{wt} x] \equiv [^0\times \ [^0\times \ ^02 \ ^0\pi] \ [^0\surd \ [^0\div \ [^0\times \ ^0m_{wt} x \ ^0l_{wt} x][^0\times \ g_{wt} x \ [^0- \ ^0m_{wt} x \ ^0m'_{wt} x]]]]]$ /10**/

The construction ^0T is in /9**/ and /10**/ provided by mutually different construction, thus we have here two different definitions for the construction ^0T. From this we draw the conclusion that we have in /9/ and /10/ two different meanings for the term "period"; the term "period" thus changes its meaning in the explanatory movement from $L^{(7)}$ to $L^{(5)}$. We can thus draw the conclusion that *explanation by gradual concretization by its very nature is not a deductive argument.*

Fourth, that heuristic aspect of explanation by gradual concretization is also important for understanding the fact why one needs at all the law(s) of the higher degree(s) of idealizations in order to derive the law(s) of the lower degree(s) of idealizations. Only via the former, which is (are) the explanatory basis of explanation and discovery, can one derive the latter. Thus, if explanation would end up with a law of the zero-degree of idealization, it still could be derived only by means of a gradual abolishment of all idealizations in a law of the type $L^{(l)}$. A scientific law of the type $L^{(0)}$ is therefore, even as a law of a zero-degree of idealization, not free from modifications conditions; the latter are given in the form of the abolished idealizations in its antecedent.

Fifth, while the D-N model views scientific explanation based on universal laws as a process in which we explain a phenomenon by i) subsuming it under a (covering) law, and then ii) deduce it from the laws and singular conditions, in explanations of a singular phenomenon-effect represented by /21/, (for example, of a concrete value of the period of swing of a real pendulum, say, in a laboratory), one has (a) to subsume the singular phenomenon of a certain type to be explained *not under a (covering) scientific law, but only under a covering kind N* and the respective covering cause C grasped via a function of the phenomenon $E^{(k)}$*, both appearing in a law of the type $L^{(l)}$ then, (b) gradually concretize the law of the type $L^{(l)}$ and, only then, (c) bringing in the respective singular conditions for the deduction of the singular phenomenon to be explained. This means, with respect to (a), that one of the important aims of explanation by gradual concretization is to prove that the phenomenon to be explained qualitatively fits the type N^* and the cause C.

Sixth, by my reconstruction of the idealized laws of mechanics, by my reconstruction of the idealized law of the types $L^{(k)}$ and $L^{(l)}$, and by providing the reconstruction of the method of explanation by gradual concretization I at the same time can overcome the one-sided logico-normative orientation of Hempel's approach to the reconstruction of scientific laws and explanation. The terms appearing in my reconstructions of laws of the types $L^{(k)}$ and $L^{(l)}$ are based on an analysis of the laws of Newtonian mechanics; they have thus a *pragmatic basis* and they show that it is possible to deal with the concept of scientific law by drawing on the real practice of science. My reconstruction shows also that in order to overcome the problems encountered by the D-N model one has to analyze and reconstruct in detail the respective laws and explanations based on them as they are given in science. With respect to this it is now clear how one should solve the above stated "shadow"-problem of S. Bromberger. One has to analyze in each particular case what the respective physical theory states – *if it states anything at all* – about the production of the shadow. Geometrical optics, mentioned by S. Bromberger, does not deal with the causal structure underlying the propagation of light[121] in the same manner as the law of the pendulum stated by Galileo, tentatively reconstructed in /14/, does not state anything about the cause of the period of the swings.

Seventh, my reconstruction of the laws of the types L_{cs}, L_{cx}, $L^{(k)}$, and $L^{(l)}$ shows that what unifies them, independently of whatever typological differences there are between them, is the fact that they are *always stated only with respect to entities of a certain kind*. Hempel's and Carnap's attempts to frame the concept of fundamental scientific law via the condition that the latter should hold for all times and spaces is thus a product of a superficial reconstruction of the structure of scientific laws that does not take into account that fact. That fact was neither taken into account by K. R. Popper. In his *Postscript to the Logic of Scientific Discovery* he, on the one hand, characterizes a scientific law already as "a universal statement, pertaining to all things of a certain kind, or to all elements of a

---

121   On this see, e.g., (Bunge 1964).

certain non-empty universe of discourse" (Popper 1983, 184) and, thus, already departs from his 1935-approach to the concept of scientific law. On the other hand, however, he gives the following compact characterization of scientific laws (1984, 184):

All things have the property P.

And so he relapses back to his 1935-approach to the concept of scientific law.

Eighth, as shown above, we have in fact *two* laws of the simple pendulum: one that antedates Newtonian mechanics and another that is derived in the framework of the latter. The same "double" existence can be identified also for the law of free fall. By drawing on Galileo's *Two New Sciences* (Galileo 1974) and on the reconstructions of the historians of science,[122] the law of free fall as stated by Galileo could have the following structure:

$$(x)(O'x \ \& \ Id_1x \ \rightarrow \ s^{(1)}x \propto t^2x) \qquad /22/,$$

where "O'" stands for an object falling with a uniformly accelerated motion from the state of rest (i.e., its initial velocity is equal to zero); "Id_1" stands for the idealization that the fall of the body occurs in a vacuum; "$s^{(1)}$" expresses the distance covered by the body under this idealization; "t" stands for the time in which it covers that distance and "$\propto$" is the sign of proportionality. Expression /22/ can be viewed as a (Humean) regularity type of scientific causal law with a structure corresponding to that of /12/. It states: Once a uniformly accelerated body is released from rest (type of a phenomenon as a type of a *cause*) in a vacuum, it will cover a certain distance (type of phenomenon as a type of an *effect*) proportional to the square of the time in which it falls. The "double" existence of the laws of physics suggests that in the development of science *types of phenomena* appearing initially in the (Humean) regularity types of scientific laws – the latter being symbolized earlier as L_{cs} and as L_{cx} – are *reinterpreted* on the basis of laws of types $L^{(k)}$ and $L^{(l)}$. This typological reinterpretation then suggests accepting into the framework of the philosophy of science Hegel's a differentiation, at the level of *categories*, between *two types of phenomena*. Phenomena that were initially cognized *prior* to the discovery of their underlying cause/ground, including that which enables one to compute the size (quantity) of the cause/ground, that is, for phenomena for which one has as yet no causal explanation, I label *phenomena as appearances*. Phenomena that are explained on the basis of their cause/ground, I label *phenomena as manifestations*. Such a change of the phenomena as appear-

---

122 See, e.g., (Naylor 1974b) and (Drake 1973).

ances E_1, E_2, ..., E_n, via the cause/ground C – the latter being discovered via an appearance in the kth degree of idealization $E^{(k)}$ – into phenomena as manifestations $E^{(l)}{}_1$, $E^{(l-1)}{}_2$, ..., $E^{(l-j)}{}_r$ can be symbolized as shown in Figure 3.13 ("Df" stands here for definition).

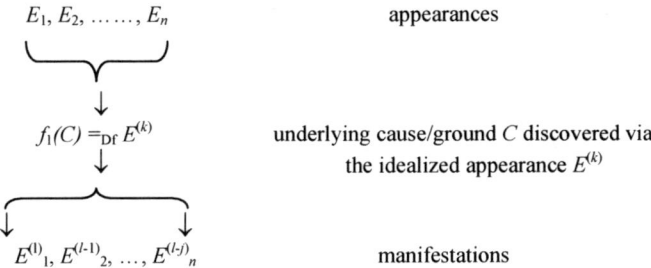

Figure 3.13 The movement of knowledge from appearances to manifestations

The *epistemic* difference between appearances and manifestations becomes even more understandable once we realize that *manifestations* are part of the meaning of the expressions in which the expression referring to their underlying cause/ground is *always* embedded, and where the source of knowledge about the underlying cause/ground is a scientific law of type $L^{(k)}$, thus, the *expression (and its meaning) referring to the cause/ground is, in the process of explanation, gradually shifted from the explanans-law to the explananda-laws*. Appearances, contrary to this, are free from any epistemic/cognitive connections with knowledge about the underlying cause/ground.

Once I differentiate conceptually between phenomena as appearances and phenomena as manifestations of a cause/ground, I also have to differentiate conceptually between two types of scientific laws pertaining to phenomena: *laws of appearances* and *laws of manifestations*. So, for example, while the law of free fall and the law of the simple pendulum as stated by Galileo and Huygens are of the former type, the law of free fall and the law of simple pendulum derived in the framework of Newtonian mechanics are of the latter type.

In order to make my conceptual differentiation as precise as possible, I differentiate further between *general appearances* – stated in the framework of laws of appearance – and *singular appearances*, which one can deductively derive from these laws together with the respective singular conditions. Finally, I differentiate also between *general manifestations* – stated in the framework of laws of manifestations – and *singular manifestations*, the latter being derived by gradual concretizations *and* deduction from these laws together with the respective modification and singular conditions. This means that the epistemic status, for example, of a concrete value of the swing of an individual pendulum deductively explained on the basis of the law for the simple pendulum as stated by Galileo, and tentatively reconstructed in /22/, differs from the epistemic status of a concrete value of the swing of an individual pendulum derived by gradual concretization *and* deduction from the law of the simple pendulum /9/, as stated within the framework of Newtonian mechanics. The former concrete value has the status of an *individual appearance*, the latter concrete value that of *an individual manifestation*. I thus enlarge Figure 3.13 as shown in Figure 3.14:[123]

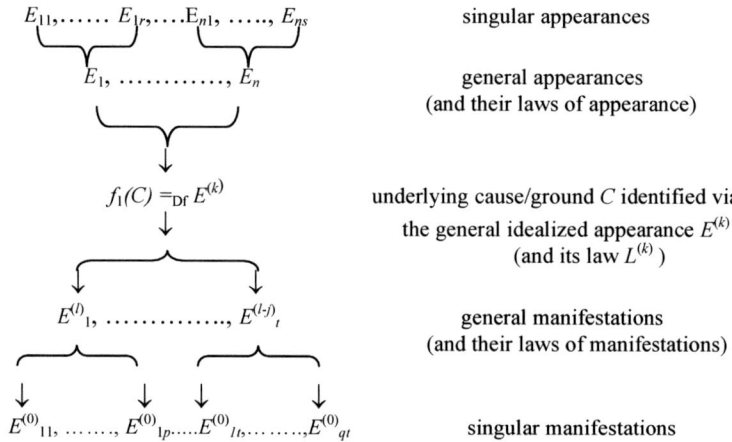

Figure 3.14 *The movement of knowledge from individual appearances to individual manifestations*

I can now give a final evaluation of Hempel's D-N model. Once it is freed it from the claim that the conditions of adequacy (R2) and (R3) are *logical* conditions, then it fits explanations based on universal scientific laws which have the status of a *law of appearance* or, as he labels them, that of an *empirical law*, and which (Hempel 1966, 54)

asserts a uniform connection between different empirical phenomena or between different aspects of an empirical phenomenon. It is a statement to the effect that whenever and wherever conditions of a specific kind F occur, then so will always and without exception certain conditions of another kind, G.

---

123 In order not to overburden this figure, I do not place the arrows at the very top to indicate the epistemic relation between the singular appearances and the general appearances with their laws of appearances. I view the singular appearances as the *presupposition* for the statement of the general appearances together with their laws of appearances and as *the explanatory consequence* of the laws of appearances.

Even if these laws very often contain idealizations, as shown above in the case of the law of free fall stated by Galileo, explanations based on empirical types of scientific laws can still be viewed as having the nature of a subsumption under these (covering) laws plus deduction, of course, under the supposition that one is not forced to abolish idealizations in the course of explanation. So, for example, if Hempel states the law as follows (1942, 232):

Below 32°F, under normal atmospheric pressure, water freezes,

then one can derive by deduction from this law and the singular condition expressed as follows: "This sample of water had a temperature below 32°F and was held at normal atmospheric pressure," the explanandum statement "This sample of water froze."

Hempel's D-N model as well as his regularity approach to causal types of scientific laws becomes, however, insufficient as an instrument for a philosophical analysis, reconstruction, and explication of the categories of scientific law and scientific explanation once explanation involves (i) the abolishment of idealizations stated in the structure of the scientific law, and (ii) the derivation of phenomena as manifestations based on causal laws identifying an underlying cause/ground.

Worth mentioning here is the fact that while Hempel's works give a quite complete reconstruction of the deductive explanation based on the universal laws of appearance (or, in his terminology, universal "empirical laws"), one can find in them also some *hints* at the existence, but *never a reconstruction*, of explanations based on laws identifying the causes underlying the phenomena, where these phenomena are initially described in universal "empirical" laws. So, for example, he states that (Hempel 1966, 70):

Theories ... seek to explain ... regularities and, generally, to afford a deeper and more accurate understanding of the phenomena in question. To this end, a theory construes those phenomena as manifestations of entities and processes that lie behind or beneath them, as it were. These are assumed to be governed by characteristic theoretical laws, or theoretical principles, by means of which the theory then explains the empirical uniformities that have been previously discovered.

He even states the following (Hempel 1970, 142):

Theories are normally constructed only when prior research in a given field has yielded a body of knowledge that includes empirical generalizations or putative laws concerning the phenomena under study. A theory then aims at providing a deeper understanding by construing those phenomena as manifestations of certain underlying process governed by laws which account for the uniformities previously studied, and which, as a rule, yield corrections and refinements of the putative laws by means of which those uniformities have been previously characterized.

Thus, I conclude that Hempel's reconstruction of deductive explanation based on universal laws of appearance is *not invalid but valid, even if only for that type of sciences which accesses its objects initially only as appearances*. Explanations based on idealized laws with the structure of /18/ and /19/ are performed in that type of sciences in which one can already transform the laws of appearance into laws of manifestation based on the grasping of the underlying cause. The method of explanation I labeled "gradual concretization" is based, as shown above, on the subsumption of a phenomenon-effect to be explained under a covering universe of discourse (i.e., under a certain kind of entities) and the corresponding cause/ ground. From this one can immediately realize *what are the limits of the method of explanation by gradual concretization*. If one wants to explain a phenomenon-effect that belongs to a universe of discourse (a type of entity) that is different from that stated in the respective idealized law of the underlying cause, then the method of explanation by gradual concretization can no longer be employed, for example, as in the case when we want to explain the distance covered by a block sliding on an inclined plane starting from the second dynamic law in Newtonian mechanics. The former pertains to a *block sliding on an inclined plane*, the latter, however, to *a mass-point given in a physical system*.

To widen my typology of explanations based on universal scientific laws, let us briefly reconstruct how Newtonian mechanics deals with the derivation of the law for the distance covered by a block sliding on an inclined plane. Starting from the second dynamic law and the law of gravity Newtonian mechanics performs a *thought-reconstruction* so that the sliding block becomes a mass-point in a physical system sliding down the inclined plane, that is, an idealized thought-object, and, then, states for this type of entity the equation $mdv/dt = mg\sin\alpha$, where the expression on right side stands for the component of the force of gravity acting on the mass-point sliding down an inclined plane. From this equation, together with its universe of discourse and the stated idealizations, it is then possible to derive the following law $L^{(4)}$ for the distance covered by the sliding mass-point:

$$(x)(O"x \:\&\: Cmod_{1,2,3,4}x = d_{1,2,3,4} \rightarrow s^{(4)}x = g^{(3)}xt^2 x \sin\alpha x) \qquad /23/,$$

where 'O'''' stands for the mass-point in a physical system sliding on an inclined plane; "$Cmod_{1,2} = d_{1,2}$" stands for the conjunction of the two idealizations given already in the second dynamic law, namely, that the considered object has a zero volume (i.e., it is a mass-point) and that it is placed into a physical system free from any impact of external forces; "$Cmod_3 = d_3$" stands for the idealization that the sliding mass-point starts its sliding from rest (i.e. its initial velocity is equal to zero); "$Cmod_4 = d_4$" stands for the idealization that there is no force of friction decreasing the accelerated motion of the mass-point along the inclined plane; "$g^{(3)}$" stands for acceleration due to the force of gravitation involving already two idealizations transposed from the second dynamic law of Newtonian mechanics reconstructed in /15/.

Based on /23/, we can explain, by abolishing idealization $Cmod_4 = d_4$, how the covered distance changes once the force of friction is at work. From $L^{(4)}$ we obtain by gradual concretization $L^{(3)}$ (the force of friction is proportional to $mg\cos\alpha$)

$$(x)[O"x \ \& \ Cmod_{1,2,3}x = d_{1,2,3} \ \& \ Cmod_4 x \neq d_4 \rightarrow s^{(3)}x = g^{(2)}xt^2x(\sin\alpha x - \cos\alpha x)] \quad /24/.$$

/23/ and /24/ can be viewed as laws of the manifestation of an underlying cause; in the former the covered distance is the effect of the force of gravity and in the latter it is the combined effect of the force of gravity and the force of friction.

From the example of the derivation of the law for the distance covered by a block sliding along an inclined plane, it is clear that the explanatory move from a law of the type $L^{(k)}$ to laws of the type $L^{(l)}$ involves a *thought-reconstruction of a situation one wants to explain by laws of the latter type in such a way that this situation after the thought-reconstruction fits already the universe of discourse of the law of the type $L^{(k)}$*. What of course unifies the law of the type $L^{(k)}$ with the laws of the type $L^{(l)}$ is *that in all of them one and the same main underlying cause/ground is presupposed*. If the cause/ground given in the explanans law and the cause/ground in explanandum law are different, then neither the method of thought-reconstruction nor that of gradual concretization can be applied. These two methods of explanation, like deductive explanation based on laws of appearance, have a certain range of application beyond which they cannot be used any more. In chapter 4 I will reconstruct the method of explanation that overcomes this limitation of the method of explanation by gradual concretization.

This limitation of the methods of explanation based on these two types of scientific laws can be traced primarily to the typology of conditions reconstructed with respect to these two types of scientific laws and that appear in scientific explanation. With respect to these two types of scientific laws I have dealt with two types conditions: the modification conditions and the singular conditions. As shown in /21/, the former are relevant for the derivation of the *general forms of manifestations* stated, via the process of concretization, in laws of the types $L^{(l-1)}, ..., L^{(0)}$, while the latter are relevant for the derivation of *singular phenomena-manifestations* of the type $E^{(0)}a$. So what is presupposed here is that both the general manifestations and the singular phenomena-manifestations are in their very existence depended on certain conditions: the former are *produced* once certain modification conditions are given, the latter are produced once certain singular conditions are given. Therefore, in the process of explanation, one can (i) via the introduction of modification conditions in the explanans containing a law of the type $L^{(l)}$ derive different explananda with various laws of the types $L^{(l-1)}$, ..., $L^{(0)}$ pertaining to different general forms of manifestations and (ii) via the introduction of singular conditions in the explanans, where the law of the type $L^{(0)}$ is already stated, to derive different explananda with various singular phenomena-manifestations. But what is supposed here to be *completely independent of any modification and/or singular conditions is the very underlying cause/ground*. All thought-operations performed on modification and singular conditions in the course of explanation are irrelevant with respect to the existence of very cause/ground; it is given as *the same both in the explanans and the explananda* when we move in the course of explanation from the former to the latter. What is thus needed is yet another enlargement of the conceptual framework, so that in addition to the concepts of modification and singular conditions, a concept of conditions relevant for the existence of the respective cause/ground would be taken into account. Such an enlargement, in order to correspond to the practice of real science, would have to draw on an identification of those scientific theories that perform thought-operations not only with singular and modification conditions, but also with conditions under which the very cause/ground of the respective objects they investigate is produced. The reconstruction of this type of conditions will be the subject matter of chapter 4.

With this limitation, I can now summarize my differentiated approach to the typology of universal scientific laws and explanations based on them as shown in Table 3.4.

3.2.2 Some Comparisons

Let us now compare my approach to scientific laws and scientific explanation with approaches appearing in more recent works in the philosophy of science.

A. Brian Ellis on Essence, Capacities and Dispositions

According to Brian Ellis (1992, 266):

we idealize for reasons which have to do with the basic aims of scientific research. Physical science, it will be argued, is fundamentally concerned to discover essential natures of kinds of things that can exist, and the kind of changes that can take place, in a world such as ours. And to achieve its aims, science must focus on the intrinsic properties and structures of the basic kinds of things and processes which are to be found existing or occurring in nature. ... The aim of science is not to describe what actually happens in nature, or

The epistemic status of phenomena in the laws of the explanans	The corresponding type of scientific law in the explanans	The corresponding type of conditions in the explanans	The corresponding type of relation between the explanans and explanandum	The corresponding type of explanandum
general appearances	law of appearance	singular conditions	L-consequence	singular appearances
idealized general appearance and idealized general manifestations	idealized law of cause underlying the idealized appearance and laws of manifestation	modification and singular conditions	construction of idealized objects, gradual concretization and L-consequence	laws of manifestation and singular manifestations

Table 3.4 A typology of universal scientific laws and explanations based on them

to systematize our knowledge of what occurs by subsuming it under a law; it is to explain what happens by showing how what occurs can be seen to arise out of the essential nature of the natural kinds and processes which constitute the real world.

A similar view is stated as follows (Bigelow and Ellis and Lierse 1992, 371, 373):

The world is one of a kind. ... Recognition that this world is one of a kind offers a new approach to the question of what a law of nature is. We argue that in general laws of nature are concerned with natural kinds. In some cases laws simply describe the essential properties of natural kinds. ... In the case of other laws, for instance, where there are interactions between things of different kinds, the laws stating how they behave are derivable from their essential natures. ... This theory of the nature of scientific laws derives from the basic idea that things behave as they do because of what they are made of, how they are made, and what their circumstances are. ... [Our] approach is to develop a theory of essences as primitive, and seek to define natural kinds in terms of essences. ... Essential properties, thus defined, could then be used to define natural kinds. Membership in a natural kind is, arguably, an essential property for each of the members of that natural kind. Consequently natural kinds could be construed, for instance, as classes of individuals which share an explanatory significant cluster of essential properties. ... Laws of nature, we claim, derive from the attribution of essential properties to things.

Let me compare these views with my reconstruction of the structure of scientific laws of the type $L^{(k)}$ as given above in /18/ as well as with my analyses of Newton's *Principia*. Even if in that reconstruction I employ terms like "universe of discourse," "kind of entities" and "cause/ground," and even if the second could be understood as "natural kind" and the last as "essence of the respective kind," still scientific laws with the structure of /18/ do not fit into the conceptual framework delineated by Ellis, Bigelow, and Lierse. This can be seen when one analyzes, for example, the second dynamic law of Newtonian mechanics. This scientific law does not put the property to have a nonzero mass into a relation with a force as the cause/ground. Neither, as I have shown in my analysis of the *Principia*, does it state what force is as a quality; it enables one only to determine the size/quantity of the force by means of the size/quantity of one of its own effects – acceleration. And last, but not least, as shown in my analysis of the *Principia*, acceleration in Newtonian mechanics has the not only the status – compared to the "essential" status of force – of an *non-essential* property, but also *exclusively the status of an appearance of force* and *never the status of an manifestation of the force*. Stated in other terms, even if Newtonian mechanics presupposes that acceleration is the effect of force as cause, this still does not explain why force (physically) necessary manifests itself as acceleration.

As shown earlier in this subchapter and in my analysis of the *Principia*, in all scientific laws derived by thought-operations from the second law of motion, as well as in all laws derived by gradual concretization from scientific laws obtained by such thought-operations, the magnitude of acceleration is permanently given, that is, *in all scientific laws that have already the status of laws of manifestations a magnitude with the status of appearance is present. Newtonian mechanics is not a science about the origin of forces, but a science only about the size/quantity of forces derived from the size/quantity of one of its effects (acceleration), and about the size/quantity of the other effects (e.g., work, momentum, turning effect, etc.) of forces derived from the size/quantity of these forces.*

Expressed in more general terms, theoretical sciences centered around laws with the structure of $L^{(k)}$ do not relate the respective natural kind to its underlying cause/ground in the sense that individuals sharing a certain essential property belong to one and the same natural kind. On the contrary, the reconstruction of the structure of $L^{(k)}$ in /18/ shows that one cannot exclude the possibility that there could be individual entities displaying the property of C as an essential property underlying all the phenomenal properties displayed by these entities, but still the latter need not belong to the same (natural) kind or universe N. And neither are theoretical sciences that are centered around laws with the structure of $L^{(k)}$ sciences the scientific laws of which express some powers, tendencies, or capacities. Once we do not know why the effect of type $E^{(k)}$ is produced by the cause/ground of type C, and when simultaneously $E^{(k)}$ is permanently present in all derivations of manifestations of C, then we have a *non-complete* derivation of manifestations. The cause/ground is then not grasped as having some tendencies, capacities, or powers to manifest itself under certain modification and singular conditions. In chapter 4 I will deal with a type of scientific law that grasps the cause/ground in such a way.

B. J. Woodward and C. Hitchcock and the typology of scientific laws

The reconstruction of the structure of laws of the types $L^{(k)}$ and $L^{(l)}$ in /18/ and /19/ enables one, in turn, to reconstruct the method of thought-construction and the method of gradual concretization as two methods of explanation clearly satisfying J. Woodward's (*f*) requirement. Even if, as shown in chapter 1, the fact that Hempel's reconstruction of explanation based on universal scientific laws cannot fulfill Woodward's (*f*) requirement has its roots in Hempel's oversimplification of the structure of scientific laws, J. Woodward and C. Hitchcock refrain from taking a step which suggests itself and that I have taken in this chapter, namely, to overcome that oversimplification and enlarge the typology of scientific laws in such a way that different models of explanations based on these types of scientific laws would already fulfill the (*f*) requirement. Instead, Woodward and Hitchcock restrict the sense of the term "scientific law" to that given in the (Humean) regularity approach and opt instead for the term "explanatory generalization."

That my enlargement of typology of scientific laws and models of explanation is a reasonable step within the framework of the philosophy of science can be demonstrated by the effects that Woodward's and Hitchcock's unwillingness to make such a revision has on the reconstruction of the structure of scientific laws and scientific explanations.

First, even if Woodward is capable of distinguishing already between explanation that can fulfill the *(f)* requirement and explanation that cannot, still he is not able to provide a more fine-grained typological differentiation between those explanations that already fulfill this requirement. This is readily seen in his examples of cases when from one set of explanatory generalizations different explananda can be derived (Woodward 1979, 47):

These generalisations are such that on the assumption that the mass of the earth had different values, a quite different value of the acceleration of a falling body could be derived. These generalizations are also such that we could use them to derive an expression for the rate of fall of a body falling from a distance which is no longer negligible in comparison with the earth's radius. Indeed these generalisations are such that we could used them to derive even more disparate explanda; for example we could use them in conjunction with other information to derive Kepler's laws and a great many other derivative laws of Newtonian mechanics.

From my typology of scientific laws and explanations already introduced here, it is clear that a deductive explanation based on a law appearance can lead to explananda with different singular appearances. For example, one can explain why a particular body falling in a vacuum (created, e.g., in a piece of equipment) covers quantitatively different distances by referring to the law of free fall /22/ and the different times it needs to cover the respective distances.

From the deductive explanation based on a law of appearance, I distinguished the type of explanation based on laws of the type $L^{(l)}$ and involving, already, thought-operations with modification conditions. These operations need not be limited only to *an abolishment of idealizations* – that is, *to a deliberate taking into account of certain modification conditions* – but could involve in a concrete case of explanation also *a deliberate introduction of additional idealizations* – that is, a *deliberate putting to zero of certain modification conditions that were before supposed to be at work* (i.e., were before supposed to be not equal to zero). To understand this type of thought-operations with modification conditions, let us return to the law of free fall, represented in /16/, as stated in the framework of Newtonian mechanics. Here g stands for acceleration due to the force of gravity and in that framework it holds $g = G \times M/(R + h)^2$. If we substitute this relation for g in the law of free fall /16/, we can understand that the law of free fall *in the form in which it was stated by Galileo* – who supposed that the body falls with a constant acceleration – is just a *special case* of the free fall derived in the framework of Newtonian mechanics – that is, once we suppose that the free fall takes place at height h which is much smaller than R, the radius of the Earth (so, $R \gg h$), then the relation for acceleration comes out as $g = G \times M/R^2$ and, thus, as a constant.

Finally, due to my enlargement of the typology of scientific laws and explanations, I was able to reconstruct yet another type of scientific explanation, where by means of thought-operations we derive from a law of the type $L^{(k)}$ various laws of the type $L^{(l)}$ type, for example, the derivation of the law for the movement of block sliding on an inclined plane from the second dynamic law in Newtonian mechanics.

The fact that Woodward does not *typologically* differentiate (i) the explanation of the different values for g given different values of R and M; (ii) the explanation of g when h is not any more negligible in comparison with R; and (iii) the derivation of such disparate explananda as, for example, the law of free fall, the law for the simple pendulum, and so forth, from the second dynamic law and the law of gravitation, is *one consequence* of his (and Hitchcock's) unwillingness to give up the reduction of all types of scientific laws to that of a (Humean) regularity type of scientific law and, at the same time, his unwillingness to enlarge the typology of scientific laws. This enlargement, as shown above, involves the enlargement of the typology of conditions appearing in scientific explanations by the concept of modification conditions. Once Woodward does not make an attempt at the enlargement of the typology of scientific laws, neither can he differentiate conceptually between modification conditions and singular conditions, and he speaks instead, very generally, about variables assuming various sets of values that he, in turn, understands as "assumptions about boundary and initial conditions" (Woodward 1979, 46).

Another consequence of that unwillingness can be found in Woodward's and Hitchcock's reconstruction of the law of free fall. This reconstruction should serve the end of comparing two explanatory generalizations G and G', so that the latter is more invariant than the former and where "G' makes explicit the dependence of the explanandum on variables treated as background conditions by G" (Hitchcock and Woodward 2003, 187). G should stand in their illustrations for the law of free fall stated by Galileo, which relates the covered distance and the time of fall. G, they claim, is invariant with respect to certain changes of the height from which the body is dropped, but (Hitchcock and Woodward 2003, 187)

it would fail to hold if the object were dropped from a height that is large in relation to the earth radius or if it were dropped from the surface of a massive body of proportions different from those of earth (such as Mars).

G' should stand for the generalization entailed by the law of gravitation and Newton's second dynamic law and that allows one to compute the time it would take an object to fall a certain distance. However, G' (Hitchcock and Woodward 2003, 187–188)

is not restricted in the way that Galileo's law is: it will remain invariant under changes in the mass and radius of the massive body upon which the object is dropped. It achieves this greater range of invariance by explicitly incorporating the mass and radius of the planet (or whatever) into the generalization as variables,

and they add in a footnote pertaining to this last quote that (Hitchcock and Woodward 2003, 199)

it will not help to explicitly incorporate various *ceteris paribus* conditions – e.g., the mass and radius of the earth as antecedents in order to render Galileo's law exceptionless. The resulting law would still say nothing about what would happen if these conditions were changed, and hence would not be invariant under *testing* interventions on the relevant variables.

From these quotes it is readily seen that because Woodward and Hitchcock do not provide an enlarged typology of scientific law involving a nonregularity understanding of the concept of scientific law neither can they grasp the profoundness of the change which the law of free fall undergoes once it is transformed from its Galilean form to the form it takes once it is derived in the framework of Newtonian mechanics. By a comparison of my reconstruction of these two forms as given in /22/ and /16/, it becomes clear that one cannot insert the clause stating the mass and radius of the Earth into the law of free fall as stated by Galileo because it would be *completely irrelevant*: in Galileo's formulation of the law of free fall, *acceleration is related neither to the mass nor to the size of the Earth, to which the body falls, because the acceleration of the falling body is here not viewed as an effect of the force of gravitation.*[124]

A similar confusion between the Galilean and the modern forms of the law of free fall, again due to the lack of a differentiated typology of scientific laws, can be found also in S. D. Mitchell's paper (2000). There she tries to illustrate the idea that the truth of scientific laws depends on certain conditions by using a continuum of contingency starting at its bottom with "All coins in Goodman's pocket are made of copper," and continuing to the law of conservation of energy at its top. Between them she locates what she labels as "Galileo's law of free fall" and about it she states the following (2000, 256):

The exact formulation of Galileo's "law" of free fall (including the acceleration due to earth's gravity) is conditional upon the mass of the earth. If the earth were of different mass, then it would no longer hold. And that the mass of the earth is what it is is a feature of the evolution of our universe – had the core been lead instead of iron, then acceleration of free falling bodies would be four times as great.

From what has been stated already in this section, it is clear that what she labels as "Galileo's law of free fall" is *neither* the law stated by Galileo – the latter did not relate acceleration to the gravitation of the Earth – *nor* the law derived inside the framework of Newtonian mechanics. In the latter the law of free fall *is stated, as not pertaining to the fall of a body toward the planet Earth with the mass it accidentally has, but to any central body of whatever mass*. Once inside the framework of Newtonian mechanics, for acceleration due to gravitation it holds that $g = G \times M/(R + h)^2$, then the law of free fall where this g appears would hold and be true even if the core of the Earth were lead.

D. Chang Liu on Idealizations

To delineate in a more precise way the concept of modification condition, which I differentiate from the concept of initial condition, let us compare my approach to idealizations[125] and gradual concretization with that of C. Liu.

---

124 Such an insertion would contradict also historical facts. The modern concept of mass appears for the first time only in Newton's *Principia*, where also, as shown above in 3.3.4, for the first time reflections on the masses of the planets, at least in their mutual ratios, are made.

125 For various approaches to idealizations in scientific theories see, e.g., (Barr 1971), (Barr 1974), (Nowak 1980), (Nowak 1992), (Nowakowa 1994), (Nowakowa – Nowak 2000), (Lind 1993), (Haase 1995), and (Teller 2004).

C. Liu, in order to deal with the issues like the realistic status of theories and their confirmation, puts the question how can one confirm a theory T, consisting of scientific laws. Once we would supply the respective initial and boundary conditions, T should yield testable predictions, but, unfortunately, "most of the obtainable initial and boundary conditions are idealized and the predictions yielded by the theory are approximate" (Liu 1999, 238). These idealized (initial and boundary) conditions he unifies into a set **I**, so that it holds that:

I = $\{I^i \mid i = 1, ..., n\}$,

here **I** is a partially ordered set against an R-scale, that is, some scale of being realistic. For all I^i's from **I** holds that T yields predictions, that is,

$T : I^i \mapsto P^i$,

where P^i is the ith prediction from a partially ordered set **P** given as

P = $\{P^i \mid i = 1, ..., n\}$,

and where any element of **P** approaches P^E, the observed datum.

On this basis, Liu then provides the following, more-generalized approach to the issue of idealizations (Liu 1999, 239). Let T^U be a nonidealized theory, that is, a theory free of any idealizations, yielding predictions, for example, $T^U \models P^E$. But, according to him, the derivation of P^E from T^U would be too complex and the introduction of idealizations is called for. Let T^n be a result of the unification of T^U with some idealizations Id, then we can derive a prediction P^n, that is,

$T^n = (T^U \ \& \ Id^n) \models P^n$,

where P^n is close but not identical to P^E, the observed datum, while for the idealizations holds:

Id = $\{Id^i \mid i = 1, ..., n\}$,

which is a set of ideal conditions such that for each element Id^i from this set we have a corresponding theory T^i entailing a P^i which is, in turn an element of the set **P** = $\{P^i\}$. Then, according to Liu, it holds that

T = $\{T^i\} \Rightarrow$ **P**,

meaning that each element from **T** yields a corresponding element in **P**. **T** stands here for a set of theories approaching theory T^U in descending degrees of idealization. Liu then gives the following notation for T^U ("C" stands for the background; "h_i" stands for variables; "w_i" stands for conditions):

$$(x)[Cx \rightarrow h_1 x = g_k(h_2 x, \ldots, h_n x; w_1 x, \ldots, w_n x)] \qquad /25/,$$

and states that "w_i" are "conditions which become ideal conditions when their values are set to zero" (Liu 1999, 239) and that "all $w_i x$'s are assumed to be non-zero values to begin with. In other words, if any $w_i x$ has a value zero before the idealization is taken, it should not be included in the original set" (Liu 1999, 253). So, for idealizations now should hold:

Id = $\{Id^i \mid Id^i = [w_i x = 0 \ \& \ w_{i+1} x = 0 \ \& \ \ldots \ \& \ w_k x = 0]; \ i = 1, \ldots, n\}$,

while for theories it holds that:

T = $\{T^i \mid T^i = [T^U \ \& \ Id^i]\}$,

where T^U is identical to /17/. For predictions derived from idealized theories it holds that:

P = $\{P^j \mid P^j = T^j [x/\alpha]\}$,

where these predictions are instantiations of T^i's, and for P^E, the datum derived from a nonidealized theory it holds that

$P^E = T^U[x/\alpha]$.

Finally, Liu uses the concept of concretization as follows: "Moving from T^i to T^U via the T^i's is called *concretization*" (1999, 239).

At least the following three points can be made against Liu's approach to idealizations and concretizations. First, he speaks about idealized conditions but adds the qualification "initial and boundary" (conditions). The term "idealized initial and boundary conditions" is, at least according to my view, an oxymoron.[126] Once we suppose that certain conditions are not at work, that is, we put them equal to

---

[126] It seems that Liu draws here on R. Laymon who speaks about the set *I* that should "represent the idealizing assumptions. Include into *I* the required parameters or initial condition values" (Laymon 1985, 148), and about a "more realistic specification of initial and boundary conditions" (Laymon 1989, 359).

zero, then they cannot be viewed as initial and boundary conditions; the latter are viewed, for example, in physics as being at work, that is, as being *not* equal to zero. Therefore, as already mentioned, I propose to differentiate conceptually within the framework of the philosophy of science between *modification conditions* which can be idealized, in the sense of being put equal to zero, and *singular conditions* which are always at work and thus can never be put equal to zero.

Second, C. Liu presupposes that one arrives at the idealized theory T^I by adding to the nonidealized theory T^U some idealization Id^i, that is, it should hold that $T^I = [T^U \& Id^i]$, or, in the terminology of his papers (2004a; 2004b), via the introduction of the idealized claim Id^i a *generic theory T is mapped on an idealized theory T^I*. But, on the other hand, he claims also that by the sequence of theories T^I one ends up in a non-idealized theory T^U. We thus have a cycle, where we *initially add* certain idealizations Id^i to T^U in order to obtain T^I, and then, by what *he* labels as *gradual concretization*, we should end up with T^U, that is, *with the same theory from which we initially proceeded*. The whole cycle is thus superfluous because *no growth of knowledge* occurs in it, contrary to the cycle of cognition given in explanation by gradual concretization. In the latter one *starts* from the knowledge of a singular phenomenon and *then* one has to find a concrete scientific law of type $L^{(l)}$ in order to subsume this singular phenomenon under the covering kind of type N^* and its cause/ground of type C given in this concrete scientific law. Then, as the *next* step, one has to find out the modification conditions, stated in the form of idealizations in the scientific law, relevant for this phenomenon and, thus, one has to take them into account by abolishing the respective idealizations and, at the same time, one has to discover the causal impact of these modification conditions. *Next*, one has to find the singular conditions of the singular phenomenon to be explained, and then one has to insert them into the already concretized scientific law and, as a last step, derive the singular phenomenon. One starts *just from the knowledge that a singular phenomenon obtained* and, via the chain of steps just mentioned, ends up with the *expanded* knowledge that the singular phenomenon is produced as an effect of a cause of type C and of the action of certain modifications and singular conditions.

Third, according to Liu, by adding idealization Id^i to T^U one should obtain the idealized theory T^I, for example, by adding the idealizations $w_i = 0 \& w_{i+1} = 0 \& \ldots \& w_k = 0$ to

$$T^U : (x)[Cx \rightarrow h_1 x = g_k(h_2 x, \ldots, h_n x; w_1 x, \ldots, w_n x)] \qquad /26/$$

we should obtain

$$T^I: (x)[Cx \& w_i x = 0 \& w_{i+1} x = 0 \& \ldots \& w_k x = 0 \rightarrow h_1 x = g_k(h_2 x, \ldots, h_n x)].$$

But, as shown above, *epistemic/cognitive* priority has the law of the type $L^{(l)}$ with the structure of /19/ together with the explanations by gradual concretization enabling to derive laws of the types $L^{(l-1)}, \ldots, L^{(0)}$. As shown in Figure 3.12, the method of explanation by gradual concretization is a method by means of which our knowledge about the causal nexuses expands and grows. Therefore, the "backward" idealization of modification conditions already introduced, that is, a reintroduction of idealizations $Cmod_i = d_i$, which have been already abolished in the s*ame concrete case of explanation* – $Cmod_i \neq d_i$ – is *formally* possible, but from the point of view of the growth of knowledge, *completely unreal*. Stated otherwise, even if it holds that ("|=" stands here for logical consequence)

$$L^{(0)} \& Cmod_1 = d_1 \models L^{(1)} \& Cmod_2 = d_2 \models L^{(2)} \& \ldots L^{(l-1)} \& Cmod_l = d_l \models L^{(l)},$$

one should *not* view the relation symbolized as "|=" as an *inverse derivation* with respect to the derivation by gradual concretization symbolized in /20/ and in /21/ as "-$_c$-|". In order to perform the operation symbolized as "|=", one has *first* to perform the operation of gradual concretization; *the latter, not the former, has – from the point of the genesis and growth of knowledge – priority.*

D. The "Noncausal" Analytic Mechanics

I reconstructed the type $L^{(k)}$ of scientific law with the structure of /19/ by drawing on Newtonian mechanics and its second dynamic law which I viewed as a (non-Humean) nonregularity type of scientific causal law.

Against this attempt to reconstruct a (non-Humean) non-regularity type of scientific causal law one could bring up the claim that classical mechanics can still be viewed as supporting the (Humean) regularity approach to scientific causal laws because it contains, not only Newtonian mechanics, but also a more-advanced apparatus, seemingly *free of any causal considerations*, suitable for the description of physical phenomena and labeled "analytic mechanics." M. Thalos seems to hold this view when she claims that the distinguishing mark of analytic mechanics is that there (Thalos 1999, 269)

are no forces or influences of any kind to bring about acceleration. Instead there are what are known as *variational principles*. Famous examples are Hamilton's principle (that the integral, over a system's path, of the difference between kinetic and potential energies is an extremum – typically a minimum).

Let us now try to find out if analytical mechanics, for example, Hamiltonian mechanics, is really non-causal; that is, whether it is free of any concept from Newtonian mechanics bringing in some aspects of causation.

Hamilton's variational principle is stated as follows:

$$\delta \int_{t_1}^{t_2} L\, dt = 0, \qquad /19/,$$

where "δ" stands for variation and "L" stands for the *Lagrangian function*. In order to state that variational principle in analytic mechanics we need to know the general structure of this function. How is the latter introduced? To answer this question we have to go back to the concept of work mentioned in my analysis of Newtonian mechanics as *force acting on a body along a certain path*.[127] If we hold to the equations $F = m\, dv/dt$ and $ds = v\, dt$, then for the integral $I = \int_{t_1}^{t_2} F\, ds$, it holds that $I = \int (m\, dv/dt)(v\, dt)$ and we obtain

$$I = [½mv^2]_{t_2}^{t_1}$$

Once we, based on knowledge from classical mechanics, view $½mv^2$ as the kinetic energy T of the particle with mass m and with the speed v, then we have

$$I = T_2 - T_1.$$

If, in addition, we suppose that work is performed in a unit of time (i.e., $dt = 1$), then I stands for the *work* performed by the force on a body.

Let me now consider the case when for the force F holds $F = -\nabla U$, where ∇ is the operator $(\partial/\partial x, \partial/\partial y, \partial/\partial z)$. For the integral $I = \int_{t_1}^{t_2} F\, ds$, already introduced, then it holds that

$$I = -\int_{t_1}^{t_2} (\nabla U)\, ds = -U_2 + U_1.$$

Taking into account that for the same integral, as shown earlier, it holds also that $I = T_2 - T_1$, we obtain the equation $T_1 + U_1 = T_2 + U_2$, which is the law of the conservation of energy E, so that $E = T + U$ for the case when $F = -\nabla U$; this latter equation characterizes the field produced by F as *conservative*.

With this background we can now introduce the Lagrangian function as follows. To the Descartian coordinate s_i we add the generalized coordinate q_k, so that it holds that the former is a function of the latter (and vice versa). We can then introduce the generalized velocity dq_k/dt via the relation

$$dq_k/dt = \sum_i (\partial s_i/\partial q_k)(dq_k/dt),$$

---

127 I draw here upon (Ter Haar 1971).

and once we have for the kinetic energy $T = \frac{1}{2} \sum_i m_i (ds_i/dt)^2$, we obtain

$$T = \frac{1}{2} \sum a_{kl} (dq_k/dt)(dq_l/dt),$$

where $a_{kl} = \sum_i m_i (\partial s_i/\partial q_k) \times (\partial s_i/\partial q_l)$. Once we move from the Cartesian coordinate s_i to the generalized coordinate q_k, then the function $U = U(s_i)$ becomes $U = U(q_k)$, and we can then introduce the Langrangian function as follows

$$L = L(q_k, dq_k/dt) = T(q_k, dq_k/dt) - U(q_k).$$

Having derived Langrangian function, we can then, finally, substitute into Hamilton's variational principle as given in /27/.

Let me now pose the following question: *Is the concept of the Lagrangian function completely independent of the conceptual apparatus of Newtonian mechanics?* I introduced it as a difference between kinetic and potential energy, where both these concepts have their origin in Newtonian mechanics. In addition, these are inside the framework of Newtonian mechanics related to the concept of force. And for the concepts of generalized coordinate q_k and (generalized) velocity dq_k/dt it holds that they are introduced as *counterparts* to the Cartesian coordinate from Newtonian mechanics. So, once I introduce the Lagrangian function using the sequence I described, then at the same time I *pull in the conceptual apparatus of Newtonian mechanics* from which the Lagrangian function cannot be separated. The meaning of the expression "Lagrangian function" in analytic mechanics cannot be separated from the meanings of terms like "coordinate," "acceleration," "mass," "force," and so forth from Newtonian mechanics. Once the latter are used in the process by means of which the meaning of the term "Lagrangian function" is introduced in analytic mechanics, then analytic mechanics cannot get rid of the meanings of those terms; they are, so to speak, "wrapped up" into the meanings of the expressions appearing in analytic mechanics. Stated otherwise: *analytic mechanics as a theory that has already been constructed cannot be separated from those meanings – having their origin in Newtonian mechanics – which were used in the construction-process of analytic mechanics*; they are not parts of a "ladder" that can be "kicked away" once classical mechanics has "climbed up" from the level of Newtonian mechanics to that of analytic mechanics. This means that from the point of view of the *construction* of analytic mechanics the latter could neither *historically* precede Newtonian mechanics, nor can it be – from the point of the *internal relations* between meanings of its own terms, like "general coordinate," "Lagrangian function," "Hamiltonian," and so forth – separated from the meanings of terms given in Newtonian mechanics.

Such a meaning-dependency of analytic mechanics on Newtonian mechanics holds, not only at the level of construction of general theory, but also at the level of the solutions to concrete physical problems. To show this, let us take the case of the movement of a simple pendulum of the length *l*, mass *m* of the suspended body, and α as the angle of deviation. Schematically, the swings of such a pendulum can be represented as follows:

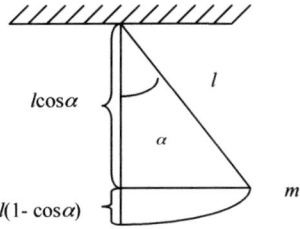

Figure 3.15 Swings of a simple pendulum according to Newtonian mechanics

Drawing on Newtonian mechanics we can state that the kinetic energy of a body rotating around its own fixed axis is $T = \frac{1}{2} J \cdot (d\alpha/dt)^2$, where "*J*" stands for moment of inertia and $d\alpha/dt$ for angular velocity, while for the pendulum it holds that $J = ml^2$. So we obtain for the kinetic energy of the pendulum $T = \frac{1}{2} ml^2 (d\alpha/dt)^2$. For the potential energy, again in Newtonian mechanics, it holds that $U = m \cdot g \cdot h$, and because for the pendulum holds $h = l \cdot (1 - \cos\alpha)$, we obtain $U = m \cdot g \cdot l \cdot (1 - \cos\alpha)$. If we suppose that the deviations are rather small then $\cos\alpha = 1 - \frac{1}{2}\alpha^2$ holds. The Lagrangian function for the simple pendulum can then be stated as follows:

$L = \frac{1}{2} ml^2 (d\alpha/dt)^2 - \frac{1}{2} mgl\alpha^2$.

If, *using the concepts of Newtonian mechanics*, the Lagrangian function is already derived for the swings of a simple pendulum, then the whole apparatus of analytic (Langrangian and Hamiltonian) mechanics can be applied to them, for example, we can state for them the Euler-Lagrange equation, the Hamilton equation, and so forth.

3.3 The Nature of Nomic Measurement

Let me now deal with the structure of measurement based on scientific laws that relate the quantity of the phenomena-appearances with the quantity of their common cause-ground and that thus enable us, by finding the size/quantity of the effects, to measure the size/quantity of the cause. I start with a reconstruction of the structure of nomic measurement, both at the level of phenomena-appearances, exemplified by the measurements performed by Mersenne and Huygens and analyzed in chapter 2, and at the level of the movement from phenomena-appearances to their common ground. Finally, as a case study of the latter type of nomic measurement, I show the main stages of Newton's measurement of the masses of the celestial bodies as given in Book III of his *Principia*.

3.3.1 Mersenne, Huygens, and the structure of nomic measurement

As shown in chapter 2, the measurements and experiments performed by Mersenne and Huygens were based on the *knowledge* and *conscious simultaneous use* of two *scientific laws*: the law of free fall stating the proportionality $s \propto t^2$ and the law for the simple pendulum stating the proportionality $t \propto l$. So as the measurements of Mersenne and Huygens remained at the level of the theory we today view as kinematics, I labeled in chapter 2 this type of measurement as the *nomic measurement of phenomena-appearances*. With respect to my reconstruction of these laws in subchapter 3.2, I can give now a more precise reconstruction of this type of nomic measurement. The idealizations involved in these two laws are, as shown in subchapter 3.2, important for scientific explanation by gradual concretization. At the same time, they are important because they enable us to view those proportionalities as two conceptually *closed* systems since we need not take into account additional concepts like friction, change of the length of the suspension cord under the impact of the force, and so forth. As shown in chapter 2, both Mersenne's and Huygens' experiments were based on an interlinking of the law of free fall and the law for the simple pendulum – where the former should serve, first, as the *explanatory* basis for the relation between the covered distance in free fall and elapsed time and should be, second, *experimentally* tested. The latter should serve as the basis for the instrument enabling *time to be measured*; and where the mediating link between the former and the latter was the magnitude of time. These two aspects of the Mersenne-Huygens experiments and measurements, namely, *conceptual closure* by means of idealizations and *conceptual linking* can be represented as shown in Figure 3.16.[128]

---

128 Here I draw on (Pawson 1989).

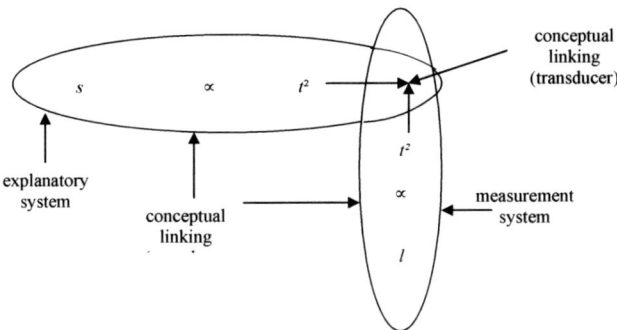

Figure 3.16 Conceptual closure and conceptual linking in Mersenne-Huygens experiments and measurements

Following R. Pawson, I label the magnitude we want to measure as the *measurand*; in Figure 3.16 it is the covered distance, while time is the so-called *transducer* functioning as the link or *transduction* between the conceptual system containing the magnitude to be measured and the conceptual system on which the measuring instrument should be based. In Figure 3.16 the transducer is the magnitude of time while the measuring instrument is the seconds pendulum.

The arrangement of the Mersenne-Huygens experiments based on conceptual linking and conceptual closure is then as follows ($s \equiv g$ expresses that for the first second s is numerically identical to g, which is the *measurand*):

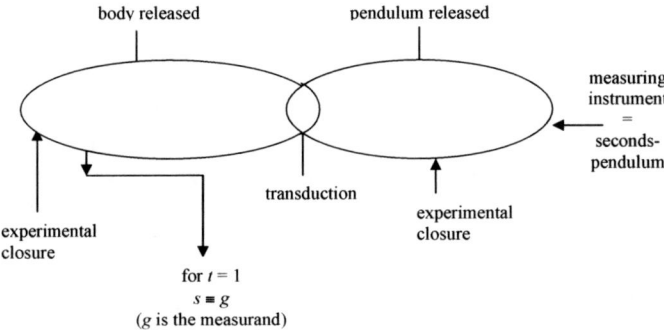

Fig. 3.17 The Mersenne-Huygens experiments and measurements from the point of view of transduction and experimental closure

From this I draw the conclusion[129] that nomic measurement at the level of phenomena-appearances is based on *transformation*, so that the *measurand* is linked – initially, conceptually and, then, practically in experiment with another property. This type of measurement is simultaneously based on an *intersection* of scientific laws – by finding which magnitudes are common to them, so that the measurement of one of them will yield the size/quantity of the other. So, for example, in the Mersenne-Huygens experiments:

(i) The magnitude l is measurable directly, while t is measurable by a mechanical clock calibrated according to the sidereal day.
(ii) Based on the knowledge of the law for the simple pendulum, the fixing of l enables one to obtain a timepiece beating certain time-intervals to be fixed initially by a "sidereally" calibrated mechanical clock.
(iii) Finally, based on the knowledge of the law of free fall, one finds by means of t, the size/quantity of the magnitude s, and for $t = 1$ one finds the numerical value of g.

The Mersenne-Huygens experiments and measurements stand for a case of the interchaining of only two laws. But once such an interchaining is accomplished and the size/quantity of the *measurand* is found out then other laws can be "hooked up" to this two-component chain. An example of such "hooking up" is Huygens' endeavor to use his newly constructed pendulum-clock which by itself (as shown in chapter 2) is the practical implementation of an initially *explanatory* system – as a *measuring device* for finding the longitudinal position of oceangoing ships.[130] The *explanatory* theory is now the knowledge that the Earth is (again, as an idealization or conceptual closure) a globe. Based on this knowledge, we can divide the surface of the Earth from the North to the South into segments using 360 principal lines of longitude (or *meridians*). And once we know that this globe rotates around its own axis once in 24 hours (yet another idealization or conceptual closure) we can view the distances between the meridians as proportional to the time the Sun (apparently) travels. So, if 24 hours corresponds to 360 degrees, 1 second of the Sun's apparent motion corresponds to 15" of space covered in this time. For any oceangoing ship, one can find out its longitudinal position, first, based on the determination of high noon (that is, when the Sun is at its highest position) by means of a sextant, and, second, by finding the time difference between the high noon on the ship and the time indicated by the ship's on-board pendulum-clock which indicates the time at the geographical location chosen as the 0th meridian.

---

129 Here I draw on (Pawson 1989).
130 For details see, e.g., (Mahoney 1980).

In the case of the ship we have a highly interesting *cognitive-practical cycle*. As shown in chapter 2, the seconds-pendulum, initially used in the Mersenne-Huygens experiments just as a measuring instrument to determine the value of *g*, is transformed by means of Huygens' *cognitive-practical* endeavor into an *explanatory* system that is then *practically* implemented in a pendulum-clock. The latter can, in turn, be used as a *measuring* instrument providing the values of the magnitude of time on which one can hook up, in turn, the *explanatory* theory for the proportionality of time and longitudinal position.

3.3.2 Newton's measurement of the masses of celestial bodies and the structure of nomic measurement

In section 3.3.1, we saw that the concatenation of several laws enables one to develop and apply a highly advanced technique of experimentation and measurement. Such a concatenation and measurement are important also for another reason, namely, that they lead to the identification of that magnitude which is viewed as *the* effect of an underlying cause and, in addition, as that magnitude the size/quantity of which can serve us as the external measure of its cause. Newton, as already shown, used the magnitude of acceleration as the measure of the accelerative force and the phenomenal characteristics of the paths of bodies orbiting a center of force to find the space-dependence of this force. Newton, however, did not stop once he found the space-dependence of the centripetal force, but used it to perform computations that – like Mersenne's and Huygens' determinations of *g* – have the character of a nomic measurement; here I refer to his endeavor in corollaries 1 and 2 in proposition 8 (Book III) to find the masses[131] of several celestial bodies.

In corollary 1,[132] he sets as his aim to find the "weights" toward the Sun and the planets Jupiter, Saturn, and Earth by using the physical characteristics of their satellites.[133] By "weight" he understands here the *weight of bodies with the same mass*, thus we will use here the notation weight*, or w^*, for short. So as according to proposition 6 (Book III) it holds that weight is a motive force and for a given mass (i.e., *m* is a constant) motive force turns into an accelerative force proportional, by Definition 7, to acceleration, w^* stands for accelerative measure in the sense of this definition. At the same time, Newton introduces the following idealizations or conceptual closures: all bodies are viewed as homogeneous (i.e.,

---

131   In Corollary 3, Newton deals with the densities of planets; see (Cohen 1998, 90).
132   I draw here on (Cohen 1998) and (Cohen 1999, 218–231).
133   For Jupiter, Newton chooses as its satellite Calisto; for Saturn, Titan, for the Sun, Venus and for Earth the Moon.

their density does not change in space) and the orbits of the satellites are circular (i.e., ellipses whose eccentricity equals zero). Based on this, he can apply the result from Corollary 2 of proposition 4 (Book I) that for the centripetal force (here gravity) acting on an object with the same mass, thus for w^*, it holds that $w^* \propto r/T^2$, where r is the radius of the circular orbit of the satellite and T is its period, while taking also into account that this proportionality holds "at the surfaces of the planets or at any distances from the center" (1999, 812).

In a next step, Newton computes the radius of the orbit of the chosen satellites on the same scale, namely, the distance scale, the unit of which is the mean distance from the Earth to the Sun.[134] Then he computes the quotient r/T^2 and obtains for each of the satellites the size/quantity of w^* with respect to the satellite's parental planet at the normal orbiting distance. Next, Newton converts these values at their respective orbital distances into the values at some common distance from the center; as such distance he chooses the radius of the orbit of Venus, drawing on proposition 8 (Book III) that "weights at the surfaces of the planets or at any other distances from the center are greater or smaller (by the same proportion) as the inverse squares of the distances", that is, $w^*_1 / w^*_2 = r_2^2 / r_1^2$ (1999, 812). On this basis Newton gives the respective values of the magnitude w^* as shown in Table 3.7.

	Sun	Jupiter	Saturn	Earth
w^*	1	1/1067	1/3021	1/169282

Table 3.7 "Weights" of the satellites in corollary 1, proposition 8

As a next step, still in Corollary 1, he computes[135] w^* at the surfaces of the four central bodies, or w^*_s for short, and obtains, based on the scale where the radius r of the Sun is 10 000, the results shown in Table 3.8.

	Sun	Jupiter	Saturn	Earth
r	10 000	997	791	109
w^*_s	10 000	943	529	435

Table 3.8 Radiuses and weights* of the Sun, Jupiter, Saturn and Earth in corollary 1, proposition 8

---

134   For the satellites of Jupiter and Saturn the computation of the radii of their orbits is given in the first edition of the *Principia* (1687/1960, 413–414); on this see also (Cohen 1998, 93–94).
135   Here Newton draws again on $w^*_1 / w^*_2 = r_2^2 / r_1^2$.

Finally, in corollary 2, the relative masses of the Sun and of the three planets are given because they are "as their forces at equal distances from their centers" (1999, 813). The results obtained by Newton in the third edition of the *Principia* are as shown in Table 3.9.[136]

	Sun	Jupiter	Saturn	Earth
mass	1	1/1067	1/3021	1/169282

Table 3.9 Relative masses of the Sun, Jupiter, Saturn and Earth in corollary 2, proposition 8

What is the moral, from the point of view of nomic measurement, of Newton's endeavor to find the relative masses of celestial objects? Newton, in his derivation of the masses of celestial objects:

a. draws on proposition 4, corollary 2 (Book I), which states that for the centripetal force F_C it holds $F_C \propto r/T^2$, and where this characteristic of F_C acting on a body moving in a circular orbit is derived on the basis of the phenomenal characteristics, first, of this orbit (arc and radius) and, second, of the characteristics of the movement of the body (its speed in corollary 1, and its period in corollary 2).
b. draws also on propositions 1 through 5 (Book III), by means of which he arrives in the scholium to proposition 5 at the conclusion that F_C "by which celestial orbits are kept in their orbits ... is [the force of] gravity" F_G (1999, 806), that is, $F_C = F_G$. As shown in 3.1.2.A and in 3.1.4.B, crucial for the derivation of these propositions are the phenomena of motion of these celestial bodies. And by supposing that F_G acts on bodies with equal mass (i.e., m is a constant) F_G is reduced to w^*.
c. From $F_C \propto r/T^2$, by a., and $F_C = F_G$ and $F_G \propto w^*$, by item b. (in this list), he obtains for the satellites' w^* at distance r from the center of the respective planets $w^* \propto r/T^2$. Based on data for r and T, Newton computes the respective values of w^*. These values he then rescales by using the radius of Venus' orbit and the equation $w^*_1 / w^*_2 = r_2^2 / r_1^2$ from proposition 8 (Book III).
d. Based on the same equation $w^*_1 / w^*_2 = r_2^2 / r_1^2$, it holds for the relation between w^* to w^* at the surface of a planet, w^*_S, for short, whose radius is r_{PL}: $w^* / w^*_S = r_{PL}^2 / r^2$. This enables him to compute the size/quantities of w^*_S for the respective celestial bodies.
e. By taking into account that (by propositions 7 and 8) $w^*_S \propto m_C/r_{PL}^2$, where m_C is the mass of a celestial body, he obtains $m_C \propto w^*_S \times r_{PL}^2$.

---

[136] For a comparison with more recent results see (Cohen 1998, 89–90).

f. Based on the equation in item d., he substitutes $r^2 \times w^*/r_{PL}^2$ for w^*_S and obtains $m_C \propto r^2 \times w^*$.

g. Finally, by taking into account that "the quantities of matter in the planets are as their forces at equal distances from their centers" (1999, 813), that is, that r is viewed as a constant, he obtains from the last relation in item f. the proportionality $m_C \propto w^*$. This explains why Newton gives for m_C and for w^* the same numerical values, as it is readily seen from the tables given in tables 3.7 and 3.9.

Newton's derivation of the masses of celestial bodies in corollaries 1 and 2 of proposition 8 (Book III) is, thus, based, first, on certain data obtained by astronomical observations and measurements. Here I mean the data for the periods of the orbital movements of the satellites and the data for the radii of their orbits relative to their parent planets.

Second, coming into play is Newton's thought-movement from the phenomena of motion of bodies to their respective forces in Book I. Here I mean the determination of the size/quantity of the centripetal force as proportional to r/T^2, based on the phenomenal characteristics of the orbits (arc and radius) of bodies and of the movements (velocity and period) of these bodies.

Third, arising from Newton's derivation of the masses of celestial bodies is his derivation (in propositions 1 through 5, Book III) of the quantitative characteristics of the force of gravity based on the phenomena of motion of the celestial bodies (radii of their orbits, periods, etc.) Newton's derivation can thus be schematically represented as shown in Figure 3.18 (I refrain here, in order not to overburden the scheme, from symbolizing both the respective idealizations/closures and the fact that the term "centripetal force" in Book III has its origin in Book I. The lower indices "S," "J," "ST," and "E" stand here for "Sun," "Jupiter," "Saturn," and "Earth;" "w^*" with these indices stand for the computed size/quantity of "weight" for the respective celestial objects; "w_S^*" with these indices stands for the computed size/quantity of "weight" at the surfaces of the respective celestial objects,; "m_C" stands for the mass of a celestial object; and "m" with those indices stands for the computed size/quantity of the mass of the respective celestial objects).

When compared with the type of nomic measurement exemplified by Mersenne's and Huygens' measurement of what we today symbolize as g, it can be readily seen that Newton's measurement of masses of celestial objects is based already on a thought-movement from the phenomena-effects to their underlying cause, so that knowing the size/quantity of the former enables one to find the size/quantity of the latter. Worth mentioning here is also the fact that the data entering into the sequence of Newton's derivation of magnitudes and enabling one to

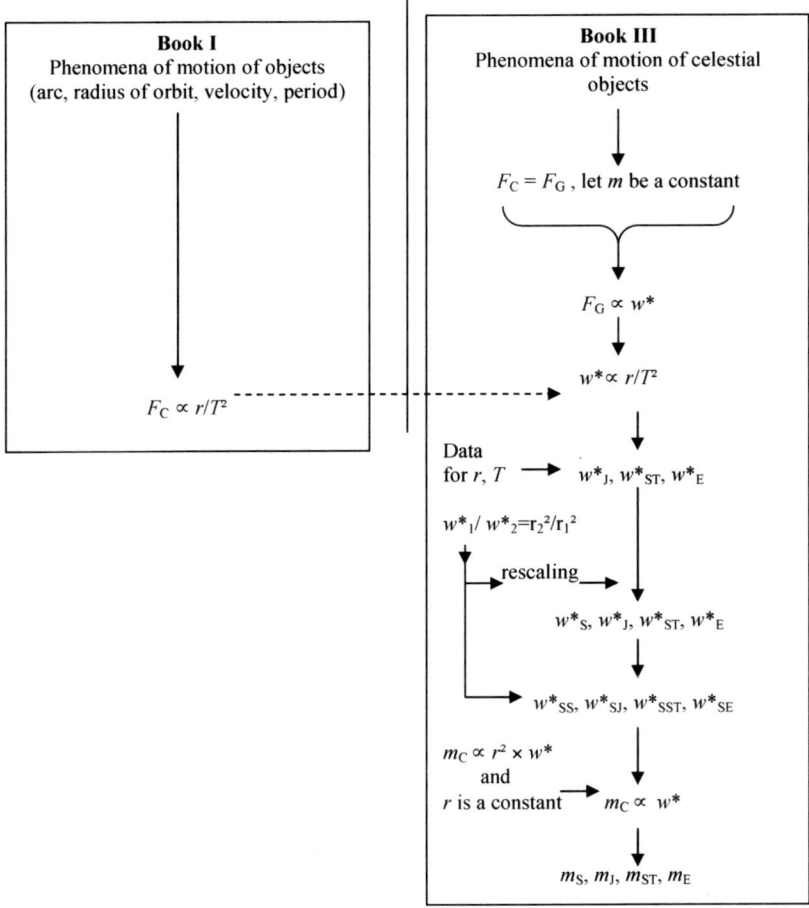

Figure 3.18 Newton's derivation of the masses of celestial objects

compute the size/quantity of "weight" are obtained *independently* of Newton's sequence of conceptual derivations in Book I and III of the *Principia*. They are thus, with respect to the latter, *but only with respect to the latter*, of a *pretheoretical* nature. This independence explains why Newton's result of the computation of the mass of the Earth is far from the mark: the input data for his computation of this mass were at the time of Newton, off the mark.[137]

---

137 On this see (Cohen 1998, 86 – 87, 93 – 94)

3.4 Three attempts at reduction

Let me now analyze three highly interesting attempts at the elimination of "metaphysically biased" theoretical concepts of science, as given in Hempel's paper "Theoretician's Dilemma," in Brian Ellis's papers (1963; 1965; 1976), and in Ernst Mach's *Science of Mechanics*.

3.4.1 Hempel and theoretical Concepts

According to Hempel (1958, 173),

> scientific research in its various branches seeks not merely to record particular occurrences in the world of our experience: it tries to discover regularities in the flux of events and thus to establish general laws which may be used for prediction, postdiction, and explanation,

and these general laws should have "the function of establishing systematic connections among empirical facts" (Hempel 1958, 177). Hempel also claims that "scientific systematization is ultimately aimed at establishing explanatory and predictive order among the bewilderingly complex 'data' of our experience, the phenomena that can be 'directly observed'" (Hempel 1958, 177). Hempel is however forced to concede that (1958, 177):

> it is a remarkable fact, therefore, that the greatest advances in scientific systematization have not been accomplished by means of laws referring explicitly to *observables*, i.e., to things and events which are ascertainable by direct observation, but rather by means of laws that speak about various *hypothetical*, or *theoretical*, *entities*, i.e., presumptive objects, events, and attributes which cannot be perceived or otherwise directly observed by us.

Hempel then, by means of these reflections, arrives at the following problem (Hempel 1958, 179):

> Why should science resort to the assumption of hypothetical entities when it is interested in establishing predictive and explanatory connections among observables? Would it not be sufficient for the purpose ... to search for a system of general laws mentioning only observables, and thus expressed in terms of the observational vocabulary alone?

This leads him to the following example which should illustrate the way how one can arrive at a scientific systematization not based on hypothetical, or theoretical, entities. Let us have the statement,

(1) Wood floats on water; iron sinks in it.

This statement has the character of an empirical generalization, but – unfortunately – it is stated only for entities from the universe of wooden objects and from the universe of iron objects, and, what is even more serious, there are entities from these universes for which (1) does not hold, for example, exotic types of woods and hollow iron spheres. It is therefore more convenient to introduce the concept of specific gravity s of a body x, defined as the quotient of its weight w and of its volume v.

(2) Def. $s(x) = \frac{w(x)}{v(x)}$.

While, according to Hempel, weight and volume are observable properties, specific gravity is not (1958, 180), and for the latter one can state, drawing on the principle of Archimedes, the following generalization:

(3) A solid body floats on a liquid if its specific gravity is less than that of the liquid.

Let us now consider the case that we want to explain[138] why a certain solid body b floats on a given body l of liquid, that is, we are searching for an explanation of a situation expressed by the following statement:

(O_1): The solid body b floats on the body l of liquid,

which, because it is stated exclusively in observational terms, we can view as an observational statement. In order to explain the state of affairs expressed in O_1 we have to find out the weight w and volume v for b and l. Let the following four statements express this information:

(O_2): $w(b) = w_1$,   (O_3): $v(b) = v_1$,
(O_4): $w(l) = w_2$,   (O_5): $v(l) = v_2$,

which, according to Hempel, are also observational statements, and where w_1, w_2, v_1, and v_2 are certain positive real numbers. From these four statements we can by means of definition (2) derive the specific gravity for b and l:

---

138 I deviate here from Hempel who considers the case of prediction and not of explanation.

(4) $s(b) = w_1/v_1$; $s(l) = w_2/v_2$.

Let us suppose now that we found out that it holds that $s(b) < s(l)$. Then it is possible to derive from law (3) the observation statement

(O_6): The solid body b floats on the body l of liquid,

by means of which, we give an explanation of the observed phenomenon expressed in statement O_1. The movement from the data given in observation to explanation of the observed phenomenon can be schematically represented as follows (here the arrows stand for deduction and the number above the arrows for the law used in this deduction):

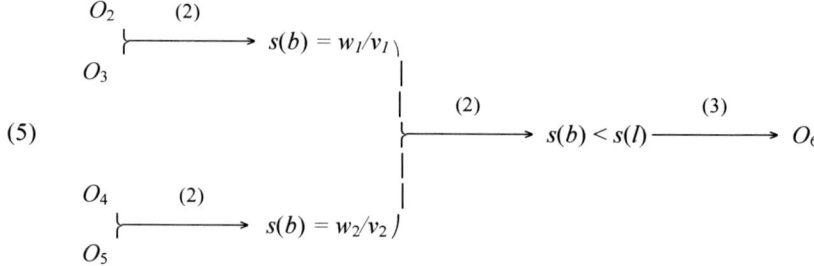

Figure 3.19 *Movement from data given in an observation to the explanation of an observed phenomenon*

According to Hempel the movement represented in this figure, on the one hand, starts and ends with observational statements but, on the other hand, contains also a theoretical roundabout via the theoretical term "specific gravity." But according to his view, it is possible to derive the statement O_6 from the statements O_2, O_3, O_4, and O_5 without such a roundabout, namely, by the following restating of the generalization (3) by means of definition (2):

(3') A solid body floats on a liquid if the quotient of its weight and its volume is less than the corresponding quotient for the liquid.

Based on such a restatement it is possible, according to Hempel, in a scientific systematization to eliminate theoretical terms.

3.4.2 B. Ellis and the Concept of Force

A similar elimination strategy was developed by Ellis for the concept of force. He claims that the characteristic feature of force is that the "distinguishing feature of forces generally is that in some sense their existence *entails and is entailed by* the existence of the effects they are supposed to produce" (1965, 31) and from this he draws the conclusion that forces are (1963, 187):[139]

> very queer scientific entities. They are not like other theoretical entities, such as atoms, or genes, since the existence of atoms and genes is not entailed by the existence of the effects they are supposed to produce. Moreover, forces are radically different from ordinary causes since causes and effects are logically independent existences. It need not surprise us, therefore, if we should find that forces are logically superfluous entities in science.

In order to prove that forces are superfluous entities in physics, he embarks on the following strategy.[140] The concept of force is, according to Ellis, introduced in mechanics in order to deal with either the persistence of a system in nonnatural state or the nonnatural way of change of the state of the system. A system is thought to be in a nonnatural state if and only if one considers that its persistence in that state requires a causal explanation, and it is considered to be changing in an nonnatural way if and only if one thinks that its changing in this way requires a causal explanation. From this, Ellis draws the conclusion that we claim that forces exist in nature simply because *we choose* to regard some states or changes of states of a system as nonnatural. But since, according to his view, there are no objective criteria for distinguishing between natural and nonnatural states or changes of states, forces have a conventional existence.

In order to prove that this element of conventionality is given in the concept of force in mechanics, Ellis constructs his own system of mechanics – the so-called *E*-system – which is based on a principle of natural motion different from that applied in Newtonian mechanics – the so-called *N*-system. The *E*-system, he claims, should as the same time be viewed as an *alternative* to the *N*-system (Ellis 1965, 50; 1976, 174). Ellis begins the construction of the *E*-system by using the law of gravity for two bodies A and B with masses m_A and m_B, respectively, in the form $f = Gm_A m_B/r^2$. From this law and Newton's second law for bodies A and B, $f = m_A a_A$ and $f = m_B a_B$, the absolute acceleration of bodies A and B, respectively is as follows:

$$a_A = Gm_A/r^2, \quad a_B = Gm_B/r^2.$$

---

139 On this see also (Ellis 1965, 47).
140 On this see (Ellis 1963, 185) as well as (Ellis 1965, 46–50).

Then the formula

$$a_A + a_B = G(m_A + m_B)/r^2$$

stands for the relative acceleration of bodies A and B toward each other. Ellis takes this formula as the basis of the first law of his E-system so that a natural change is a change that is in accord with this formula. This law states (1965, 49):

> every body has a component of relative acceleration toward any other body in the universe directly proportional to the sum of their masses and inversely proportional to the square of the distance between them – unless it is acted upon by a force.

The second and the third laws of the E-system are identical in formulation but not in application with the second and third laws of the N-system; for the term "acceleration" has here the meaning "acceleration relative to the natural accelerated motion" rather than "acceleration relative to the natural non-accelerated motion", and there is no law of gravity in the E-system. In this way, according to Ellis (and similar to Hempel's practice of elimination of the term "specific gravity") we get rid of the concept of force because the laws of the E-system (1976, 173)

> enable us to calculate directly the magnitudes of the effects to be explained from the magnitudes of those other quantities whose measures are used to determine the magnitude of the forces we say combine to produce these effects.

3.4.2 E. Mach's Elimination of the Concept of Force

Mach attempted to purge mechanics from the concept of force as something causal and replace it by an understanding of a kind that would allow it to be derived as a purely mathematical expression in the form $F =_{Df} ma$. He puts forward the following statement: "All those bodies are bodies of equal mass, which mutually acting on each other produce in each other equal and opposite acceleration" (Mach 1960, 266). He generalizes this case and shows that if bodies A and B receive respectively, as the result of their mutual interaction, the acceleration with the absolute value of a_A and a_B, then body B has a_A/a_B times the mass of body A. Then "if we take A as our unit we assign to that body the mass m which imparts to A m-times the acceleration that A in reaction imparts to it" (Mach 1960, 266). Finally, he claims that "my definition is the outcome of an endeavor to establish the interdependence of phenomena and to remove all metaphysical obscurity" (Mach 1960, 267).

3.4.3 How to evaluate philosophical attempts at reduction?

Let us now analyze Hempel's, Ellis's and Mach's elimination strategies. In the case of Hempel's strategy, it is possible to distinguish, at least according to my view, two levels of the elimination strategy: the first is the level of *philosophico-methodological reflections* which should serve as an introduction to the attempt at the elimination of the term "specific gravity," and the second is the very attempt at this elimination. Let me start with the first level. When one gets acquainted with the introductory philosophico-methodological reflections, he or she can notice the contradiction that Hempel expresses as the contradiction between two *really given "states of affairs" in empirical sciences*. He claims, on the one hand, that the aim of science is to discover regularities in the flux of events given to us in experience and, thus, to state empirical laws enabling us to express systematic relations among empirical facts. On the other hand, however, Hempel is forced to concede that empirical sciences state, not only laws remaining at the level of the so-called empirical generalizations, but also laws pertaining to nonobservable entities, that is, laws that are already part of the so-called level of *theory-formation* (Hempel 1958, 177–178). How should we deal with this contradiction? According to my view one faces here not a contradiction between states of affairs really given in empirical sciences, but in fact a confusion, on Hempel's part, between the descriptive and the normative, as well as an imputation of certain philosophico-methodological views on scientific theories to the very empirical sciences.

Hempel's claim that the aim of science *is* to discover regularities in the flux of events given to us in experience and, thus, to establish *empirical* laws with the aim of expressing systematic relations between observable facts represents, in fact, a *normative* statement formulated in the framework of a phenomenalistically oriented philosophy to which Hempel holds. This normative statement can be expressed as follows:

Science should, according to the representatives of the so-called logical empiricism, make attempts to discover regularities in the flux of events and, thus, establish general laws the function would be to express systematic relations between observable facts. The aim of scientific systematization should therefore be to find an explanatory and predictive order among the bewilderingly complex "data" of our experience, the phenomena that can be "directly observed."

With the background of such a restatement of the original, seemingly descriptive statements of Hempel, one can also evaluate in a different manner Hempel's attempt at eliminating the term "specific gravity" in favor of terms that should have, according to Hempel, the character of observational terms. First, it is *not* an attempt that should realize tendencies that are somehow inherent to the very empirical sciences, but in fact an attempt to prove the validity of a certain philosophical view, namely, a phenomenalistic view, of empirical sciences. Second, it is an attempt that, with respect to empirical sciences, is part of *metascientific* reflections that appropriate and interpret the results of these sciences from a *specific*, philosophico-methodological point of view. Thus, these reflections are of a *post festum* nature; they are accomplished only *after* empirical sciences have arrived at certain results.

Third, by differentiating between the normative and descriptive components in Hempel's views in his paper "Theoretician's Dilemma" one faces the question of how one should evaluate the former. One option is to confront the content of the statement characterize above as a (masked) normative statement with the descriptive statement. If science states not only empirical laws and theories but also laws and theories referring to nonobservable entities, then it is possible to declare that Hempel's normative prescriptions do not correspond to the actual state of empirical sciences in the sense that they are too simplifying or even reductive; they do not take into account the higher level of the endeavor Hempel labeled as "scientific systematization."

Against such a claim, a phenomenalistically minded philosopher could object that empirical sciences subject themselves to an excessive high risk of pseudo-scientific metaphysics if they use terms referring to nonobservable entities. A second option for evaluating Hempel's normative aspect of his paper "Theoretician's Dilemma" is to leave the level of philosophico-methodological reflections and to move to an evaluation of the very attempt on his part to eliminate the term "specific gravity." If it would be possible to prove that his attempt fails, then one could view this, if not as refutation of the (hidden) normative aspect of the paper then, at least, as casting doubt on it.

Discounting Hempel's view that his reduction of the term "specific gravity" is successful, one can state at least the following three objections. First, Hempel accomplishes his reduction in such a way that he restates, by means of the expression (2) and statement (3) into statement (3'). This means that he draws on expression (2), which he introduced in order to define the term "specific gravity." Hempel in his endeavor for a reduction uses the quotient $\frac{w}{v}$ but, the same time, claims that he can detach it from the term "specific weight." But then the *very physical sense of the introduction of this quotient is being completely lost.*

Second, the elimination of the term "specific weight" via a "return" to the quotient $\frac{w}{v}$ does not represent, as Hempel claims, a complete elimination of hypothetical, nonobservable entities. Even if we would understand the term "weight" as a common sense term not as yet shaped by Newtonian mechanics, that is, *not* as a force, still, the term "volume" cannot be applied to objects without the use of geometry, which guides our practice of its measurement and of scale-construction for its measurement.

Third, even if Hempel differentiates in scientific systematization between the level of empirical generalization and the level of theory formation, still, he claims that the process of explanation and prediction enabling the derivation of a singular statement from laws that are already part of the level of theory formation is, at the same time, a movement from this level to a level where theoretical terms are already absent, that is, to a level that is of a *non*theoretical nature. This, as shown already for the case of (Kepler's) harmonic law in the *Principia*, does not hold. Once we derive a singular statement from a theoretical law, then the derived singular statement already contains theoretical terms. From this point of view Hempel's scheme (5) is not precise enough. Even if the aim of explanation is to explain the process as described in statement O_1, the very process of explanation does not end in the formulation of the statement O_6, but in the statement, "The solid body *b* floats on the body *l* of liquid because the density of *b* is less than that of *l*." The term "density" is thus moved from the law (3) into this statement. Stated otherwise: The statement that was the initial starting point of our endeavor for explanation, and that expressed the state of affairs we want to explain, and the statement we derive in the process of explanation from theoretical laws are not *identical*; the latter contains additional theoretical terms from those theoretical laws. They differ mutually by their respective *epistemic* statuses: while the former has the status of *appearance*, the latter, that of *manifestation*.

Let me now analyze Ellis's proof of the superfluousness and eliminability of the concept of force from Newtonian mechanics. Three principal objections can be raised against his proof. First, he claims that the *E*-system is an alternative to the *N*-system. But the opposite is true. In order to derive the first law of the *E*-system, he needs the *N*-system (both its second law and the law of gravity) as a *basis for its construction*. In addition, he needs the *N*-system also as a *permanent background* for the *E*-system in order for the latter to be intelligible at all. The constant *G* appearing in the first law of the *E*-system acquires its meaning only by means of the law of gravity from the *N*-system, and the sense of the expression "acceleration relative to the natural acceleration" is codetermined by the sense of the expression "acceleration" which has its origin in the *N*-system. So, Ellis's system is *not an independent alternative* to the *N*-system. In order to be such, it would have to be constructed "from scratch" completely independently from the *N*-system.

Second, because the *E*-system requires the *N*-system, the former represents a superstructure, while the latter represents a base structure. But such a superstructure hopelessly contradicts its own basis. In Newtonian mechanics one starts from the supposition that accelerated motion (in Newton's notation: $f \propto a$) is a nonnatural (caused) type of motion and then one states the three laws of motion plus the law of gravity. But then one should not, as Ellis does, suddenly, switch the conceptual system and claim that only acceleration with respect to "natural" acceleration *a* requires causal explanation by means of the concept of force. Expressed otherwise: Because the first law of the *E*-system explicitly depends on the law of gravity from the *N*-system which in turn depends on the second law of motion, but at the same time the *E*-system changes the meaning of the term "natural motion," the *E*-system not only draws on the *N*-system but at the same time *contradicts it*.

Third, so as the *E*-system requires the *N*-system as a permanent basis and background, the latter has to be already given in advance, thus, the whole idea of constructing the *E*-system has the character of a philosophical *ad hoc* construction that is *parasitic upon an existing system of knowledge already given in an empirical science*.

The same *ad hoc*–ness can be discovered in E. Mach's attempt to dispose of the concept of force when one compares his approach to Newton's treatment, analyzed earlier, of the relation of two bodies in the third law after he introduced the concepts of accelerative, motive, and absolute forces. Once Newton, in the third law and its commentaries starts from the equation $f_A = f_B$, he can, by bringing in the concepts of accelerative and motive force, claim that $dp_A/dt = dp_B/dt$ holds, that is, that $m_A a_A = m_B a_B$ holds and, then, finally, state that "changes in velocities ... are inversely proportional to the bodies because the motions are changed equally" (1999, 417). Mach accepts this result of Newton's process of thought, that is, that $a_A/a_B = m_B/m_A$ holds, but disposes of both the notion of absolute force, that is, [f], as well as $f_A = f_B$. What is left over is, *seemingly*, only a statement concerning the relation between the phenomena a_A and a_B, as it is given in the last of Mach's quotations presented earlier. But one has to bear in mind that this is only a *post festum* disposal of Newton's approach to the concept of force. Mach's rejection of the concept of force as a concept pertaining to causation can be performed *only afterwards* and *only because Newton had already accomplished the cyclical thought-movement*. Newton starts from the phenomena of motion as appearances (change of speed and of motion in time) and, by means of them, derives the concepts of accelerative force and of motive force. He then introduces the concept of absolute force and arrives, even if in an incomplete way (as we have seen), by means of $f_A = f_B$, at $a_A/a_B = m_B/m_A$. One can express this in even stronger terms. Mach accomplishes his antimetaphysical endeavor at the categorial level of *phenomena as manifestations*, even if he *claims* that he is moving exclusively at the level of *phenomena as appearances*.

Mach's attempt at reduction can be criticized also from another point view. Mach, as we have seen, presupposes that bodies interact; only then can he apply his "antimetaphysical" attitude. But the fact that bodies interact was already realized by Kepler for celestial bodies and by Galileo and Huygens for terrestrial objects. Newton, starting from the knowledge produced by them – that bodies interact in certain proportions and due to these interactions acquire certain properties in a certain size/quantity – pushed further, impelled by the ambition, first, to discover the common ground of these interactions and of these proportions and quantities and, second, to derive the former from the latter. Mach's ambition, contrary to Newton's, is just to remain at the level of knowledge that fixes the "fact" that bodies interact in certain proportions. From the point of view of scientific knowledge it represents a retreat from, and a regression with respect to, the level of knowledge attained in the *Principia*.

Chapter 4: The Cyclical Method of Theory Construction II. The Retreat into and Coming Out from the Real Ground

The aim of this chapter is to reconstruct yet another type of the cyclical method of theory construction, namely, that given in Marx's *Capital*. I start with a reconstruction of the three contexts of Marx's theory of value as given in Chapter 1 of *Capital*, Volume I, and reconstruct at the same time epistemic categories relevant to this chapter together with the method employed in its construction. Then I compare those categories with categories employed by Ricardo in his theory of value. Finally, I reconstruct the structure of measurement based on Marx's approach to the concepts of value, surplus-value, production price and average rate of profit as given in the *Manuscript 1861–63* and in Volume I and the manuscript of *Capital*, Volume III.

4.1 The Three Contexts of Marx's Theory of Value

In this subchapter, I reconstruct the epistemic, the sociological and politico-enonomic dimensions of Marx's theory of value of value as given in Chapter 1 of *Capital*, Volume I, in comparison with the views of C. J. Arthur and G. Reuten. I start with a brief presentation of the views of Reuten on Marx's approach to value in that chapter. Next I reconstruct those three dimensions and, when analyzing the epistemic dimension, I compare Hegel's and Marx's approach to epistemic categories with the views of Arthur. Then, by comparing these three dimensions with the views of Arthur and Reuten, I try to locate the roots of their misunderstanding of Marx's approach to value as given in Chapter 1 of *Capital*, Volume I. So as these misunderstanding are partially rooted, at least in my view, in wrong translations of Marx's German terminology into English, I translate quotes from Marx directly from the German originals reprinted in the second MEGA-edition. Then I provide a critique of Reuten's views. Finally, I will give a comparison of the German MEGA-original with Ben Fowkes's English translation in (Marx 1976a) and explain in what respect they are not precise enough or even highly misleading.

4.1.1 G. Reuten

Reuten provides in his papers (1993) and (2005) a very detailed analysis and interpretation of Marx's approach to concepts like labor, value, and money. In (Reuten 1993) one can find three possible approaches to Marx's Chapter 1 of *Capital*, Volume I: the concrete-labor-embodied approach, the abstract-labor-embodied approach, and the systematic-dialectic-view approach; G. Reuten holds to the last one.

The *concrete-labor-embodied* approach takes, according to Reuten, as its point of departure, the following claim of Marx: "For the sake of simplification, henceforth any kind of labor-power counts for us directly as simple labor-power; by this we save ourselves the effort of reduction" (Marx 1872/1987b, 79; 1976a, 135). Reuten introduces the following magnitudes and their relations. Let "L" stand for "abstract labor" or "labor," for short, in my terminology and "W" for "concrete labor" or "work" in my terminology.[141] If W_i and W_j are two[142] different kinds of work, then one can state, *according to Reuten*, the following equation:

$$\alpha_i W_i + \alpha_j W_j = L \qquad /1/.$$

Here α_i and α_j are the so-called discounting coefficients, which, according to Reuten, *can be put equal to one*. The principal problem of the concrete-labor-embodied approach he views in the fact "that we need a procedure to quantify the discounting coefficients. It is hard to see how this could be done prior to the market" (Reuten 1993, 98). So as W_i and W_j are not homogeneous, the following equation does *not* hold:

$$W_i + W_j = W \qquad /2/,$$

or, to be more precise, this equation *does not make any sense at all* (Reuten 1993, 108).

The *abstract-labor-embodied* approach is based on Marx's claim that commodities are "a mere coagulation of human labor devoid of any differences, i.e. … the expenditure human labor-power without any regard to the form of its expenditure" (Marx 1872/1987b, 72; 1976a, 128). Therefore, the value of any of these commodities is measured "by the size of the 'value-forming substance', the labor" (Marx 1872/1987b, 72; 1976a, 129).

---

[141] Here I change Reuten's notation; instead of his symbol "Λ" I use the symbol "L" and instead of his symbol "L" I use the symbol "W." The sociological, epistemic, and politico-economic aspects of the use of the terms "labor" and "work" will become apparent later.

[142] Following Reuten I restrict myself here to the case of just two kinds of work.

Marx's claims that "human labor-power in liquid state, or human labor, creates value, but is not value. It becomes value in coagulated state, in objectified form" (Marx 1872/1987b, 84; 1976a, 142) is interpreted by Reuten in such a way that some labor is embodied in the produced commodities and that both *labor* and its crystallization in a product – namely, *value – exist prior to exchange*; "value is identified with the (reduced) abstract labour L, insofar it is objectified or expended" (Reuten 1993, 99).

So as for any L_i and L_j it holds that they are mutually homogeneous, it holds as well that

$$L_i + L_j = L \qquad /3/.$$

Based on /1/ and /3/ one obtains

$$\alpha_i W_i = L_i; \; \alpha_j W_j = L_j \qquad /4/.$$

So as from /4/ follows $\alpha_i = L_i/W_i$ and $\alpha_j = L_j/W_j$, Reuten interprets these coefficients as *the value productivity of works* W_i and W_j (Reuten 1993, 100).

Reuten views the abstract-labor-embodied approach as Marx's approach, but – as in the case of the concrete-labor-embodied approach – he states the following objection to it (Reuten 1993, 99):

It relies on an abstract entity, the reduced abstract labour, but it is also given – already at this level – a fairly concrete meaning, especially because [of] what is added on measurement. It is not made clear, however, how to undertake this measurement ... prior to the market because we are left in doubt about the actual discounting to simple labour,

and "at a certain stage a procedure has to be developed for getting to the discounting coefficients. It has not been shown how this can be done prior to the market" (Reuten 1993, 103).

Finally, Reuten's own *systematic-dialectic-view* approach introduces the magnitude *l* standing for the magnitude *L* in Marx's understanding, because neither *l* nor *L* as *objectified* is value; at the same time *l* should contradict W.[143] This contradiction gets its expression in the market in terms of money and the following holds:

$$m_i l_i + m_j l_j = ml \qquad /5/,$$

---

[143] In Reuten's notation this contradiction is expressed as $W \bullet \times \bullet L$, using my symbols for labor and work.

where m_i, m_j, and m stand for the monetary expression of labors l_i, l_j, and l. According to Reuten: "In the market, labour takes the value form. Thus labour is actually converted (transformed) into an abstract entity" (1993, 108). This means that contrary to Marx's abstract-labor-embodied approach, here "the simple labour discounting ... pertains to a process that actually takes place in the market (m_i, m_j). Of course, the current theory maintains that value has no existence prior to the market" (Reuten 1993, 108).

In his 2005 paper Reuten further developed his views on Marx's approach to labor, value and money. Worth to be mentioned here are the following four points. First, he states that in his paper (Reuten 1993) "I adopted for abstract labour the composite mW, where m is the monetary expression of labour; and W is in fact added-up concrete labour. As an interpretation of Marx this is wrong" (Reuten 2005, 83).

Second, he now assigns directly to Marx the view that (Reuten 2005, 80)

money is one *constituent* of value (he does not use exactly this formulation). The immanent or introversive *constituent* of value is undifferentiated "abstract labour" (chapter 1), its extroversive (außer) constituent is money (chapter 3); but these two *inseparably belong together*. Money is the necessary form of expression of value (Ausdruck). That is, *value has no existence without money*.

Third, Reuten quotes Marx's view from Chapter 3 of *Capital*, Volume I, which states "money as the measure of value is the necessary form of manifestation of the immanent measure of value of commodities, of labor-time" (Marx 1872/1987b, 121; 1976a, 188). He then analyzes this view in such a way that while its "first line is an obvious reference back to the chapter 1, simple-abstract 'immanent' or introversive notion of value with its immanent measure, namely labour-time" (Reuten 2005, 82), the second line "posits that [m]oney ... is the 'necessary form of appearance' of that immanency" (Reuten 2005, 82). From such an analysis he then draws the conclusion that for Marx "commodity, and hence value, has *no existence* without money" (Reuten 2005, 82). According to Reuten, this is in fact an outcome of Chapter 1; "Its section 3 presents the *formation* of the form of money, or one could say it posits the *form* of extroversion (Veräußerlichung) which is the starting point for chapter 3" (Reuten 2005, 81).

Fourth, so as abstract labor cannot be quantified prior to the market, "abstract labour has no determinate existence" (Reuten 2005, 85), and thus Marx does not present to us in Chapter 1 "'a labour theory of value' ... in any quantifiable sense" (Reuten 2005, 86).

4.1.2 Chapter 1, Volume I of Capital: The Three Dimensions

Let me now present the following three quotes from Marx, in which I emphasize by means of italics certain words; at the same time I suppress Marx's emphasizes in order to avoid any possible confusions.

1) (Marx 1872/1987b, 69–71; 1976a, 125–126):

> The *wealth* of societies in which the *capitalist mode of production* dominates, *appears* as an 'immense collection of *commodities*' ... The *usefulness* of a thing makes it a *use-value*. ... *Use-values* constitute ... the material carrier of – *exchange-value* ... *Exchange-value appears initially* as the *quantitative relation*, the *proportion* in which *exchange-values* of one kind exchange against *exchange-values* of another kind ... The exchange-value *appears* therefore as something *accidental* and purely *relative*, an *exchange-value inherent to value*, an *immanent exchange-value* ... thus a contradiction in terms.

2) (Marx 1867/1983, 41–42):

> Really, all *use-values* are *commodities* because *products* of mutually independent *private labors* ... The standard (*Maßstab*) of "sociality" has to be taken from the relations proper to each *mode of production*, not from representations alien to it.

3) (Marx 1872/1987b, 72; 1976a, 128):

> Let us now look at the residue of the *labor-products*. Nothing is left of them but ... a mere coagulation of human labor devoid of any differences, i.e., the expenditure of human labor-power ... As crystals of this social substance common to them they are values ... In the very *exchange-relation* of *commodities appeared* to us their *exchange-value* as something completely independent from their use-values. If we now really abstract from the *use-value* of the *labor-products*, then we obtain their *value* as it just has been determined. The common what *represents itself* in the *exchange-relation* or *exchange-value* of *commodities*, is thus their *value*. The *progress of investigation* will *lead us back* to *exchange-value* as the *necessary mode of expression* or *form of manifestation* of *value*, which, however, has to be considered independently from this form.

If one looks at these three quotes, one can discern in them the following types of terms. First, we see terms like "to appear," "to appear initially," "quantitative relation," "inherent/immanent," "common substance," "the progress of investigation," "proportion," and "form of manifestation," which are important from the point of view *how scientific knowledge proceeds*. Second, we see terms like "use-value," "product," "mode of production," "capitalist mode of production," and "mutually independent private labors," which characterize the most general social determinations providing the basis and framework for the production of material means satisfying human needs. Third, terms like ""commodity," "exchange-value," "value," "human labor devoid of any differences," and "human labor-power" used to characterize the historically specific way how these needs are satisfied. Each of these three types of terms stands for one dimension of Chapter 1, namely the *epistemic*, the *sociological* and the *politico-economic* dimensions. As we will see below, the first dimension is the subject matter of a discipline I label *theory of scientific knowledge*, or *epistemology*, for short.[144] Let me now start with the epistemic dimension of Chapter 1.

4.1.2.A Categories: Hegel and Marx vs. C. J. Arthur

The terms like "appears," "appears initially," "form of manifestation," and so forth I view as terms whose meaning stands for categories in the sense explicated above in the Introduction. I claimed in the Introduction that epistemic categories come into being only where the relation of the subject and object is given – but only on a relatively high stage of the development of society – do they become the subject matter of a special investigation in the framework of philosophy that expresses them as philosophical categories. I will label them in this book – with respect to epistemic categories – as *metacategories*.

In the European philosophical tradition, and of course with respect to Marx's *Capital*, an outstanding place has to be assigned to Hegel's attempt at the creation of a *system of metacategories*. This attempt displays certain *positive* as well as certain *negative* aspects. The former can be best understood in comparison to *Kant's* approach to categories. The latter subjected categories, namely concepts of understanding (*Verstand*), to a special investigation and claimed that we cannot think a single object of our sensory experience without categories (Kant 1965, 147, 160; A125, B143), that is, "we cannot think an object save through categories" (Kant 1965, 173; B165). At the same time Kant however claims that they are inherently restricted to the empirical realm and we are not allowed to use them beyond the realm of the empirically given (Kant 1965, 239, A219). In addition, according to Kant, there exists an unbridgeable gap between the *phenomenon* and the *noumenon*. Even if categories enable us to synthesize phenomena given to us at the level of sensuous representations, where these phenomena are some-

---

144 These terms are of course primarily the subject matter of theory of knowledge or *gnoseology*, for short. But because I deal here only with a *scientific theory*, namely Marx's *Capital*, I restrict myself here only to theory of scientific knowledge or epistemology, for short.

how related to the object of cognition, and so enable us to *think this object at all*, still we should not be able to recognize what the object is beyond the sphere of the phenomena. Expressed otherwise: "though we cannot *know* ... objects as things in themselves, we must be yet in the position to *think* them as things in themselves" (Kant 1965, 27; Bxxvi). This means that we face here a fundamental dichotomy. On the one hand, Kant grounds his approach to what I label here as theoretical (speculative) reason by accepting the difference between the phenomenal determinations of the object and its *noumenon*. On the other hand, however, he does not provide any further metacategories which would make the category *noumenon* more specific and thus turn it in fact into a *cluster of metacategories*.

It was Hegel who undertook this endeavor by taking a new approach to epistemic categories. He targets the central problem of epistemology by means of the following question: "How do we, as subjects, get over to the objects" (Hegel 1842, §246, Zusatz, 13; 1970, 8), that is, the "difficulty lies in the passage from the thing to the cognition which is accomplished by means of reflection (*Nachdenken*)" (1840, xiv; 1991, 6). Hegel gives the following solution to this problem. To appropriate a thing means to think it; "intelligence familiarizes itself with things ... by thinking them, it sets their content into itself" (Hegel 1842, §246, Zusatz, 16; 1970, 9). By thinking *real singular objects*, we "add to them, so to say, the form of universality ... This universal of things is not a subjective belonging to us, but rather a noumenon opposed to the transitory phenomena; the true, the objectivity, the real of the things themselves" (Hegel 1842, §246, Zusatz, 16–17; 1970, 9). Hegel claims that via the phenomena, understanding is capable of penetrating into the inside (*Inneres*) of things (Hegel 1928, 110–111; 1977, 86–87).

This, then, enables one to understand why in his *Science of Logic* he introduces the cluster of metacategories labeled by him as "Essence". Given his view – that we can think the noumenon, penetrate it by means of thinking, and to acquire knowledge what is given beyond the level of the phenomena – he inevitably replaces Kant's category of noumenon by the cluster of metacategories he labels as "Essence." Thus in turn leads Hegel to a reinterpretation of Kant's category of phenomenon which is of a profound importance, as we will see below, for the internal structure of Chapter 1 of Marx's *Capital*, Volume I. By replacing Kant's metacategory of noumenon by the cluster labeled "Essence" he in turn reinterprets Kant's category of phenomenon, so that it becomes either the cluster "Schein," which I translate as "appearance," or the categorial cluster "Erscheinung", which I translate the as "manifestation," depending on the epistemic/cognitive status of the phenomenon with respect to the knowledge of the essence. In case when the knowledge of the phenomena serves as the starting point for the production of knowledge about the essence, he labels these phenomena as "appearances." In case when the knowledge of the very essence becomes the starting point for the production of knowledge about the phenomena, he labels the latter "manifestations."

Marx accepts Hegel's anti-Kantian differentiation between the category of appearance and that of manifestation. So, for example, when evaluating David Ricardo's approach to concepts of political economy he explicitly states the latter "has reduced the apparent relativity which these things, e.g., diamonds and pearls, possess as exchange-values, to the true relation hidden behind this appearance, to their relativity as mere expressions of human labor" (Marx 1872/1987b, 113; 1976a, 177). At the same time, *after* he accomplished the thought-analysis of the internal contradiction of labor/work given in commodity, he declares: "We started ... from exchange-value or the exchange-relation of commodities in order to track down their value hidden in it. Now we have to return to this form of manifestation of value" (Marx 1872/1987b, 80; 1976a, 139).

Let me now turn to the *negative aspect* of Hegel's approach to categories. The fact that Hegel was able to overcome Kant's paradox that "we cannot *know* the noumenon but we can *think* it" is based on an overcoming of Kant's epistemological mode, which can be represented as follows:

Human beings – Thoughts of human beings – Objects

This means that according to Kant "we put thoughts between us and the objects as a medium (*Mitte*) in such a way that this medium separates us from the objects" (Hegel 1923, T. 1, 15; 1969, 36). Hegel criticizes Kant because, by postulating the thing-in-itself, "the logical determinations remained burdened with (*behaftet*) the object ... and a thing-in-itself, as an infinite impulse (*Anstoß*) was left on them as a beyond" (Hegel 1923, T. 1, 32; 1969, 51). The fact that we as humans are able to penetrate in our thinking beyond the phenomenal characteristics of objects and grasp their inside is, according to Hegel, just the result of the fact that *what penetrates*, the thinking, and *what is penetrated*, the object, are, so to say, made of the same "stuff" – *thoughts*. This becomes readily seen from Hegel's characterization of the case when consciousness, via the play of forces, looks into the true background of things. According to him (1928, 128; 1977, 103):

reason experiences ... in the inside of the appearances only *itself*. Raised above perception, consciousness reveals itself united with the supersensuous world via the medium of appearance. The two extremes, the one that that of the pure inside, the other that of the inside looking into the pure inside ... merge together.

Hegel transforms everything into the objective content of thinking – the "objective thought is the inside (*Inneres*) of the world" (1840, §24, Zusatz 1, 45; 1991,

56). Hegel's replacement of Kant's category of noumenon by the cluster "Essence" is thus based on a *thoroughly accomplished idealistic construction*.

Now what are the consequences of this idealistic construction for Hegel's *Science of Logic*? There is, first, a "macroscopic" consequence of it. Hegel views categories of thinking as the medium through which the self-development of the absolute concept is realized in its various stages that are described by him in the *Speculative Logic* (*Science of Logic* and *Minor Logic*), *Philosophy of Nature* and *Philosophy of Spirit*; therefore, he views his *Science of Logic* as "the delineation of God as he is in his eternal essence before the creation of nature and the finite mind" (Hegel 1923, T. 1, 131; 1969, 50). In addition, there are also profound "microscopic" consequences of Hegel's idealistic construction of his system of metacategories. Here I mean that the idealism of his construction penetrates into the very clusters of metacategories in his *Science of Logic* and determines the respective understanding and position of individual categories, as well as the location of clusters of metacategories within it.

That this is so will be evident when we analyze Hegel's approach to categories *measure* and *substance* which, as we will see later, reappear in Marx's approach to value, labor, money, and price, and where they are assigned a fundamental epistemic/cognitive role. Hegel introduces the category of substance into the subcluster "Wirklichkeit"[145] based on his own reflections from previous subclusters, where he dealt with issues like the relation of the inside of things to their outside (external manifestation), the relation of the essential to the nonessential and so on. Unfortunately, according to Hegel, everywhere throughout this territory one faces unresolved problems. Instead of self-relation only relations are dealt with; the essential is not properly understood because it is essential only with respect to the non-essential and the substrate is just a mere "also" ("*Auch*") of apparent qualities (Hegel 1923, T. 2, 118; 1969, 495) and forfeits its independence from them, though it ought to have an independence somehow related to the essence. In order to solve all these problems Hegel introduces the category of substance. This category should lead cognition to the totality proper: to the understanding of self-relation and self-movement. Here Hegel also introduces the characterization of causation as *mutual interaction* (*Wechselwirkung*) and stresses that in order to solve the problems mentioned earlier, one has to broaden the category of causation so that it encompasses, not only unidirectional causal action, but also mutual action and reproduction, that is, he in fact acknowledges the need to arrive at the category of *causa sui*. But Hegel understands this broadening of the categorial apparatus at the same time as a development from the cluster "Essence" to that of "Notion," at which point he makes the explicit ideal-

---

145 Translated by A. V. Miller in (Hegel 1969) as "Actuality."

istic claim that "the *dialectical movement* of the substance through causality and mutual interaction is ... the immediate genesis of the *notion* by means of which its *becoming* is expressed" (Hegel 1923, T. 2, 214; 1969, 582). Even if Hegel treats the category of substance as both a part of the cluster "Essence" and the cluster "Notion," the category does change depending on which cluster subsumes it, for when it is part of the *former* cluster, substance *should lack the character of a self-related, self-moving, internally differentiated entity*. Thus, because of Hegel's idealism there exists in the *Science of Logic*, with respect to its clusters, "Being," "Essence," and "Notion" – a *fundamental trichotomy*. Wherever (in the cluster "Being") he mentions, *at least indirectly* as we will see later, the category of immanent measure, he relates it neither to the category of essence (in the cluster "Essence") nor to the category of substance-subject (in the cluster "Notion"). Wherever he deals with the essence as essence (subcluster "Essence as Reflection in Itself" as part of the cluster "Essence"), he relates it neither to the category of immanent measure nor to that of the substance-subject. Finally, when he treats in the cluster "Notion" substance-subject as something internally active, differentiated, self-related, and self-moving, categories like measure and essence are viewed by him, with respect to the cluster "Notion," as already overcomed and untrue categories.[146]

This explains why Hegel – contrary to Marx – in the *Science of Logic* provides *no positive characterization of the category of immanent measure*. He only *terminologically* distinguishes between the external and the immanent measure by claiming, for example, that (Hegel 1923, T. 1, 341; 1969, 331)

the development of measure which ... starts with the immediate, external measure ... should ... indicate the connection between this determination of measure and the *qualities* of natural object, at least in general,

and when dealing with the specific quantum, the rule, and the standard, he makes this claim about the conventionally chosen unit of measurement: "Such a unit ... insofar as it is also used as a standard for other things, is in regard to them only an external measure, not their original measure" (Hegel 1923, T. 1, 343; 1969, 334).

Hegel thus *positively* knows only the meaning of the term "external measure" and he has *no positive knowledge* about the meaning of the terms like "original measure" or "groundmeasure" (*Grundmaß*);[147] nowhere in the cluster labeled by him as "Measure" does he define the meaning of either of these two terms.

---

146 Categories of this cluster, Hegel claims, replace "forms such as categories and determinations of reflections, whose finitude and untruth has demonstrated itself in logic" (Hegel 1923, T. 2, 231; 1969, 592).
147 On Hegel's approach to this term see (Hegel 1923, T. 1, 344; 1969, 334).

What conclusion can be drawn from my analysis of Hegel? *First*, Marx inherits Hegel's anti-Kantian distinction between the category appearance and the category manifestation, and at the same time assigns to them a *realistic* interpretation, namely, that they are categories framing thought-movements in the process of theory-construction.

Second, Marx's critique of Hegel's idealism starting from *Zur Kritik der Hegelschen Rechtsphilosophie* up to the introduction in the manuscripts *Grundrisse*, leads, from the point of view of his economic theory, to a cancellation of Hegel's *Philosophy of Nature* and *Philosophy of Spirit* and, at the same time, to a profound restructuring of the categorial "microstructures" in Hegel's *Science of Logic*. There exists a profound difference between Hegel and Marx, not only in the understanding of the role of epistemic categories, but, *third*, also in a profound shift in the very meaning of the terms standing for those categories. This will become readily seen later when I will deal with Marx's understanding of categories as they are given in his economic theory.

It is my view that when these just-stated differences between Marx and Hegel are not taken into account, one develops a distorted understanding of the epistemic dimension of Marx's economic theory. This holds, for example, for the works of philosophers like P. Murray and C. J. Arthur. Let me start with the former. He states that[148] "under [the] ... dialectical (or internal logical) conception of essence and appearance, science is no longer a one-way street that externally relates appearances to the essence, but works both from the appearances to the essence *and* from the essence to the appearances" (Murray 1990, 133–134). But this would mean – if we view categories as ways of the epistemic/cognitive appropriation of the world, that knowledge/cognition, after it accomplishes the cycle appearances → essence → appearances ends up with same knowledge. Phenomena are known at the beginning and at the end of the cycle as only appearances and, thus, *no valuable extension (growth) of knowledge has taken place* and the whole cycle is completely superfluous.

What here seems just as an empty game with name-labels fully displays its negative impact when economists like F. Moseley, drawing on the works of philosophers analyzing Marx, try to correlate the epistemic framework of Marx's chapters 1 and 3 in *Capital*, Volume I, with the concepts of political economy employed in these chapters. Moseley in his critique of Reuten provides the diagram shown in table 4.1 (here "SNLT" stands for "socially necessary labor-time" and "abs. labor" for "abstract labor") (Moseley, 1999, 37):[149]

---

148 The same claim appears again in (Murray 1993, 40).
149 I slightly change F. Moseley's diagram so that it becomes clear that it unifies in itself epistemic categories and concepts of political economy from Marx's *Capital*, Volume I.

Capital, Vol. I Epistemic categories	Chapter 1 Sections 1-2	Chapter 1 Section 3	Chapter 3
Form of appearance	Exchange- values ↓	Money Prices ↑	Money Price
Substance/ Magnitude	abs. labor SNLT →	abs. labor SNLT ⟶	abs. labor SNLT

*Table 4.1 F. Moseley on epistemic categories and concepts of political economy in Chapters 1 and 3 in **Capital**, Volume I*

Here it becomes clear that this diagram contains a *contradiction between the employed epistemic categories and the employed concepts of Marx's politico-economic theory*. From the point of view of categories one *starts* and *ends up* with appearances and thus ends up with *no growth of knowledge at all*; but from the point of view of the concepts of Marx's politico-economic theory, Moseley correctly shows that one ends up with a *profound increase of it*; one is able to derive, based on the concepts of abstract labor and SNLT, concepts like exchange-value, money, and price – and not just presuppose the first of the concepts as given at the very beginning of the derivation of concepts in section 1 of Chapter 1. One has to be, of course, tolerant; Moseley is an *economist* who relied on the work of *philosophers* who, unfortunately, have not done their homework properly.

    This holds foremost for *C. J. Arthur*. Against views he expressed in (Arthur 2004) and (Arthur 2005), the following three objections can be raised. *First*, he wrongly collapses Hegel's distinction between *Schein* and *Erscheinung* into one term, namely, "appearance," because he does not take into account that philosophers translating Hegel clearly distinguished between *Schein* and *Erscheinung*; for example, W. Wallace uses in (Hegel 1975) for *Schein*, the English term "appearance" and for *Erscheinung*, the term "manifestation," while A. V. Miller uses in (Hegel 1969) for the former the term "illusionary being" and for the latter the term "appearance." As a consequence, Arthur accepts (2004, 37) for Marx's statement about commodity "Sie stellt sich dar als dieß Doppelte was sie ist, sobals ihr Werth eine eigne … Erscheinungsform besitzt" (Marx 1872/1987b, 92), B. Fowkes' *wrong* translation: "It appears as the twofold thing it really is as soon as its value possesses its own particular form of manifestation" (Marx 1976a, 152). So as Marx builds here already on the knowledge of value, labor, SNLT, and the differentiated unity of labor/work and value/use-value, the subsection where this quote is given is *not part of the thought-movements at the level of appearanc-*

es but already of *manifestations*; therefore Marx uses at the end of this sentence the expression "Erscheinungsform" *correctly* translated by B. Fowkes as "form of manifestation." But once the latter term is used, one cannot translate Marx's expression "sich darstellen," given in the same paragraph and pertaining to the same subject matter, as "it appears as"; the proper translation would be either "it represents itself" or "it manifests itself."

Second, because Arthur does not take into account that Marx restructures the relations and positions of categories from Hegel's *Science of Logic*, he provides the correlation shown in Table 4.2 between Hegel's categories and Marx's sequence of concepts of political economy (Arthur 1993, 87).

Logical categories (Hegel)	**The Value-Form** (Marx)
Doctrine of Being	*Commodity*
Quality	Exchangeability
Quantity	The Bargain
Measure	Value in exchange
Doctrine of Essence	*Money*
Ground of Essence	Value in itself
Appearance	Forms of value
Actuality	Money
Doctrine of the Concept	*Capital*
Subjectivity	Price
Objectivity	Metamorphoses of commodities
Idea	Self-valorization of capital

Table 4.2 Arthur on the relations between Hegel's categories and Marx's concepts of political economy

That such correlation is highly misleading becomes readily seen once we realize that Hegel in the category-cluster labeled as "Idea" provides categories like truth, good, volition, means, purpose, practical idea, etc., that is, categories that are central for the creation of thought-projects of changes of the external world; the sequence of concepts of political economy in *Capital* is, however, placed into the framework of what Marx labels as theoretical (speculative) reason and *not* of practical reason (*praktische Vernunft*).

Third, so as Hegel does not, as shown previously, provide any positive account of the term "immanent measure," Arthur is not capable of reconstructing what Marx means at all by the category of immanent measure and how he in fact overcomes Hegel. Therefore, once he deals with Marx's term "immanent measure," he just holds to such superficial claims as "value has an identity with itself and thereby grounds some immanent measure" (Arthur 1993, 75). He also introduces, *based on examples from physics*, a three-fold distinction for measure: *immediate comparative measure*, where the measure shares the same inherent dimension with the measured (e.g., the measurement of length of an object by the length of a ruler); *indirect measure*, here the measure is foreign to what is measured (e.g., the measurement of heat by a thermometer); *measurement of complex magnitudes* (e.g., the measurement of accomplished work by the measurement of force, time and distance). What strikes us here is the fact that he relies on examples only from physics and does not provide a reconstruction of the correlation of the category of measure and of the concepts of political economy used by Marx himself in Chapter 1 of *Capital*, Volume I. In Part 3, we will see the negative consequences of Arthur's attempt to reconstruct Marx's approach to measure based on these examples from physics.

So as philosophers like Arthur do not provide a reconstruction of that correlation, Reuten as an *economist* has to rely on common sense examples like the reduction of tables to their common substance, namely, timber (Reuten 2005, 84), and, finally, he arrives at the conclusion that "Marx's immanent measure of value in Chapter 1 – time of abstract labour – is *very* abstract. It does not provide a measure of value in the sense that we (nowadays) usually use the term measure" (Reuten 2005, 85).

4.1.2.B The Sociological Dimension

Let me now turn to Marx's description of the most general social determinations which provide, according to him, the basis and framework for the production of material means satisfying human needs. I will, first, analyze Marx's approach to determinations necessary for the production of use-value and then of surplus-value. Based on this analysis, I will then provide the starting point for the reconstruction of how Marx understands the determinations necessary for the production of value as given in Chapter 1.

Marx uses the term "concrete labor," in my terminology "work" (or W for short),[150] to designate the source in which originate, due to the process of produc-

---

150 Here I draw on the resources of languages like English and the Slavic languages which, unlike the German used by Marx, still have the ability to differentiate be-

tion, *use-values*. He characterizes work as "purposive activity for the appropriation of the natural (*das Näturliche*)" (Marx 1980b, 115). In addition to the term "purposive activity" he employs also the following terms to describe work leading to the production of a use-value, say, a coat: "To produce it, requires a certain kind of *purposive productive activity*. It is determined by its purpose, mode of operation, object, means and result" (Marx 1867/1983, 22). Drawing on these terms, I now describe the three *essential (existential)* conditions under which work can exist at all: the existence of *intentionality*, under which I subsume Marx's terms "purposive activity," "mode of operation," "object," "means," and "result"; the existence of a *propositionally structured language* that enables one to refer to the state of affairs in the real world and last, but not least, the presence/givenness of the (human)[151] body[152] enabling to operate with certain objects as means of action on other objects and to make of the latter the result of action. The conjunction of these three essential conditions I will write as C_E^A.

The term "essential conditions" in fact appears in Marx's reconstruction of the conditions for the creation of surplus-value from surplus-labor. In the *Grundrisse* he lists them as follows (Marx 1981a, 371–372; 1973, 463–464):[153]

(1) On the one side, the existence of the living labor-capacity as purely *subjective* existence, separated from the moments of its objective reality; therefore separated just as much from the *conditions* of living labor as from the *means of existence*, the *means of subsistence*, the means of self-maintenance of *living labor-capacity*, the living possibility of labor on one side in this complete abstraction.
(2) The value or objectified labor given on the other side, must be an accumulation of use-values, sufficiently large to provide the objectified (*gegenständlichen*) conditions, not merely for the production of products or values, necessary to reproduce or maintain the living labour-capacity, but to absorb surplus-labor – to provide the objective material for the latter.
(3) Free exchange relationship – circulation of money – between the two sides; a relation between the two extremes based on exchange-values, not on the lord-subject relationship, i.e., production which does not supply the means of subsistence directly to the producer, but is mediated by exchange; as little as it can directly command alien labor directly, but has to buy it from the worker himself, has to exchange it; finally

---

tween concrete labor and abstract labor by means of one single expression: "work" and "labor" in English, "работа" and "труд" in Russian, "робота" and "праця" in Ukrainian, "robota" and "praca" in Polish, etc.
151 I put here the term "human" into brackets because it remains an open question which species other than *homo sapiens* in the group of hominids fulfilled these three conditions.
152 R. Bellofiore characterizes the capability for work as: "inseparable from the living body of human beings" (Bellofiore 2004, 173).
153 The same four essential conditions are given also in the *Manuscript 1861–63* (Marx 1982a, 2287).

(4) the side – which represents the objectified conditions of labor in the form of independent, for themselves existing values – has to present itself as *value*, and has to regard value-positing, self-valorization, money-making as the ultimate aim – not immediate enjoyment or creation of use-value.

Worth mentioning here is the fact that Marx lists as the fourth condition the conscious aim of making money, which has – with respect to Marx's views – the character of a transposed, false form of consciousness. Below we will see how this can contribute to the reconstruction of Marx's thought-movements inside Chapter 1 as well as in his thought-movement from Chapter 3 to Chapter 5 in *Capital*, Volume I.

Let me now turn, finally, to Marx's characterization of the conditions essential for the existence of abstract labor, or, in my terminology, labor which creates commodities endowed with value. According to him, the producer of the commodity has to be an "independent private individual" (Marx 1980a, 51); "indeed, all use-values are commodities only because they are products of mutually independent private labors" (Marx 1867/1983, 41), and "only the products of independent (*selbstständige*) and mutually independent (*voneinander unabhängige*) private labors face each other as commodities" (Marx 1872/1987b, 75; 1976a, 132). So, the first essential condition for the existence of labor leading to the production of a commodity as a carrier of value is the existence of producers privately owning the means of production. I thus introduce the expression "C_E^{B1}" for this essential condition, so that "$C_E^{B1}(x, y)$" stands for "x is the private producer of y," where the individual variable x ranges over producers while the individual variable y ranges over commodities. In Section 1 of Chapter 1, where Marx introduces the condition C_E^{B1}, he does not, however, introduce the form of consciousness which is essential for the production of value from labor. It is introduced by him, as we will see later, in the subsection of Section 3 where he deals with commodity fetishism. The conditions C_E^A and C_E^{B1}, as we will see, play a central role in Marx's theoretical grasping of the law of value in Chapter 1.

4.1.2.C Unification: Political Economy, Sociology and Scientific Knowledge in Chapter 1

Let me now bring into unity the sociological, politico-economical and the epistemic/cognitive dimensions of Chapter 1 by providing a reconstruction of Marx's thought-movement in it.

Marx starts from a fact that he regards as apparent, namely, that capitalist economy is an "'immense collection of commodities'" (Marx 1872/1987b, 69;

1976a, 125) and then he chooses any one commodity C here initially presupposed to stand in an exchange relation to any other commodity C'; this I symbolize as $C \rightarrow C'$. Next, Marx views C as, first, a carrier of a use-value from which he temporarily abstracts, and, second, as a carrier of an exchange-value, once we take it as related to commodity C'; "Exchange-value appears first as the quantitative relation, the proportion in which use-values of one kind exchange for use-values of another kind" (Marx 1872/1987b, 70; 1976a, 126). In a next step, proceeding from the empirically given fact of an exchange relation between C and C', for example, 1 quarter of wheat = a lbs. iron (Marx 1872/1987b, 71; 1976a, 127), he abstracts, again temporarily, from C' and subjects the commodity C itself to a thought-analysis. He claims that such an equation expresses just that (Marx 1867/1983, 19):

the *same value* exists *in two different things*, in 1 quarter of wheat and likewise in a lbs. of iron. Both are therefore equal to a *third*, which in and for itself is neither the one nor the other. Each of both, as far as it is exchange-value, has to be, independently of the other, reducible to this third ... Independently of their exchange-relation or the form in which they appear as exchange-values, the commodities are to be considered simply as values ... Their being as *values* makes up their ... unity. This unity arises ... from society.

Then Marx traces the origin of value back to its substance: "*The common social substance*, which represents itself in different use-values only differently, is – *labor*. As *values* are commodities nothing than *crystallized labor* ,,, only *socially necessary labor-time* counts as value-creating" (Marx 1867/1983, 19–20). From now on I will signify under labor, or L, *socially necessary labor*, and I introduce the symbol SNLT for socially necessary labor-time which Marx characterizes as "labor-time required to produce any use-value with the given socially normal conditions of production and the social average degree of skill and intensity of labor" (Marx 1872/1987b, 73; 1976a, 129).

Based on the concept of SNLT he then gives a more precise characterization of the *quantitative* determination of value; "It is thus only the amount of the socially necessary labor or the labor-time socially necessary for the production of a use-value which determines the size of its value" (Marx 1872/1987b, 73; 1976a, 129). This, in turn, enables him to state the conditions of the *quantitative identity* for the value of two commodities: "Commodities which contain the same amounts of labor, or which can be produced in the same labor-time, have the same size of value" (Marx 1872/1987b, 73; 1976a, 130). Marx, finally, gives the following summary of his approach to value and labor; "Now we know the *substance* of value. It is *labor*. We know the *measure of its size (Größenmaß)*" (Marx 1867/1983, 21).

Let me now try to approach the just-reconstructed thought-movement of Marx at the very beginning of Chapter 1 from the point of view of a unity of *epistemology*, *sociology*, and *political economy*.

Marx starts from a certain *appearance*; an immense accumulation of commodities and the exchange relation between them: $C \to C'$, so it holds that, in Marx's notation, *a* commodities $C = b$ commodities C', where "*a*" and "*b*" denote the amount of the respective commodity (e.g., two pieces, three kilograms). Then C is picked out and a use-value and an exchange-value is discerned on it; the latter initially in relation to C'. Then he abstracts from both the use-value-determination of C and its exchange-relation to C' and moves in thought to value, V, for short, which he understands as a result of labor expended on its production and where the latter is measured or quantified by Marx on the basis of SNLT. As shown previously, for Marx L and, thus, SNLT, can exist only under the specific social condition that the producers of the products are *private* producers.

Worth mentioning here are the conceptual resources on which Marx consciously draws in the very beginning of Chapter 1. As quoted already, he positively evaluates Ricardo's reduction of exchange-value to labor and claims also: "So it was only the analysis of the commodity-prices that led to the determination of the magnitude of value, only the common money-expression which led to the establishment of their value-nature" (Marx 1872/1987b, 106; 1976a, 168).

I can now express the result of Marx's thought-movements from the concept exchange-value/price to that of value and from the concept of labor to that of value as follows. Let P_g stands for the price as a certain weight-amount of gold, V for value and $f_i(L)$ for a function of labor; I then write:

$$P_g(y) =: kV(y) \qquad /6/$$

$$V(y) := f_i(L(y)) \qquad /7/$$

Here y stands for an individual variable ranging over commodities, and k stands for the constant mediating between the right and left side of the equation. $V(y)$ then stands for the value of y, and $L(y)$ stands for the labor socially necessary to produce y. In order to indicate Marx's *direction* of the thought-movement *from* the concepts of exchange-value/price *to* that of the value, I add the sign ":" to the right side of the sign of equality and, thus, obtain in /6/ the sign "=:"; and in order to indicate Marx's thought-movement *from* the concept of labor *to* that of the value, I add the sign ":" to the left side of the sign of equality and obtain in /7/ the sign ":=".

Let me now try to determine in a more precise way, *first*, the nature of the function f_i in /7/. Here I can draw on Marx's claim that commodities are (Marx 1872/1987b, 72, 84; 1976a, 128, 142) a mere coagulation of human labor power ... As crystals of this social substance common to them they are – values ... Human labor-power in liquid state, or human labor, creates value, but is not value. It becomes value in coagulated state, in objectified form.

Once we translate the German expressions "Gallerte" as "coagulation," and "geronnener Zustand" as "coagulated state," and "gegenständliche Form" as "objectified form," we can pick up the following three statements we regard as given in Section 1 of Chapter 1:

(1) Value is coagulated labor.
(2) Value is crystallization of labor.
(3) Value is objectified labor.

If we view "is" in these expressions as the sign of equality "=," and if we at the same time view the term "value" as being defined by the expressions standing on the right side of "is," then, we can restate the statements (1) through (3) as follows (here "$=_{Df}$" stands for the sign of definition and "y" is a sign of the argument ranging over commodities):

(4) Value of $y =_{Df}$ Coagulated labor in y.
(5) Value of $y =_{Df}$ Crystallized labor in y.
(6) Value of $y =_{Df}$ Objectified labor in y.

We can then restate the last three statements as $V(y) =_{Df} f_i[L(y)]$ ($i = 1, 2, 3$) so that the expression "V" stands for "value of," "f_1" stands for the "coagulated," "f_2" stands for "crystallized," "f_3" stands for "objectified," and "L" stands for "labor in." In order to grasp[154] the generality of the function f_i referred to by the expression "f_i", I indicate the generality of the function V referred to by the expression "V" and of the function L referred to by the expression "L" by removing from both expressions the symbol "y" and by replacing it by the Greek vowel "ε". We thus have

$$V(\varepsilon) =_{Df} f_i[L(\varepsilon)] \qquad (i = 1, 2, 3) \qquad /8/$$

Now what is the relation between the functions $L(\varepsilon)$ and $V(\varepsilon)$? From the equality between $V(\varepsilon)$ and $f_i[L(\varepsilon)]$ as expressed in /8/, it becomes evident that in my approach the value-function $V(\varepsilon)$ and the *very* labor-function $L(\varepsilon)$ are *not identical*; thus /8/ can be viewed as an explication of Marx's claim that "human labor-power in liquid state or human labor ... is not value" (Marx 1872/1987b, 84; 1976a,

---

154 Here I follow G. Frege's approach to functions in (Frege 1984a).

142).[155] But at the same time I regard the *course of values* for the function $V(\varepsilon)$ and the *range of values* for the function $f[L(\varepsilon)]$ as *identical*. By means of this I can take into account the fact that *by quantifying labor by means of labor-time, one can quantify also value by means of labor-time* and thus explicate Marx's claim that "labor as process, *in actu*, is ... measure of value" (Marx 1982a, 2099).[156] And what are the values from the course of values for the functions $V(\varepsilon)$ and $f[L(\varepsilon)]$? Once we take into account Marx's claim that the "quantity of labor ... is measured by its duration, and the labor-time has in turn its standard of measurement in certain parts of time, like hour, day etc." (Marx 1872/1987b, 72; 1976a, 129), then these values are just certain time-intervals, for example, 5 hours, 3 days and so on. If we apply to the expression which stands for names of functions, Frege's ideas from (Frege 1984a) and those presented in subchapter 1.4 of this book, then we can represent both the *value* of a commodity as well as the *labor* expended on it by means of one and the same ordered pair[157] ($\{t\}$, $[t]$), where "$\{t\}$" denotes a variable acquiring values from the set of positive numbers, while "$[t]$" stands for the time-dimension; the whole expression "($\{t\}$, $[t]$)" is the generalized reconstruction of expressions like "5 hours" and "3 days.". The notations "$V(\varepsilon)$" and "($\{t\}$, $[t]$)" stand for two different representations of one and the same *magnitude* of political economy, namely, value. If we speak about the concrete *size* of this *magnitude*, for example, "5 hours" or "3 days", we use the notation "($\{t_i\}$, $[t]$)" or "$\{t_i\}[t]$," where "$\{t_i\}$" denotes a positive number. Later I will give a more precise representation of the magnitude of labor by means of an ordered triple and that of value by means of an ordered quadruple.

Let me now try to determine in a more precise way, *second*, the nature of the constant k in /6/. Here we can draw on *categories* which Marx employs in order to characterize the movement from price, exchange-value and money to value and labor. He views, as quoted already, exchange-value, price and money as the phenomena-*appearances* of a certain underlying structure. That this is really his view can be also seen from the following claim of him: "Historically it is quite right that the search after value leans itself initially against the *appearing* expression of commodities as value, against money" (Marx 1979, 1332), and at the same

---

155 In the *Manuscript 1861–63* he states in the same vain that "labor as process, *in actu* is ... not value" (Marx 1982a, 2099).
156 This, at least in my view, removes the objection of P. Ruben, to phrases like "measurement of value by means of labor-time," which he regards as "expressions which are in respect to theory of measurement absurd ... Measurement is comparison under the supposition of use of measurement-units which have the qualities (dimensions) of the measured objects ... A measurement always mutually relates things of the same quality, so that masses measure only masses, lengths only lengths, labors only labors, etc." (Ruben 1997, 53).
157 In representing magnitudes by means of ordered pairs I draw on (Berka 1982).

time he classifies here, from the point of view of epistemic categories, money "as the external measure of values" (Marx 1979, 1330).

This means that $P_g(y)$ in /6/, as the (gold) price of y is a certain weight-amount of gold; thus we can express it as an ordered pair. Its first element $\{w_0\}$ is a variable acquiring values from the set of positive numbers, while the second element, $[w_0]$ gives the weight-dimension. We thus can write instead of "$P_g(y)$" the expression "$(\{w_0\}, [w_0])(y)$." If we now take into account the above given explication of the expression "$V(\varepsilon)$" (given earlier) by means of an ordered pair, then we obtain

$$(\{w_0\}, [w_0])(y) =: k(\{t\}, [t])(y) \qquad /9/.$$

Once we take /9/ as an expression of equality, then we can view the constant k introduced in /1/ as given by the ordered pair

$$k = (\{w_0\}/\{t\}, [w_0]/[t]) \qquad /10/.$$

In addition to the characterization of *measure* at the level of *appearances* as *external measure*, Marx provides, *contrary to Hegel*, an additional characterization of it at the level of the (social) *structure underlying these appearances*, and it pertains to the characterization of the *quantitative* determination of value by means of the quantitative characterization of labor, that is, by means of SNLT. Marx labels this new determination of the category of measure as "immanent measure" and states – with respect to labor-time – that "labor-time is the living existence of labor ... it is its living existence as a quantitative simultaneously with its immanent measure" (Marx 1980b, 109–110). Later he claims: "The size of value expresses ... a relation to the social labor-time which is immanent to its creation-process" (Marx 1872/1987b, 128; 1976a, 196). This means, from the point of view of epistemic categories, that the *quantitative determination of the substance in which the respective magnitude originates provides the immanent measure of this magnitude*. For example, with respect to the given analysis of value, this means that the ordered pair $(\{t\}, [t])$ enables one to inherently measure the quantitative determination of the magnitude of value.

Marx, however, does not stop at the level of the knowledge where only the *quantitative* characterization of value is given. In a next step he targets the *qualitative* determination of the substance of value and, thus, of value as well. So as the essential conditions symbolized previously as "C_E^A" and "C_E^{BI}" are *together* necessary for labor to exist, while the former is *sufficient* for work to exist, it holds what Marx states, namely, "that in the commodity indeed are not contained two kinds of labor but still the *same* labor which is differently and even contradictory determined, depending on whether it is related to the *use-value* of the commodity as its *product*, or to the commodity-value as its merely *objectified* expression" (Marx 1867/1983, 26–27). Thus Marx widens the concept of labor so that now he views it as a *differentiated unity of work and labor*, which I symbolize as "*L/W*"; I therefore widen my representation of labor from the ordered pair ($\{t\}$, $[t]$) into that of an ordered triple ($\{t\}$, $[t]$, L/W). So, as labor is the substance of value, and the former can exist only in unity with work, the latter can exist as well only in unity with use-value; together they set up in the commodity a *differentiated unity*; according to Marx, "commodity is use value ... and 'value'" (Marx 1872/1987b, 92; 1976a, 152). If we introduce the symbol "*UV/V*" for the differentiated unity of use-value and value, then we can represent Marx's understanding of the value of a commodity by the ordered triple ($\{t\}$, $[t]$, L/W). And if we take into account also that value has its origin in the substance that is the differentiated unity of labor and work, then we can finally represent the value of a commodity as the ordered quadruple ($\{t\}$, $[t]$, UV/V, L/W). This quadruple can be viewed as a restatement of the value-function $V(\varepsilon)$ and the ordered triple ($\{t\}$, $[t]$, UV/V) as a restatement of the labor-function $L(\varepsilon)$. The fact that the first and second elements from the ordered quadruple correspond to those from the ordered triple is just the expression of the fact that the value-function and the labor-function have one and the same course of values. The difference between these two functions is expressed by the fact that the number of elements differs from one "tuple" to the other one.

Marx moves in a next step to a conceptual grasping of the way how the inherently differentiated unity of value/use-value in any *one* commodity *manifests itself* in the *mutual exchange-relations between commodities.*

So, as any human product has, from the point of sociology and political economy, the social property of value only if it is a product of a private producer, value of a product can manifest itself only in the social relation between producer accomplished by means of exchange of their products, that is, commodities; (Marx 1979, 1317)

Value implies ... exchanges, [and] exchanges are exchanges of things between men; exchanges do not concern absolutely the things as such ... Exchanges of products as commodities is a certain method of exchanging labour, and of the dependence of the labour of each upon the labour of others, a certain method of social labour or social production;

and (Marx 1872/1987b, 80; 1976a, 138–139):

Let us remember ... that commodities possess value-objectification (*Werthgegenständlichkeit*) only in so far as they are expressions of the same social unity, human labor, thus their value-objectification is purely social. Then it a matter of course also that it can manifest itself only in the social relation between commodities.

This means that after Marx temporarily abstracts from the exchange relations (in my notation $C \to C'$; 1 quarter of wheat = a lbs. of iron in Marx's example) and conceptually grasped the inner, social determinations of commodity means of the concepts of political economy like value, labor, SNLT, value/use-value, and labor/work, he now brings in the social relation given in the exchange relations between commodities. And this exchange can be accomplished only by the private producers as owners of their products/commodities. This, in turn, means that while up to Section 2 of Chapter 1, Marx spoke about social conditions characterizing the producers of commodities and did not speak about the *conscious* aspect of their production activities, now he prepares the basis in order to bring in their action as guided by their aim-setting *consciousness*. The basis for this is that Marx presupposes that the various types of the value-form relate products that (1) are products of *exclusively* private producers, that is, they are always value-objects (*Werthgegenstände*), and (2) they *contain the same amount* of SNLT, that is, *their values are of the same size*.

What does this mean from the point of view of the *epistemic categories*? After Marx moved *from the phenomena as appearances to their essence and determined the quantity and quality of the very essence*, he now consciously moves "back" the phenomena in order to derive them, based already on the knowledge of their essence, as *manifestations*: "We started ... from exchange-value or the exchange-relation of commodities in order to track down their value hidden in it. Now we have to return to this form of manifestation of value" (Marx 1872/1987b, 80; 1976a, 139). In addition, the derivation of the manifestations is based here on Marx's move to presuppose that into exchange of commodities, *consciously* planned in advance by their producers, enter entities (1) that are *qualitatively* identical, that is, they all have the property value/use-value, and (2) in such *sizes* that they are *quantitatively* identical, that is, their *quantitative determination by means of SNLT is of the same size*.

Then, Marx, by analyzing the various types of the value-form (simple/singular, total/unfolded, the general, and the money-form), shows by means of thought-operations that in order for exchange to work – that is, that any product fulfilling the already stated qualitative and quantitative characterizations, can express its qualitative identity with and quantitative proportion to any other commodity – gold as money has to grow out of the world of value-things.[158]

From the point of view of *epistemic categories*, Marx evaluates his already accomplished derivation of gold as money as a *thought-movement leading to the phenomena as manifestations of the immanent measure of their underlying essence*: "money as the measure of value is the necessary form of manifestation

---

158 For a detailed analysis of this derivation see (Moseley 2003).

of the immanent measure of value of the commodities, the labor-time" (Marx 1872/1987b, 121; 1976a, 188).

Let me now try to show, again from the point of view of *epistemic categories*, how Marx derives the concept of the (golden) *price* as a phenomenon with the epistemic/cognitive status of a *manifestation*, and how it differs from the concept of price I reconstructed previously, where it had as yet only the epistemic/cognitive status of an *appearance*.

We suppose now, following Marx, that we have two products, namely, a certain weight-amount of gold and any other commodity y. So as we now view that weight-amount as a commodity produced by a certain amount of labor, we can represent the labor expended on this weight-amount by the ordered triple $(\{t_1\}, [t], L/W)$, where "$\{t_1\}$" denotes a positive number, while for the labor given expended on any other commodity y we have the ordered triple $(\{t_2\}, [t], L/W)$, where "$\{t_2\}$" denotes a positive number. If we accept the conditions (1) and (2) for these two concrete commodities, then we can represent the relation between them as follows (here "g" stands "gold"):

$$(\{t_1\}, [t], L/W)(g) = (\{t_2\}, [t], L/W)(y) \qquad /11/$$

So as for the two commodities holds the condition (2), $\{t_1\} = \{t_2\}$ holds for them as well. This then means that for g and y, it holds, with respect to their values, that

$$V(g) = V(y) \qquad /12/$$

It holds also, as shown earlier, that the weight-amount of gold is given by the ordered pair $(\{w_0\}, [w_0])$. We can then suppose that the following holds:[159]

$$V(g) = nV_g(w_0)(g) \qquad /13/,$$

where $V_g(w_0)$ is the value of the chosen weight-unit for gold we represent as "$1w_0$" and "n" denotes a positive number. By substituting the right side of /13/ for the left side of /12/, we obtain the expression

$$nV_g(w_0)(g) = V(y) \qquad /14/.$$

By multiplying it with $1/V_g(w_0)$ and simultaneously with $1w_0$, we obtain

$$nw_0(g) := [1w_0/V_g(w_0)]V(y) \qquad /15/.$$

---

[159] Here I draw on (Quaas 1984) and (Quaas 2001, 171–172).

I interpret /15/ in such a way that the left side stands for a certain weight-amount of gold given by a certain multiple of the chosen weight-unit represented as "$1w_0$," and, thus, for the (*golden*) *price of commodity y*. And because /15/ was derived based on the thought-movement *from* value *to* price/money, I add the sign ":" to the left side of the sign of equality. Based on Marx's derivation of the concept of (gold) price, we then have the equation

$$P_g(y) := [1w_0/V_g(w_0)]V(y) \qquad /16/$$

Here the concept of price is already derived from the concepts of value. And if we take into account also equation /7/, both for the value of the weight-unit w_0 of gold, $V_g(w_0)$, and the value of the commodity y, we obtain the equation /17/, where $L_g(w_0)$ stands for the labor expended on the production of the weight-unit w_0 of gold:

$$P_g(y) := \{1w_0/f_i[L_g(w_0)]\} f_i(L(y)) \qquad /17/.$$

Thus the concept of price is derived in /17/ already on the basis of the concepts labor and SNLT. By comparing /17/ with /6/, we find out that the constant k is now, due to Marx's thought-movement from the concept of labor to that of price, understood not just as a constant simply mediating between the two sides of equation /6/, but as

$$k = 1w_0/f_i[L_g(w_0)] \qquad /18/,$$

that is, as the ratio of the chosen weight-unit of gold to a function of the amount of SNLT expended on the latter's production. It thus becomes apparent that the movement from the level of knowledge given in /6/, and expressed by the category of *external measure*, to the level of knowledge given in /17/, and expressed by the category of *manifestation of immanent measure*, yields an *increase* of scientific knowledge.

What makes up the *epistemic/cognitive* difference between scientific knowledge given at the level of phenomena as appearances and scientific knowledge given at the level of phenomena as manifestations? While scientific knowledge at the former level expresses the habitualized, common sense knowledge of certain phenomena, scientific knowledge at the latter level involves in a noneliminable manner – counterintuitive, as compared to common sense knowledge – concepts referring to the essence of these phenomena. In the process of conceptual derivation of the phenomena from their essence, the concepts referring to the latter are *shifted* to the statements referring to the former. Stated otherwise: *if the explanans contains concepts referring to the essence of the phenomena to be explained, then the explanandum of the phenomena contains concepts referring not only to the phenomena, but also concepts referring to the essence of these phenomena.* So, for example, while the statements at the very beginning of Section 1 in Chapter 1 containing the term "exchange-value" refer exclusively to the state of affairs given in the sphere of exchange of commodities ($C \to C$'), at the end of Section 3 statements containing that term also contain terms like "value," "labor," and "SNLT" and, thus, refer to a differentiated unity of exchange and production. So, for example, while the statements at the very beginning of Section 1, Chapter 1, containing the term "exchange-value" refer exclusively to the state of affairs given in the sphere of exchange of commodities ($C \to C$'), at the end of Section 3 statements containing that term contain also terms like "value," "labor," and "SNLT" and, thus, refer to a differentiated unity of exchange and production.[160]

That we have here really an increase in scientific knowledge becomes apparent when we restate /18/ as $\{1\}[w_0]/\{t'\}[t]$, where $\{t'\}[t]$ is the amount of SNLT expended on a chosen weight-unit of gold. For k as an ordered pair holds:

$k = (\{1\}/\{t'\}, [w_0]/[t])$ /19/.

What above in /10/ seemed to be a ratio $\{w_0\}/\{t\}$ of only externally related quantities $\{w_0\}$ and $\{t\}$, is now understood to be the inverse of the quantity $\{t'\}$, that is, of the amount of SNLT expended on the production of the amount $1w_0$ of gold.

What has to be emphasized is, *first*, that $nV_g(w_0)$ in /14/ stands for the *measure of value* of any and all other commodities y; it is the *immanent measure of them as it manifests itself* in one exclusive commodity – gold. In terms of epistemic categories the measure of value can be understood *as the manifestation of the immanent measure of the essence of a certain type of phenomena in this type of phenomena.* Be relating the category of essence to the category of *phenomena of a certain type*, we can bring in an additional expression for an epistemic category, namely, "ground" which I view as synonymous with the expression "essence," because essence is the ground of phenomena of a certain type (or of several different types). Here I draw on Hegel who introduced the category of ground into

---

160 By such a reconstruction of the difference between the epistemic status of appearances and that of manifestations, it is possible to find a solution at the level of metacategories to the problem indicated by Ira Gerstein, namely, that "the Marxian transformation problem is usually thought of as bridging the transition from 'essence' (value) to phenomena or surface (prices). This paper shows that such a conception is incorrect. The transition is actually between two theoretical levels of the economic region of the capitalist mode of production. The first of these levels *is production in itself* ... while the second is the *complex unity of production and circulation*" (Gerstein 1986, 45).

the cluster "Essence" in his *Science of Logic*, as well as on Marx, who employs this category in his economical works. For example, he claims: "The process by which the values within the money-system are determined by labour-time does not belong into the examination of money and drops outside circulation; it stays behind it as an acting ground and presupposition" (Marx 1981a, 662; 1973, 794), and he characterizes both the status of the labor expended on a commodity to the latter's price (Marx, 1979, 1324) and the status of the rate of surplus-value with respect to the rate of profit (Marx, 1992, 57) by means of the category of the *inner ground*.

By the transformation of /14/ into /15/ we obtained on the left side of the latter equation the expression for a certain weight-amount of gold, understood here as a multiple of a chosen weight unit of gold, and thus this transformation led us from the *concept of measure of value to that of the standard of prices*. Again, we started, drawing on Marx's commentaries on his predecessors in political economy as science, from the concept of price understood as a certain weight-amount of gold and, finally, returned to it by explaining how it originates in the transformation of the (golden) measure of value into the (gold) standard of prices. About this transformation and, thus, about this return to the concept of price introduced in /6/ just as a weight-amount of gold, that is, to the phenomenon of price as an *appearance*, Marx states (Marx 1872/1987b, 120–121; 1976a, 187):

We saw how even in the most simple expression of the exchange-value x commodity A = y commodity B, the thing, *in which* the size of the value of another thing is represented, appears to posses its equivalent form independently from that relation, as a social *natural property*. We followed the consolidation of this false appearance. It is accomplished as soon as the general equivalent form is tied to the natural form of a particular kind of commodity or is crystallized into the money-form. A commodity appears not to become money because the other commodities completely represent their values in it; but, on the contrary, they appear to represent universally their values in it because it is *money*. The mediating movement disappears in its own result and leaves no traces behind. Without any contribution on their part the commodities find their own value-form ready-made as a value-body outside and alongside them. These things, gold and silver, as they come out of the entrails of the Earth are at the same time the immediate incarnation of all human labor.

If we now look at the concept of standard of prices appearing on the left side of /6/, /16/, and /17/, we find out that it is Janus-faced. In /16/ and /17/, it is derived as a concept of political economy from concepts value, labor, and SNLT. But in /6/ it antedates as a concept those concepts of political economy and once isolated from them it is just a common sense concept applied by the very actors of production, namely, just a multiple of a chosen weight-unit. These two "faces" of the concept price in Chapter 1 enable me now to deal with Marx's conclusion of Section 3 in this chapter, namely, with his concept of commodity-fetishism as well as with his introduction of the concept of a consciously acting actor of production into the description of the production and exchange of commodities.

The price of any commodity y appears on the surface of social relations, that, in the interaction between the producers, one selling and the other purchasing y, just as a certain, say, physically measurable amount of a specific commodity (e.g., a certain weight-amount of gold). The latter is for the seller of y the *subjective, consciously given aim* he or she wants to achieve when selling y, and it is the (*subjective*) *mean* for the buyer to obtain y. So what the seller and the purchaser as actors *consciously* perform is *not the exchange of values as the crystallization of a certain amount of SNLT, but the exchange of things*. What they perform *objectively* is, of course, something else; they exchange them in *exactly this proportion* only because *objectively* (*and behind their backs*), in both, the same amount of SNLT is objectified. That they can *consciously* plan and then *knowingly* exchange *at all* certain weight-amount of gold against the thing y has its objective *basis* in the fact that in both the same amount of SNLT is objectified. But, and here comes in the crucial point, in order for the production of commodity y to be *initiated* and then to be *renewed*, y has to be *consciously* and *knowingly* exchanged, by means of that weight-amount of gold, for all those commodities that enter into the production process of y. The *subjectively* (*consciously*) and the *objectively (behind the back) given aspects* of the production of commodity y, even if the former is *not a consciousness about and knowledge of the latter, cannot be separated at all*. According to Marx (1872/1987b, 104–105; 1976a, 166–167 [emphasis is mine]):

Humans ... mutually relate their labor-products as values not because these things are *valid* for them as mere thing-like envelopes of the same human labor. On the contrary: by mutually equating their various products in exchange as values, they mutually equate their different labors as human labor. *They do it without knowing it.*

So, only by the mutual exchange of commodities – *objectively* coming out of production *as* crystallizations of labor and, thus, *as values*, but *subjectively* (i.e. *consciously*) exchanged by the producers as *things*, where the exchange-relations between these commodities are *objectively* (*and independently of the consciousness producers*) regulated by the mutual ratios of the SNLT expended on them – can the objective determination of the value of each commodity by SNLT in the process of production and thus their objective mutual exchange relations exist at all. This is my interpretation of the following two claims of Marx: "In fact, the value-character of the labor-products is stabilized only by their activation as magnitudes of value" (Marx 1872/1987b, 105; 1976a, 167), but at the same time it holds as well that "the latter change continuously, independently from the will, foreknowledge and action of the exchangers" (Marx 1872/1987b, 105; 1976a, 167).

From this I draw the conclusion that for the objective social relation between value and labor to exist at all, not only the essential objective condition C_E^{B1} reconstructed earlier has to be given, but also the *subjective, consciously given aim* of actors of production to acquire by exchange *things* or *use-values*, I view it as yet another *essential condition* for the existence of objective social relation between value and labor. Marx gives the following description of this second condition (Marx 1872/1987b, 105; 1976a, 167):

What the produce-exchangers practically concerns first is the question how many products they obtain for their own product, thus in what proportions the products are exchanged. As soon as these proportions have acquired a certain habitual stability, they seem to come from the nature of the labor-products.

This essential condition is thus completely *topsy-turvy with respect to the objective determinations of commodities by value, labor, and SNLT;* it has with respect to these objective determinations the status of a *false consciousness, but still it cannot be separated from them.*

Let me now express this essential condition, which is of a subjective nature, as "the aim of the producer x of commodity y is to obtain use-values in exchange for it," or as $C_E^{B2}(x, y)$, for short. The importance of the reconstruction of C_E^{B2} lies in the fact that together with the essential condition introduced previously as C_E^{B1}, it is possible to reconstruct the structure of the scientific law of value as it is given at the very end of Section 3 of Chapter 1. Let C_E^B stand for the conjunction of C_E^{B1} and C_E^{B2}, then the law of value there can be stated as follows:

$$(x)(z)\{C_E^B(x, z) \xrightarrow{n} V(z) = f[L(x, z)]\} \qquad /20/$$

Here the individual variable x ranges over the set of humans fulfilling the two essential condtions together symbolized as "C_E^B," "z" ranges over the set of products produced by x, "$V(z)$" stands as before for the value of z, "$L(x, z)$" for the labor expended by x on the production of z, and "$\xrightarrow{n}$" is the sentential connective whose reading in English is "if ..., then necessarily __". The elements given in the law of value as a *scientific* law appearing in Chapter 1 of *Capital*, Volume I, are distributed, so to speak, among the various sections of this chapter. The essential condition C_E^{B1} and the equation $V(\varepsilon) = f[L(\varepsilon, \phi)]$ are given already in Section 1 while the conceptual grasping of the essential condition C_E^{B2} is performed only at the very end of Section 3. C_E^{B1} by itself, as well as C_E^{B2} by itself, is *necessary* and together they are *sufficient* for labor to exist. One can state this also as follows: Once these two conditions are given together, work acquires the new determination of labor, products become commodities and their use-value acquires the new determination of value.

A closer look at the condition C_E^{B2} discloses that even if it is an essential condition with the status of a *false* consciousness with respect to the condition C_E^{B1}, it still *correctly* reflects the fact that commodity-production is at the same time, *use-value* production. In addition, as a type of *consciousness*, it depends on the existence of both *intentionality* and *language*. Thus, for the condition C_E^{B2} to exist, the three essential conditions, we expressed previously as C_E^A have to be given as well. This is just a restatement, in the language of the essential conditions, of the fact noticed already, that labor is a differentiated unity of work and labor, and value, a unity of use-value and value.

How can we characterize the law of value as reconstructed by Marx in Sections 1 through 3 of Chapter 1? *First*, it is *social law* that is at work only and only where the conditions C_E^A, C_E^{B1}, and C_E^{B2} are given. *Second*, it is a historically specific way of *action*, namely, the way of action of beings fulfilling the conditions C_E^A, C_E^{B1} and C_E^{B2}. *Third*, it *regulates action* and, because it essentially, that is, in its very existence depends also on the givenness of condition C_E^{B2} – and thus on the givenness of *false* consciousness – it is at work without being acknowledged by the actors performing that action; it operates behind their backs. This is my interpretation of Marx's characterization of value "as a regulative law" (Marx 1872/1987b, 106; 1976a, 168) to which he adds in a footnote the following statement of Engels: "It is just a natural law based on the lack of awareness of the participants" (Marx 1872/1987b, 106; 1976a, 168). *Fourth*, because it is at the basis of the exchange of commodities between private producers, it regulates their mutual socioeconomic relations.

Fifth, it is a *law of reproduction*. Let me explain this feature of the law of value. Earlier I claimed that on the basis of the concepts of value, labor, and SNLT pertaining to a commodity one can derive the concept of its (gold) price. But the (gold) price of a commodity is the expression of its relation to a chosen commodity – gold – and *via* the latter to all other commodities. Let now y stand for a certain *type* of commodity (say, a coat), let $C_1, \ldots, C_n$ stand for individual commodities of this type all produced by a producer x; and let $C'_1, \ldots, C'_r$ stand for individual commodities of another *type* (say, linen) produced by another producer. And let us suppose also that the aim of x is to acquire by means of exchange for money (hereafter, M) the commodities $C'_1, \ldots, C'_r$, which are the means of production (MP for short) required as input for the production of $C_1, \ldots, C_n$. This exchange we express as follows:

$$(C_1, \ldots, C_n) \to M \to (C'_1, \ldots, C'_r)_{MP}$$

M stands here for the point at which the producer of $C'_1, \ldots, C'_r$ enters the scene with this commodities; while both $C_1, \ldots, C_n$ and $C'_1, \ldots, C'_r$ are the results of production regulated by the law of value. How can we relate this exchange-process with the process of production? Let L_v stand for the law of value and L_p for the law of price, then this relation can schematically be represented as shown in Figure 4.1:[161]

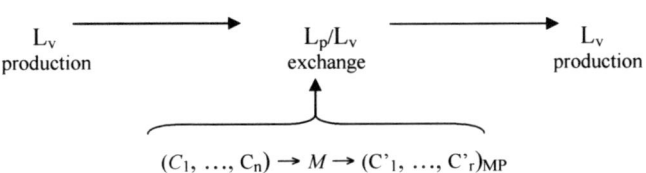

Figure 4.1 The law of value as a law of reproduction

Here L_p/L_v expresses the fact that the law of value is at work also in exchange and regulates the latter by means of the law of price for which it provides its objective ground. $(C_1, ..., C_n)$ stands for the commodity-output of the production process represented on the left side of the diagram, while $(C'_1, ..., C'_r)_{MP}$ stands for the means-of-production-input of the production process to be renewed and represented on the right side of the diagram. Thus, the law of value is a law of reproduction because it enables the renewal of the production process and, thus, the reproduction of the social structures in those societies where the law of value is at work and, also, the reproduction of the social being as commodity-producers of those persons who are at work in the production of $C_1, ..., C_n$.[162]

How can we characterize, from the point of view of *epistemic categories*, the law of value as given in Section 3 of Chapter 1? This *scientific law* conceptually grasps social relations *inherently* given both in production and (via price *based on* or *grounded in* value) in exchange; it is thus an *inherent* type of scientific law. It enables one also to theoretically grasp the unity of production and exchange so

---

161   For the scientific law of price, conceptually derived by Marx from the scientific law of value in Section 3 of Chapter 1, should hold the equation similar to that given in /17/. The internal structure of this scientific law would be much more complex than that of the scientific law of value given in /20/ because its completely reconstructed structure would have to involve both the description of two commodities – gold and any other commodity – and the description of two producers: the one producing gold and the second producing the other commodity. I therefore refrain in this book from the reconstruction of the structure of the scientific law of price as given in Section 3 in Chapter 1.

162   For a detailed analysis of this aspect of the law of value see (Quaas 1997).

that it explains why the process of production, once mediated by exchange, is continuously renewed, and why the exchange-process, once mediated by production, is renewed as well. This unity can be viewed as a whole to which one can assign, following Marx's explications in the introduction to the *Grundrisse* (1976b, 35; 1973, 100), the term "organic whole." We thus label the law of value as given in Section 3 of Chapter 1 as the *inherent organic type of scientific law*.[163]

Let me now provide in Table 4.3 a summary of Marx's unification of the sociological, politico-economic and epistemic approaches in Sections 1 through 3 in Chapter 1.

From the point of view of Chapter 1	concepts of political economy	concepts of sociology	the corresponding epistemic categories
Section 1	exchange of commodities, commodity, use-value, exchange-value, work, labor, SNLT	the existence of private (separate) producers, necessary objective social conditions	appearances, external measure, quantity and quality of appearances, essential conditions, essence/ground, the substance of the essence/ground, immanent measure of the substance
Section 2	differentiated unity of labor/work; differentiated unity of use-value/value	the differentiated unity of the objective social conditions	inherent contradiction in the substance of the essence/ground; the essence/ground as inherently differentiated

---

163 By this characterization I explicate Marx's phrase "inherent organic laws of political economy now at work in every civilized town" (Marx 1984, 252).

Section 3	exchange-value, money, price, measure of value, standard of prices, commodity fetishism, law of value	necessary subjective (consciously given) social conditions, jointly sufficient social conditions, false consciousness social law, social structure, reproduction of social structure, social being of a person	manifestations, manifestations of the inherent contradiction, manifestation of the immanent measure, scientific law, scientific inherent organic law

Table 4.3 A summary of Marx's unification of politico-economic, sociological and epistemic dimensions in Chapter 1.

4.1.3 What Went Wrong? Some Clarifications

Let me now return to my critique of G. Reuten and C. J. Arthur. As shown already, the latter introduces, after distinguishing between the *immediate comparative* and the *indirect* (*external*) measure, the third category of measure, namely, that (Arthur 2005, 116)

some dimensions are 'rock bottom' (e. g. extension and mass), but some others are complex (e.g., work). Work is a function of the force moving across distance in time; and if we could measure these three can calculate the magnitude of work units.

Based on this third type of measure he then goes over to the concepts of political economy and states the following (2005, 116):

There is an *immediate* measure of value in commodity money. This is distinct from value's determinants but, if we have a theory which determines the immanent magnitude of value by that of labour time, it can be measured by calculation ... it may be said that labour time is an indirect immanent measure of value and that money is an indirect external measure of labour.

Unfortunately, Arthur does not grasp the *fundamental difference* between what he views as the third type of measure as given in classical mechanics and the category of measure as given in Marx's approach to the concepts of value and labor. In the case of classical mechanics, force is measured by acceleration, that is, only by its effects as *appearances*; thus, here we have again a case of what I labeled previously as *external* measure. Therefore, also in the measurement of work as a magnitude in classical mechanics one has to employ the magnitude of acceleration and thus quantify work by means of an *external measure*. Contrary to this, Marx assigns to value as a magnitude of his economic theory its own inherent measure, namely, the *quantity of the substance in which it originates*. If I use the method employed earlier to represent magnitudes by means of ordered n-tuples, then the difference between the magnitude of force in classical mechanics and the magnitude of value in Chapter 1 of Capital, Volume I, can be expressed by putting side by side the representation of the former and the latter. For the magnitude of value we have, as shown previously, $(\{t\}, [t], U V/V, L/W)$, while for force, F, for short, we have $(\{f\}, [f], 0, 0)$, where $\{f\}$ stands for the quantitative determination of the magnitude of force and $[f]$ for its quality; the first zero expresses the fact that classical mechanics does not view force as internally differentiated, the second zero expresses the fact that in classical mechanics we do not know the origin of forces.

Here lies the principal difference between the categories one can assign to classical mechanics as a scientific theory and to Marx's concepts of political economy. So as classical mechanics is, as shown in chapter 3 of this book, *not a theory about the origin of forces*, it remains at the category-level of external measure. Marx's political economy as given in Chapter 1 of *Capital*, Volume I, is a theory about the *origin of value in labor*;[164] it thus does not remain stuck at the level of external measure but moves from the latter to that of the immanent measure. Therefore, I view also Arthur's term "indirect immanent measure" which, *he claims*, should correspond to the determination of the size of value by the size of labor as an *oxymoron*; *immanent measure is a direct measure*.

Let me now apply the difference, stated earlier, between classical mechanics and Marx's political economy to Marx's well-known example of how the actors consciously aiming at determining the weight of a sugar-loaf take pieces of iron weights and express the weight of the former as a multiple of the weight of the latter (Marx 1872/1987b, 89; 1976a, 148–149). *On the one hand*, what makes this example of Marx similar, from the point of view of *epistemic categories*, to his reflections on the genesis of the money-form, is that he presupposes that the actors of weighing do not know, say, from classical mechanics, that the force gravity (*Schwere*) is the basis of weight, in the same way that the actors performing the exchange of their products know nothing (say, on the basis of Marx's analysis from Sections 1 and 2) about the nature of value and quantitative proportions

---

164 Below I will show that Marx's political economy in chapters following after Chapter 4 of *Capital*, Vol. I, is a theory about the *origin of surplus-value in surplus-labor*.

of labor crystallized in these products. *On the other hand*, however, there is a profound dissimilarity between Marx's analysis of the genesis of the money-form and his example of the weighing of the sugar-loaf. Not only, as Marx notes, is gravity a natural property while value is a social property, but there exists a profound difference between the *epistemic context* into which he places that analysis of genesis of the money-form and the *epistemic context* into which he could have placed at all his example of weighing in the 1860s when he worked on the first volume of *Capital*. While in Marx's approach the analysis of the genesis of the money-form *follows after* his reconstruction of the *origin of value in labor* and of the *determination of the quantitative characteristic of value by SNLT*, what Marx could have placed *before* his example of the weighing of the sugar-loaf was just the knowledge of classical mechanics as the sole scientific theory about the force of gravity in the 1860s, which – as shown in chapter 3 – even till now is neither *a theory about the origin of gravity* as a force in some specific substance nor a theory that would be able to derive the quantitative determination of gravity from the quantitative determination of this substance.

That profound difference between the epistemic context of Marx's analysis of the genesis of the money-form and the epistemic context of his sugar-loaf-example indicates the *potential limitations* of, and even *potential dangers* inherent in attempts (performed, e.g., by Arthur) to approach the categories given in Chapter 1 of *Capital*, Volume I., by means of categories corresponding to the epistemic context of other scientific theories.

Arthur's attempt to assign to the concept of money the category he labels as "indirect external measure" fairs even worse. So, as he does not, as shown earlier, distinguish between appearance and manifestations as two types of knowledge of the phenomena, neither can he distinguish between the two types of measure related to these two types of knowledge: the *external measure* related to phenomena with the epistemic status of *appearances* and the *manifestation of the immanent* measure related to phenomena with the epistemic status of *manifestations*. Therefore, he misses the crucial point that money, as shown already in quotes from Marx, (a) is the *direct external measure* by means of which the pre-Marxian political economy set out in the direction of the concepts of value and labor and, (b) is for Marx in Chapter 1 the *indirect*, that is, the *derived manifestation of the inherent immanent measure* (SNLT) given in the exchange-relations between commodities and, thus, not any more an *external* measure, as claimed by Arthur.

We can now evaluate Arthur's comparison of Hegel's logical categories with Marx's concepts of political economy as given in Table 4.2. As shown above in my Table 4.3, to Marx's thought-movement from exchange-value to exchange-value in Chapter 1 can be assigned the sequence of cluster of categories appearance → essence/ground → manifestation. So, as Arthur in his comparison brings in also Marx's concept of self-valorization of capital, we face the following question: What categories can be assigned to those chapters of *Capital*, Volume I, where Marx deals with concepts like surplus-labor, surplus-value, variable capital, and so on?

First of all, let me note that Marx's point of departure in his thought-movement to these concepts is the formula $M \to C \to M'$, that is, a *paradoxical* or, in Marx's own terminology, "contradictory"[165] formula (Marx 1872/1987b, 173–182; 1976a, 258–269) expressing the *phenomenon* of "making money" to which he assigns the epistemic category of *appearance*; "$M \to C \to M'$ is the general formula of capital as it immediately appears (*unmittelbar erscheinen*) in the sphere of circulation" (Marx 1872/1987b, 173; 1976a, 257). From here Marx moves to the origin of surplus-value (hereafter, *SV*) in surplus-labor (hereafter, *SL*).

One has to realize here, *first*, that according to Marx, surplus-labor is produced only and only where three *new essential conditions are given*: the existence of producers privately owning the means of production and purchasing, in addition to instruments and objects of labor, also labor-power; the existence of a group of free people selling their labor-power; the subjective (conscious) aim of the producers being to make money (money $\to$ commodity $\to$ money'). *Second*, these three essential conditions, which I represent as C_E^C, coexist with the essential conditions expressed above as C_E^B. Based on this, we can realize that Marx's movement from value and price as given in Chapters 1 through 3 to value in the later chapters is a thought-derivation of a new *essential* (social) determination of commodities and value and at the same time a profound transformation of the sociological understanding of the social structure. While in Chapters 1 through 3, this structure was set up only by private producers, in the later chapters the structure is set up by the private producers *and* the owners of the labor-power. And if we, *third*, realize that in Marx's approach surplus-labour is viewed as the *substance* of surplus-value, and socially necessary surplus-time as the *immanent measure* of surplus-value, then we come to the conclusion that the third column in my Table 4.3 states categories that correspond, not only to Marx's concept of political economy in Chapter 1, but also to concepts in the later chapters of *Capital*, Volume I, but with one *important restriction*. While Marx derives in Chapter 1 the price as the manifestation of value, he derives in Volume I, in chapters that follow Chapter 3, *neither* the manifestations of surplus-value (profit, rent, interest) *nor* the manifestations of the capitalistically determined law of price (the law of the production price). This is performed by Marx in the manuscripts of Volume III of *Capital*, where the phenomenon of making money from which Marx ini-

---

165 Marx devotes both in both the first and second editions of *Capital*, Volume I, a subchapter to the analysis of the conceptual contradictions given in this formula.

tially departed as an appearance is derived as the manifestation of surplus-value and named as "profit," "interest," "and rent."

The structure of law of value given in Chapter 1 of *Capital*, Volume I acquires in Chapter 5 the following form:

$$(x)(y)(z)\{Ces^C(x, y) \ V_{c+v+m}(z) = f[LSL(x, y, z)]\} \qquad /20*/$$

Here "$V_{c+v+m}(z)$" stands for the value of z composed of constant capital, c, variable capital, v, and surplus-value, m; "$LSL(x, y, z)$" stands for labor and surplus-labor performed by x for y in the production of z," while the individual variables "x" and "y" range over a set of humans fulfilling the condition Ces^C.

Marx's movement from the law of value, L_v, with the structure of /20/, to the law of value with the structure of /20*/ (hereafter, $L_{v(s)}$), can serve as the basis for the solution, at the level of categories, of the problem indicated already in chapter 3 of this work, namely, that neither the method of gradual concretization nor that of thought reconstruction can be applied once the ground/cause/essence, given in the law of the explanandum, differs from that, given in the law of the explanans. In chapter 3, I traced this limitation back to the fact that in the laws of the types $L^{(k)}$ and $L^{(l)}$, the essence/cause/ground is viewed as *completely unconditioned*. Thus, by introducing the category of essential conditions, one can reconstruct a type of explanation where the explanans-law and the explanandum-law differ mutually by their very essence/ground/cause.

Marx's thought-movement from the law of value L_v, given in Chapters 1 through 3 of *Capital*, Volume I, to the law of value $L_{v(s)}$, as given in Chapters 5 through 23 of this Volume I, represent as follows (here "$=_{tr}=|$" stands for transformation):

$$L_v \ \& \ Ces^C \ =_{tr}=| \ L_{v(s)}$$

How does L_v differ from $L_{v(s)}$? In the case of the former, according to Marx, value is viewed so far only as the objectification of a day's labor performed by the "producer" (Marx 19872/1987b, 78; 1976a, 135), and who this "producer" is is as yet undetermined (Marx 1988, 33; 1976c, 954). Thus value V is a function of the labor L performed during that time; in symbols $V = f(L)$. What is objectified in a commodity in the case of $L_{v(s)}$ is prepaid labor – variable capital, unpaid labor, that is, surplus-value and value of constant capital; in symbols $V = f(c+v+m)$.

At the same Marx views the result of his thought derivation of $L_{v(s)}$, given in the explanandum as pertaining not to a single commodity, as in the case of L_v, but to a whole mass of commodities (Marx 1988, 33; 1976c, 954). This, thus, means that while in /20/ the individual variable z ranged over a set whose elements were *singular commodities*, in /20*/ this variable ranges over a set whose elements are *sets of commodities*. Already in the framework of the first volume of *Capital* one faces, therefore, not only a shift of meaning of the term "exchange-value," as shown earlier, but also of the term "value." And in both cases the basis of these meaning shifts is the *cyclical nature* of theory construction. In the case of the meaning of the term "exchange-value," it is the cycle *exchange-value → value → exchange-value**, and, in the case of the meaning of the term value, it is the introduction of the terms like "surplus-labor" and "surplus-value" which leads to the cycle *value → surplus-labor/surplus-value → value**. Marx, in fact, explicitly relates the meaning shift of the terms "commodity" and "value" to the "circular movement of our representation" (Marx 1988, 24; 1976c, 949).

What becomes readily apparent here, and what comes close to what was stated previously in chapter 3 for the *Principia*, is that *a social science theory, once it is built by a cyclical method, can fulfill a critical function*. Marx, by using this method, is able, first, to derive *anew* and, thus, explain the origin of exchange-value by grasping the internal differentiation of commodity into use-value/value. Second, by deriving $L_{v(s)}$ from L_v at the very end of *Capital*, Volume I, he prepares the explanatory basis for the derivations given in the manuscripts of Volume III of *Capital*. At the end of *Capital*, Volume I, the law of value regulates, not the (re)production of a new exemplar of the same single type of commodity,[166] but the reproduction of the whole mass of commodities of different types; and, then, in the manuscript of Volume III of *Capital*, he is able to *return* to the explanation of the conditions under which a single type of commodity is produced anew – once it is sold for a price that equals its *production price*; the latter is according to Marx's derivation "in the long run ... a condition of supply, of reproduction of commodities in every industrial sphere of production" (Marx 1992, 272; 1981b, 300).

In analogy with Figure 4.1, my Figure 4.2 expresses Marx's integration of the concept of surplus-value into the law of value as a law of reproduction by the following figure:

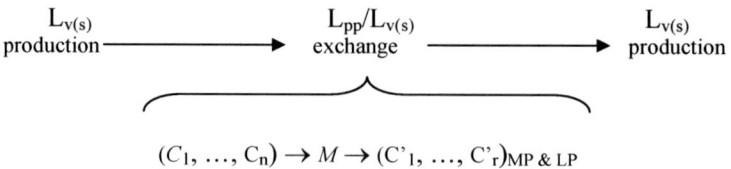

Figure 4.2 The law of value as a law of reproduction

---

166 In Marx's thought-movement, the law of value, *initially*, expresses the socially necessary labor "to produce a new exemplar of the same commodity under generally given production-condtions" (Marx 1980b, 111)

Here $L_{pp}/L_{v(s)}$ expresses the fact that the law of value is at work also in exchange and regulates the latter by means of the law of the production price, L_{pp}, for which it provides the essence/cause/ground. $(C_1, ..., C_n)$ stands for the commodity-output of the production process represented on the left side of the diagram, while $(C'_1, ..., C'_r)_{MP \& LP}$ stands for the means-of-production-input and labor-power-input of the production process to be renewed and represented on the right side of the diagram. Commodities $C_1, ..., C_n$ coming out of the production process have, according to Marx's explanation as given in Volume I of *Capital*, their value determined by the sum $c + v + s$.

I can now evaluate Ben Fowkes' translation of *Capital*, Volume I from the point of view of the following three aspects.

The first is Marx's terminology, which pertains to terms "Größe" and "Quantum" and which is relevant for the category of measure. The problem Fowkes's translation faces when dealing with the first of these terms is that the term "Größe" has in German *two different meanings* that can be read off from the linguistic pun "die Größe einer Größe." *On the one hand*, "Größe" refers to a property of an object, where the property has both *quantitative* and *qualitative* characteristics. The quantitative characteristic of the commodity, for example, labor, was expressed earlier by means of the symbol "$\{t\}$" while "$[t]$" referred to the dimension and, thus, to the *quality* of the characteristic. "Größe" then stands in general for a unity expressed by the ordered pair $(\{q\}, [q])$ and *not* for a (*concrete*) *size*. The term "Größe," once understood in such a way, should be translated, at least in my view, as "magnitude." *On the other hand*, the term "Größe" can refer also to the *size of a certain magnitude*; for example, earlier we used for it, in the case of labor, the notation "$(\{t_1\},[t])$." In this case we prefer to translate the term "Größe" as "size," and the general notation here is "$(\{q_1\},[q])$." The seeming paradox given in the German expression "die Größe einer Größe" can thus be dissolved by its translation into English as "the size of a magnitude."

That the reconstructed two meanings of the German term "Größe" are really present in Marx's texts, as can be seen from the following text of Marx and its comparison with Fowkes' translation, who does not differentiate between these two meanings as given at its very beginning and its very end and emphasized by me; I include in Table 4.4 my proposed translation.

(Marx 1872/1987b, 82)	(Marx 1976a, 140–141)	Proposed translation
... die *Größen* verschiedner Dinge erst quantitative vergleichbar werden nach ihrer Reduktion auf dieselbe Einheit. Nur als Ausdrücke derselben Einheit sind sie gleichnamige, daher kommensurable *Größen*.	... the *magnitudes* of different things only become comparable in quantitative terms when they have been reduced to the same unit. Only as expressions of the same unit do they have a common denominator, and are therefore commensurable *magnitudes*.	... the *sizes* of different things become quantitatively comparable only after their reduction to the same unit. Only as expressions of the same unit are they of the same denomination and therefore mutually measurable *magnitudes*.

Table 4.4 *Marx on the term "Größe" in Chapter 1 of Capital, Volume I*

If one accepts the translation proposed by me, then Marx's terms "Größe eines Werthes" and "Werthgrösse" (Marx 1872/1987b, 72, 73) should be translated as "size of the value" and not as "magnitude of value" (Marx 1976a, 129) and the term "Größenmaß" that Marx applies to value (Marx 1867/1983, 21) should be translated as "measure of the size" of value.

Marx uses, in addition to the ambiguous term "Größe," also the term "Quantum," for example, when dealing with the question of how to measure the size of the value of a commodity; he gives the following answer: "Durch das Quantum der in ihn enthaltenen 'werth-bildenden Substanz', der Arbeit" (Marx 1872/1987b, 72). Thus, the meaning of the German term "Quantum" is close to the first of the earlier given meanings of the term "Größe."[167] I therefore propose to translate the term as "amount" or simply use for it the term "quantum"; Fowkes, contrary to this, does not differentiate between the German terms "Quantum" and "Quantität" and translates both as "quantity" (Marx 1976a, 129).

The second aspect is Marx's terminological differentiation between "Schein" and "Erscheinung." So as Fowkes does not thoroughly base his translation on that differentiation on Marx's part, he provides in those places where Marx uses one of these terms *confusing* translations, at best, and *inconsistent* ones at worse. Let me start with the confusing ones. On the one hand, in those places where Marx uses the noun "Schein" or the verbs "erscheint zunächst" and "scheint," Fowkes correctly translates the former by the noun "appearance" and the latter, by the verb "to appear." For example (here I emphasize the relevant terms):

---

[167] This interpretation corresponds to Hegel's approach to the category Quantum: "Under the expression *size* (*Größe*) ... is understood *quantum* and not quantity, therefore this name from foreign language has to be used" (1923, T. 1, 179).

1) (Marx 1872/1987b, 70–71) Der Tauschwerth *erscheint zunächst* als das quantitative Verhältnis, die Proportion worin sich Gebrauchswerthe einer Art gegen Gebrauchswerthe anderer Art austauchen ... Der Tauschwerth *scheint* daher etwas Zufälliges und rein Relatives ...	1) (Marx 1976a, 126) Exchange-value *appears* first at all as the quantitative relation, the proportion, in which use-values of one kind exchange of use-values of another kind. ... Hence exchange-value *appears* to be something accidental and purely relative ...
2) (Marx 1872/1987b, 113) [Ricardo] hat die *Schein*relativität, die diese Dinge, Diamenten und Perlen z.B., als Tauschwerthe besitzen, auf das hinter diesen *Schein* verborgene wahre Verhältnis reducirt ...	2) (Marx 1976a, 177) [Ricardo] has reduced the *apparent* relativity which these things (diamonds, pearls, etc.) possess to the true relation hidden behind the *appearance* ...

On other hand, however, Fowkes uses very often the English expression "form of appearance" for Marx's term "Erscheinungsform" in those positions where Marx derives concepts of political economy, such as exchange-value, value-form and money, on the basis of the concepts of labor/work; use-value/value, and SNLT, that is, when the former concepts have already the epistemic status of *manifestations* of the latter concepts, for example (here I emphasize again the relevant terms):

1) (Marx 1872/1987b, 72) Der Fortgang der Untersuchung wird uns zurückführen zum Tauschwerth als der nothwendigen Ausdrucksweise oder *Erscheinungsform* des Werths ...	1) (Marx 1976a, 128) The progress of our investigation will lead us back to exchange-value as the necessary mode of expression, or *form of appearance*, of value.
2) (Marx 1872/1987b, 80) Wir gingen in der That vom Tauschwerth oder Austauschverhältnis der Waaren aus, um ihrem darin versteckten Werth auf die Spur zu kommen. Wir müssen jetzt zu dieser *Erscheinungsform* des Werths zurückkehren.	2) (Marx 1976a, 139) In fact we started from exchange-value, or the exchange relation of commodities, in order to track down the value that lay hidden within it. We must return to this *form of appearance* of value.

Thus, these last translations are wrong. Stated in a more general way: Fowkes's translations here violate the principle: *To two (or several) interchainings of concepts appearing at two (or several) different places in the structure of one and the same scientific theory, where they stand for different levels of scientific knowledge, one should not assign epistemic categories standing for one and the same level of scientific knowledge.*

Let me now turn to the *inconsistent* translations, for example, as discussed earlier (here I emphasize again the relevant terms):

(Marx 1872/1987b, 92) [Die Waare] ... *stellt sich dar* als dieß Doppelte was sie ist, sobald ihr Werth eine eigne ... *Erscheinungsform* besitzt ...	(Marx 1976a, 152) [Commodity] ... appears as the twofold thing it really is as soon as its value possesses its own particular *form of manifestation* ...

This means that to *one and same the level of scientific knowledge* attained by Marx in the construction of *Capital*, Volume I, namely, to the scientific knowledge given at the level of the subsection "*Das Ganze der einfachen Werthform*" (Marx 1872/1987b, 92; 1976a, 152) are – due to Fowkes's wrong translation – in fact, assigned two categories standing for *two different levels of the development of scientific knowledge*, namely, *appearance* and *manifestation*. This case of an inconsistent translation is, unfortunately, not the only one in the subsection "*Das Ganze der einfachen Werthform.*" While in the earlier given quote the German term "Erscheinungsform" was translated by Fowkes as "form of manifestation," the same term German term given *in the same subsection* just four paragraphs farther, as part of the following statement (the term under discussion emphasized by me) "Die einfache Werthform einer Waare ist also die einfache *Erscheinungsform*" (Marx 1872/1987b, 93), is translated by Fowkes in completely opposite way, namely as follows (the term under discussion emphasized by me): "Hence the simple form of value of a commodity is the simple *form of appearance*" (Marx 1976a, 153).

Fowkes's *inconsistent* translations violate the following general principle: *To the interchaining of concepts appearing at a certain place in the structure of one and the same scientific theory, where they stand for a certain level of scientific knowledge, one should not assign epistemic categories standing for different levels of scientific knowledge.*

The third aspect is Marx's terminology that pertains to the reconstruction of the issue of commodity fetishism. In the sentence that follows after the passage quoted previously in my analysis of Marx's approach to commodity fetishism, he states: "In der That befestigt sich der Werthcharacter der Arbeitsprodukte erst durch ihre Bethätigung als Werthgrösse" (Marx 1872/1987b, 105). Fowkes's translation of this last sentence is as follows: "The value character of the products becomes firmly established only when they act as magnitudes of values" (Marx 1976a, 167). I interpreted Marx's views in the subsection on commodity fetishism in such a way that they pertain to the *action* of the producer who consciously performs something very different from what goes on, however, only due to their action guided by the conscious aims, behind their back. If one accepts this interpretation, then one cannot accept the claim that *very products of labor act*. Marx's German phrase "Bethätigung der Arbeitsprodukte als Werthgrössen" one should not, therefore, translate as "products act as magnitudes of values."

Those who act are the *producers themselves*; in their consciously planned and accomplished exchange of products, they take the results of their own production activities – products endowed behind their back with value – and then, again consciously, renew their production activities. As has been already stated: only due to the *conscious activity* of exchange of products between their producers can the process of production, objectively (i.e., behind their back), regulated by the law of value, take place at all. The translation of Marx's German phrase "Bethätigung der Arbeitsprodukte als Werthgrössen," thus should be: "activation of the products as magnitudes of values." For this same reason, one should not translate Marx's German sentence "die Privatarbeiten bethätigen sich ... erst als Glieder der gesellschaftlichen Gesammtarbeit durch die Beziehung, worin der Austausch die Arbeitsprodukte ... versetzt" (Marx 1872/1987b, 104) as Fowkes does, as "the labour of the private individual manifests itself as an element of the total labour only through the relations which the act of exchange establishes between the products" (Marx 1976a, 165); again, as above, the expression "bethätigen sich" should be translated as "are activated."

That Fowkes's understanding of Marx's approach to commodity fetishism, which, in turn, determines his translation here, misses the crucial aspect of Marx's action-approach becomes readily seen when one compares Marx's quote of Engels on the nature of the law of value with the translation of this quote used by Fowkes. Engels's claim, in German, is as follows: "Es ist eben ein Naturgesetz, das auf der Bewußtlosigkeit der Betheiligten basiert" (Marx 1872/1987b, 105); Fowkes's translation is: "It is just a natural law which depends on the lack of awareness of the people who undergo it" (Marx 1976a, 168). This translation is wrong, from the point of view of *sociology, first*, because it suppresses the action-nature of the law of value; Marx quotes Engels because he speaks about the "Betheiligten", that is, about producers as *participants* and not about those who just "undergo it,", that is, are just *subject to it*. It is wrong from the point of view of sociology, *second*, because it translates the German phrase "basiert auf der Bewußtlosigkeit" as "depends on the lack of awareness" and, thus, suppresses Marx's view on the law of value as a *social law the very existence of which* depends on the lack of awareness on the side of the actors; it can be at work only if there is no awareness of it.

Let me now turn to Reuten's interpretations of Marx. Against Reuten's view, expressed in (Reuten 2005), namely, that, according to Marx *himself,* value has no existence without money, I bring in the following quote from Marx's *Urtext* of his *Zur Kritik der politischen Ökonomie* of 1859: "*What, once we look at the form of circulation itself, becomes, originates, is produced in it, is money itself, nothing further.* Commodities are exchanged in circulation but they do not originate in it" (Marx 1980a, 72); and he goes on as follows: "Circulation does create neither exchange-value nor its size. In order for a commodity to be measured in money, money and commodity have both to behave mutually as exchange-values, i.e., as objectification (*Vergegenständlichung*) of labor-time" (Marx 1980a, 72).[168]

As to his claim that Marx does not provide in Chapter 1 a labor-theory of value in any quantifiable sense (Reuten 2005, 86) my Table 4.3 gives the answer. The term SNLT, so as it pertains to the quantitative determination of the substance in which value originates, is permanently present in all conceptual derivations in this chapter; from this point of view F. Moseley's diagram given in Table 4.1 is correct.

Against Reuten's paper (1993) the following objections can be stated. First, the whole idea of introducing the "discounting" coefficient of the type α_i is, at least according to my view, wrong because it should, on the one hand, mediate between two *different magnitudes* – work and labor – but, on the other hand, it could be, according to Reuten, put equal to the *pure* number 1. Thus α_i should be devoid of any dimension, but in order to mediate between two different magnitudes, it should have a dimension and, thus, it should *not* be a pure number; only then could the equation $\alpha_i W_i = L_i$ be dimensionally meaningful.

That work and labor are, at least in Marx's conceptual system of political economy as science, really *different magnitudes* can be seen once we compare them mutually. They differ at least in the following four crucial aspects. First, according to Marx their existential conditions are, as shown earlier, different. Second, according to him (Marx 1872/1987b, 79; 1976a, 136–137), the laws by which they are related to the production of commodities are different. Marx provides the following sequence of thought-operations. Let us suppose that we hold the *time-span* of production constant and increase the productive power (*Produktivkraft*). This enables *work* to produce an increased amount of use-values, but, because the total labor-time expended is the same as before, the total value of this amount of commodities is still the same. Let us now suppose a *fixed amount* of commodities and suppose an *increase* in the productive power. This requires a decreased time-span to produce this amount of commodities and, thus, also a decreased amount of labor expended and, therefore, also the value of this amount of commodities decreases; but the use-value of this amount is constant. *Third*, in the process of production, the *value*, but *not the use-value*, of the means of production used in the production of a commodity is *passed over to* and *preserved in* that commodity. *Fourth*, according to Marx, "the use-value is realized only in use or in consumption" (Marx 1872/1987b, 70; 1976a, 126). So it holds that what concrete use-value a certain product has depends, not only on the type of work

---

168 Here Marx's term "Tauschwerth" refers to what Marx's term "Werth" refers to starting from the second edition of *Capital*, Vol. I.

used for its production, but also on the concrete type of its usage or consumption; usage gives to the use-value its final "touch." If, for example, a person's products are produced, on the basis of a blueprint, for and with a conscious aim of, making furniture, but then are always burned to produce heat, then the work performed is not that of a carpenter but that of a producer of heating material. In Marxian term: "A railway on which no trains run, which thus is not consumed, is only a railway δυνάμ, not in reality" (Marx 1976b, 28; 1973, 91). Contrary to this, as shown already, labor is expended in the production of a commodity, and, thus, it has value, even if it is not afterwards exchanged for money and, instead of being used and consumed, falls apart in a warehouse.

One could state an even stronger objection against Reuten's equation $a_i W_i = L_i$, namely, that even if he claims that it *expresses the view of Marx on the reduction of work to labor*, in fact, it *contradicts it fundamentally*. That this is so becomes readily apparent when Reuten transforms $a_i W_i = L_i$ into $a_i = L_i/W_i$ and interprets the latter as the *value productivity of work W_i*. But under this interpretation, *work* and *not labor, would be the substance of value*. But according to Marx, the property of the production-activity he labels as "konkrete Arbeit" ("work" in my terminology) leads to the creation of the use-value-properties of the product, the property of value of the product originates in the property of the production-activity he labels as "abstract labor" ("labor" in my terminology). I therefore view, with respect to Marx's concepts from Chapter 1, Reuten's expression "value-productivity of concrete labor (work)," as an *oxymoron*.

4.2 Marx, Ricardo, and Bailey on Value: A Comparison

In this subchapter, I first analyze David Ricardo's approach to value. I then show, from the point of view of epistemic categories, its relation to Marx's approach to value,[169] as well as to Newton's approach to the concept of force. Then I pay special attention, from the point of view of the philosophy of science, to Samuel Bailey's antimetaphysical attitude to the concept of absolute value; the latter is, as we will see, comparable to Ernst Mach's antimetaphysical attitude to the concept of force analyzed in chapter 3. In order to avoid in advance any possible misunderstanding of my aims, I emphasize that from various approaches of Ricardo[170] to the concept of value, I am here interested in only that approach in which *he relates the value of commodity to the amount of labor expended on*

---

169 The idea of a comparison, from the point of view of *epistemic* categories, of Ricardo's and Marx's theories of value was suggested to me by M. J. Carlson's article (Carlson 1995) which puts Marx's theory value into a category with that of Ricardo.
170 For an excellent analysis of these approaches see chapters 4 and 5 in (Peach 1993).

its production; only this approach on his part will be analyzed here, by means of categories.

The reconstruction of categories, presented earlier, in which, as a medium, Marx's thought-movement is accomplished in Chapter 1 of *Capital*, Volume I, enables us to reevaluate Marx's critique of D. Ricardo's *Principles*, as given in *Manuscript 1861–63*, from the point of view of epistemic categories. Such a reevaluation is, not only possible, but also necessary due to the *availability of two new types of data*. I have in mind, first, Ricardo's *Notes on Malthus*, his extensive correspondence on the issues of absolute value and relative value, as well as his two manuscripts on absolute and relative value. Second, we have, finally, at our disposal the original manuscripts of what, due to Engel's editorial work, became known as the third volume of *Capital*. Even if Marx in his *Manuscript 1861–63* claims that "one has to reproach Ricardo ... that he very often forgets ... 'real' or 'absolute value' and adheres only to 'relative' or 'comparative values'" (Marx 1978, 825), neither in the fist volume, nor in the third volume of *Capital* does he himself seem to show any allegiance to the concept of absolute value. The more recent publication of one of the manuscripts of the "future" third volume of *Capital* shows that Marx explicitly operates with the term "absolute value" (Marx 1992, 118, 310, 334) and that it disappeared from that third volume only due to Engel's editorial work.[171]

4.2.1 Ricardo on Relative, Absolute Value, and Measure

From the point of view of the aims pursued in this chapter ,the most important terms in Ricardo's works are "relative value," "absolute value" and "measure of value."

Ricardo starts his *Principles* from the concept of the value of a commodity, which he initially characterizes as "the quantity of any other commodity for which it will exchange" (Ricardo 1951-1973, Vol. I, 11). Then, following the view of A. Smith that the former type of value stands for "the power of purchasing goods" (Ricardo 1951-1973, Vol. I, 11), labels it more precisely as "exchange value."

So as this type of value expresses the proportion or relation between commodities, Ricardo also labels it as "relative value," "proportional value" (Ricardo 1951-1973, Vol. IV, 398), or "comparative value" (Ricardo 1951-1973, Vol. I, 373). He introduces the first of them in cases where the mutual exchange proportion

---

171 It is possible that Engels' removal of the concept of absolute value from Vol. III of *Capital* is related to his view expressed in *Anti-Dühring*, where he claims that the term "'absolute value', ... so far as our knowledge goes, has never had currency in political economy" (Engels 1988, 386).

between commodities has varied "Two commodities vary in relative value, and we wish to know in which the variation has really taken place" (Ricardo 1951-1973, Vol. I, 17). What he has in mind is to locate the source of that variation in the *variation* of quantity of labour expended for the production of one (or both) of them, because the value itself of a commodity is, according to Ricardo, determined by "the relative quantity of labour which is necessary for its production" (Ricardo 1951-1973, Vol. I, 11).

While Ricardo grounds the relation between commodities in the amount of labor expended on each during the process of its production, he also uses the term "relative value" in a sense different from that just mentioned. Because each type of commodity is produced by the expenditure of a certain amount of labor, he claims that there exists a value determined by that quantity, and because the latter varies as we move from one type of commodity to another, the value, which has its *origin in labor*, can also be labeled "relative value." He claims: "It is the comparative quantity of commodities which labour will produce, that determines their present or past relative value" (Ricardo 1951-1973, Vol. I, 17).

For Ricardo, then, the term "relative value" has *two meanings*. One, comparative value, pertains to the mutual relation of *already given, produced* commodities, while the other pertains to the *process of production* of commodities, to their *origin in labor*. This is readily seen when one compares his understanding of value expressed at the very beginning of the *Principles* quoted earlier, namely, the "power of purchasing other goods" with his later contention that he "cannot agree with Mr. Say, in estimating value of a commodity by the abundance of other commodities for which it will exchange" (Ricardo 1951-1973, Vol. I, 284). Ricardo labels the type of value pertaining to its origin in the labor producing it, not only as "relative value," but also as "positive value" (Vol. IX, 339, 351), "natural value" (Vol. IV, 375), "real value," (Vol. II, 65) or "absolute value."

It is worth noting that Ricardo's double meaning of the term "relative value" reappears in Marx's *Zur Kritik der politischen Ökonomie* of 1859 as well in the first edition of his *Capital* from 1867. In these works, Marx assigns to the term "exchange-value" (*Tauschwert*) both the meaning of the *internal* quantitative determination (*Wertgrösse*) of a commodity as well as the meaning of the *external, apparent* quantitative determination of the relative value (*Grösse des relativen Wertes*) of a commodity. Only later, by means of a manuscript of 1871–1872 (Marx 1987a), did he change his terminology in such a way that, starting from the second German edition of the first volume of *Capital* (1872/1987b), he used the term "value" to refer to the internal determinations of a commodity, while restricting the use of the term "exchange-value" to describe the external (both apparent and manifest) determination of value generated in the relation between commodities.

It is especially with respect to the "absolute value" that one can see that Ricardo views value as something intrinsic to commodities, as given to them by the very process of their production or in their origin. Accordingly, he claims that "commodities ... have an absolute value directly in proportion to the quantity of labour bestowed upon them" (Vol. IV, 382), while arguing, on the other hand, that "by exchangeable value is meant the power which a commodity has of commanding any given quantity of another commodity, without any reference whatever to its absolute value" (Ricardo 1951-1973, Vol. IV, 398).

It is precisely here, with respect to the distinction between relative (comparative, proportional) value and absolute (real, positive, natural) value, that the issue of the so-called "invariable measure of value," haunting Ricardo to the last days of his life, comes in. He starts his deliberations as follows (Ricardo 1951-1973, Vol. I, 17–18):

Two commodities vary in relative value, and we wish to know in which the variation has really taken place. If we compare the present value of one, with shoes, stockings hats, iron, sugar, and all other commodities, we find that it will exchange for precisely the same quantity of all these things as before. If we compare the other with the same commodity, we find it has varied with respect to them all: we may then with great probability infer that the variation has been in this commodity, and not in the commodities with which we have compared it ... If I found that an ounce of gold would exchange for a less quantity of all the commodities above enumerated ... and if, moreover, I found that by the discovery of a new and more fertile mine, or by the employment of machinery to great advantage, a given quantity of gold could be obtained with a less quantity of labour, I should be justified in saying that the cause of the alteration in the value of gold relatively to other commodities, was the greater facility of its production, or the smaller quantity of labour necessary to obtain it.

Therefore (Ricardo 1951-1973, Vol. I, 17):

If any commodity could be found, which now and at all times required precisely the same quantity of labour to produce it, that commodity would be of an unvarying value, and would be eminently useful as a standard by which the variation of other things might be measured.

Expressed otherwise: "I endeavour to measure the variations in the real value of commodities by comparing their value at different times with another commodity which I have every reason to be believe has not varied" (Ricardo 1951-1973, Vol. II, 32).

To have an invariable measure of value is, according to Ricardo, even more important if one wants to trace back the change of money-price of a commodity to the variation of its absolute (positive, real, natural, real) value. In such a case "we should carefully distinguish between those variations which belong to the commodity itself, and those which are occasioned by a variation in the medium in which value is estimated, or price expressed" (Ricardo 1951-1973, Vol. I, 48).

If such an invariable measure or standard of values, Ricardo claims, would be given to him, Ricardo claims "the advantage is, that I shall be enabled to speak of the variations of other things, without embarrassing myself on every occasion with the consideration of the possible variation in the value of the medium in which price and value are estimated" (Ricardo 1951-1973, Vol. I, 46). In this case of measurement, *according to him*, one has the situation which is identical to measurement of magnitudes of physics such as length, weight, or capacity. He claims: "Length can only be measured by length, capacity by capacity and value by value" (Ricardo 1951-1973, Vol. II, 33). He claims also that (Ricardo 1951-1973, Vol. IV, 361):

the only qualities necessary to make a measure of value a perfect one are, that it should itself have value, and that that value should be itself invariable, in the same manner as in a perfect measure of length the measure should have length and that length should be neither liable to be increased or diminished; or in a measure of weight that it should have weight and that such weight should be constant.

But Ricardo is well aware of the fact that with respect to value there is no commodity with a nonvarying value. The idea of a commodity with a constant absolute value was set forth with the purpose of reflecting upon the changes of the absolute value of other commodities that express their value in that commodity. But "if it be admitted that one commodity may alter in absolute value, it must be admitted that 2, 3, 100 and million may do so, and how shall I be able with certainty to say whether the one or the million had varied" (Ricardo 1951-1973, Vol. IV, 401). This means that Ricardo's requirement that "the only qualities necessary to make a measure of value a perfect one are, that it should itself have value, and that that value should be itself invariable" (Ricardo 1951-1973, Vol. IV, 361) is *inherently contradictory*. A commodity can function as a measure of absolute value of other commodities only if, at least according to Ricardo, its own absolute value is constant. But then it has no absolute value, because the latter always changes due to the changes of the amount of labor expended on its production. This means that if all commodities have a variable absolute value, then none of them is capable of fulfilling the function of a measure of value, even though it is obvious that, for example, money as a commodity functions as such a measure.

How it is possible to solve this problem and abolish that contradiction? Let me return to the idea of Ricardo that we encountered earlier (Ricardo 1951-1973, Vol. I, 18):

If I found that an ounce of gold would exchange for a less quantity of all the commodities above enumerated ... and if, moreover, I found that by the discovery of a new and more fertile mine, or by the employment of machinery to great advantage, a given quantity of gold could be obtained with a less quantity of labour, I should be justified in saying that the cause of the alteration in the value of gold relatively to other commodities, was the greater facility of its production, or the smaller quantity of labour necessary to obtain it.

Three facts are readily seen here. First, Ricardo is well aware of the fact that his thought-movement consists of *two mutually interconnected, but nevertheless different phases*. He *initially* starts from a change in the exchange ratios between an ounce of gold and other commodities and *then traces* this change back to a change in the very process of production of gold. Second, he is well aware of the fact that he starts with certain phenomena-effects and from them he infers their cause. This is what I labeled already in chapter 3 – following Hegel – as the *retreat into the ground*. Third, what, however, completely escapes Ricardo's attention is that when he speaks about the invariable measure of value, he presupposes that he already knows that what makes up the essence/ground of the relation between commodities (of their exchange value, proportional value, comparative value) is absolute value; the mutual relation between the absolute values of the commodities and where these values are of course variable.

From all this it becomes obvious that Ricardo does not realize that when one passes from the phenomena-effects to their ground/essence as something *as yet unknown* (*what has to be discovered*), and when one passes in mind from the already *cognized* (i.e., *known*) *essence/ground to its effects*, one moves in *two different processes of the production/creation of new scientific knowledge*. The production of knowledge in these two phases is guided by two different clusters of epistemic categories. In the *first phase* it is that of *external measure*, and here the category of scale (standard) comes in. Of course, what is required here in order to pick up the entity that generates the changes of the phenomena, is that the standard (scale) does not change. Otherwise one would not be able to pick up the entity into which one has to investigate in order to determine the essence/ground of that variation. This is what Marx had in mind with his claim about the idea of an invariable measure of value: "it has much to do with first finding value ... finding in what way the values in use ... fall under the common category and denomination of *values*" (Marx 1979, 1343).

In the *second phase*, after the essence/ground itself is already known, one explains how the mutual relations, the mutual proportions between the entities are generated. Here one already knows the origin of their mutual relations and the cause of the possible changes in their mutual ratios. The idea of invariability is here already *explained and derived (reinterpreted) from the point of view of the measure of the essence/ground*, that is, *the manifestation of the immanent measure of the essence/ground*. Ricardo, however, has not distinguished properly between the two types of knowledge of the measure of the phenomena-effects: *phenomena as appearances related to the external measure together with the invariable scale (standard)* and *phenomena as manifestations related to the manifestation of the immanent measure of the essence/ground*. In fact, he *wrongly merged* the categorical cluster guiding the production of the knowledge of the phenomena as *appearances* with that guiding the production of knowledge of the phenomena as *manifestations*. This is the key to understanding the seemingly surprising fact that he permanently confounds two terms: "the invariable standard of prices," appearing in economic knowledge when one proceeds from the money-price to discover labor/value as its essence/ground, and the "variable measure of value," appearing in economic knowledge when one proceeds from labor/value to the relation between commodities and to the genesis of the money-expression of value. Stated otherwise, he wrongly merged these two terms into an antinomic unity labeled "invariable measure of value."

The product of such an erroneous "synthesis" is then itself involved in an additional confusion. Even if Ricardo, as shown earlier, clearly distinguishes between two phases of the thought-movement from the phenomena-effects to their essence/ground, his work shows that he is incapable of differentiating between category clusters providing the framework for the knowledge of the phenomena (having here, of course, in mind that there two such clusters and not one!) and the category clusters providing the framework for the knowledge of the very essence/ground, namely, of the inner ground of the essence/ground. Here we draw on Marx who, as shown already, characterizes the *relation of labor* expended on a commodity to its value by means of the category *inner ground* (Marx 1979, 1324).

This incapability goes well beyond Ricardo's endeavor as a political economist; it pertains, so to speak, to the very category "outfit" of his thinking. To substantiate this claim of mine let me analyze his reflections on the measurement of physical magnitude of length. He states (Ricardo 1951-1973, Vol. IV, 399):

All measures of length are measures of absolute as well as relative length. Suppose linen and cloth to be liable to contract and expand, by measuring them at different times with a foot rule, which was neither liable to contract or expand, we should be able to determine what alteration had taken place in their length. If at one time the cloth measured 200 feet and at another 202 feet, we should say it had increased 1 per cent. If the linen from 100 feet in length increased to 103 we should say it had increased 3 per cent, but we should not say the foot measure had diminished in length because it bore a less proportion to the length of the cloth and linen. The alteration would really be in the cloth and linen and not in the foot measure.

So, according to Ricardo, the foot rule (measure) enables one to measure the ratio as well as the changes in the ratio of lengths (e.g., of cloth to linen and vice versa). In the example chosen by Ricardo, cloth increased three times less in length than (relative to) linen, and linen increased three times more in length than (relative to) cloth. But, and here Ricardo's confusion between the measure of the phenomena and the measure of the very essence/ground becomes apparent, he claims that the invariable foot rule should be able to express the *absolute length*. The truth, however, is that, if by means of an invariable foot rule (measure), one finds that cloth initially had a length of 200 feet and then increased by 1 per cent, one still has not arrived at an understanding of the "absolute" determination of length of that piece of cloth. In order to obtain that understanding, one has to find the mechanism that generates the magnitude labeled "length" which, in turn, would enable one to understand the mechanism causing changes of length. The invariable foot rule in Ricardo's example can fulfill only the function of an *external measure* guiding knowledge into the direction of a discovery of the *internal measure* of the magnitude of length.

The confusion between the measure of the phenomena (having again in mind that there are *two* measures pertaining to phenomena: the *external measure* at the level of *appearances* and the *manifestation of the immanent measure* at the level of *manifestations*) and the measure of the very essence/ground with respect to magnitudes of political economy is apparent already in Ricardo's *Principles*. He claims about the exchange of products of fishing and hunting (Ricardo 1951-1973, Vol. I, 27–28):

> If with the same quantity of labour a less quantity of fish, or a greater quantity of game were obtained, the value of fish would rise in comparison with that of game. If, on the contrary, with the same quantity of labour a less quantity of game, or a greater quantity of fish was obtained, game would rise in comparison with fish. If there were any other commodity, which was invariable in its value, we should be able to ascertain, by comparing the value of fish and game with this commodity, how much of the variation was to be attributed to a cause which affected the value of the fish, and how much to a cause which affected the value of the game.

This means that Ricardo, on the one hand, clearly distinguishes the *relative* (comparative, proportional) value of fish with respect to the value of the game and vice versa, from the *absolute* value of each of them, having its origin in the labor used to produce each of them. But, on the other hand, he claims that there is a need for a third type of commodity with an invariable absolute value in order to investigate the very changes of the process of obtaining game and fish, without being aware of the fact that the knowledge of the *measure of the phenomena (appearances*, to be precise) *only points to the immanent measure of the cause/ground and guides cognition to the latter but cannot substitute it in our knowledge.*

The fact that he confounds the category of the measure of the phenomena with the category of the measure of the very ground (immanent measure) is also obvious in his *Notes on Malthus* and in his two manuscripts on absolute and relative value where he claims the following: "I endeavour to measure the variations in the real value of commodities by comparing their value at different times with another commodity which I have every reason to believe has not varied" (Ricardo 1951-1973, Vol. II, 32). This means that if the exchange-value or price of commodity A changes, ΔP_A, with respect to the medium in which the price is expressed, while this medium has an invariable absolute value, then the absolute value of commodity A has varied, ΔV_A. If simultaneously the price of the commodity B has varied, ΔP_B, with respect to the same medium, then the latter's absolute value has varied as well, ΔV_B. Then, by comparing ΔP_A and ΔP_B one is able to determine the ratio ΔV_A of ΔV_B. It holds that:[172]

$$\Delta P_A / \Delta P_B =: \Delta V_A / \Delta V_B$$

Ricardo's mistake here is that he does not realize that when he deals with changes expressed by ΔP_A, ΔP_B, and $\Delta P_A / \Delta P_B$, he moves at the level of *phenomena*. But when he deals with expressions like ΔV_A, ΔV_B, and $\Delta V_A / \Delta V_B$ his thinking is moving at a level at which one completely abstracts from phenomena, because one is dealing with the quantitative changes in *the inner ground of the essence/ground*, that is, with the *"pure" essence/ground independently from any phenomena.*

A similar situation is encountered in the first manuscript on absolute and relative value. Initially, he claims that "any commodity which continued uniformly to require the same quantity of labour would be an accurate measure of value" (Ricardo 1951-1973, Vol. IV, 364). He then asks: "Have we no standard by which we can ascertain the uniformity in the value of a measure?" (Ricardo 1951-1973, Vol. IV, 381). But his answer to this question indicates that he, in fact, *moves well beyond the level of phenomena and that his thinking moves already in the "pure" essence/ground and its inherent measure* (Ricardo 1951-1973, Vol. IV, 381–382):

It is asserted that we have [a standard by which we can ascertain the uniformity in the value of a measure], and that labour is that standard ... A commodity produced in a given time by the labour of 100 men is double the value of a commodity produced by the labour of 50 men in the same time. All then we have to do it is said to ascertain whether the value of a commodity by now of the same value as a commodity produced 20 years ago is to find out what quantity of labour of the same length of time was necessary to produce the commodity 20 years ago and what quantity is necessary to produce it now. ... Having discovered this standard we are in possession of a uniform measure of value. ... Commodities would then have an absolute value directly proportional to the quantity bestowed upon them.

---

172 In this equation I draw on (Mongin 1989).

Now it is possible to compare mutually the categorical structures of Newton's and Ricardo's thought-movements. The fact that Ricardo is able not only to introduce the term "absolute value," but also to pass from absolute value to its substance shows that his thinking is embedded within category clusters that are richer and more advanced that those of Newton. While Newton arrives only at the discovery of the quantitative determinations of force on the basis of the quantitative determination of the phenomena-effects as appearances, Ricardo proceeds further by showing that the *quantitative* determination of value has its origin in the *quantitative* determination of the substance of value, that is, labor. But, on the other hand, because Newton realizes that he knows nothing about the force before it appears, *he views his own understanding of it as principally incomplete.* Contrary to this, because Ricardo realizes that he knows absolute value in itself before it appears as comparative (proportional) value and that he knows its origin in labor, he does *not* view his own understanding of the absolute value as incomplete, as lacking the understanding of the *quality* of the substance of value. The absence of investigation into the qualitative aspect of value/labor on Ricardo's part is related to the following feature of his understanding of absolute value and proportional (comparative) value. Ricardo simply presupposes the existence of the proportional (comparative) value as a phenomenal form of absolute value, and it does not occur to him that he should explain *why* value acquires at all this phenomenal form. Had he made an attempt at such an explanation, he would probably have realized that his knowledge of the quantitative determination of value/labor is an *insufficient explanatory basis* for a thought derivation of the concept of money-price. This thus means that while Newton, as shown in chapter 3, *consciously applies* the cyclical thought-movement phenomena → cause/ground → phenomena, *this cyclical thought-movement is altogether lost in Ricardo's reflections on value.*

By comparing Newton's, Marx's, and Ricardo's categorical structures I arrive at two conclusions. *First*, Ricardo has a more advanced categorial structure or "outfit" of thinking as compared to that of Newton, but less advanced as compared to that of Marx. *Second*, these structures stand to each other in a relation that can be viewed as developmental in its nature. The categorial structure guiding the construction of the *Principia* represents the first type of theoretical reason, aiming at passing from the phenomena as appearances via essence/ground to the phenomena as manifestations; therefore, it enables a *cyclical method* of theory construction. Its *ultimo ratio* with respect to that essence/ground is represented by the categories of external measure and formal ground.

The categorial "outfit" of Ricardo's works in political economy represents a more advanced type of theoretical thinking. It incorporates the "Newtonian" category of the measure of phenomena (external measure) and subordinates it to the category of the quantitative determination of the essence/ground – the latter being its *ultimo ratio*. Finally, the categorical "outfit" of Marx's economical works incorporates both the "Newtonian" categories of formal ground and external measure, and the originally "Ricardian" category of the quantitative determination of the essence/ground. It reinterprets them by subordinating them to a completely new category – missing in both Newton's and Ricardo's categorial clusters – namely, to that of the qualitative determination of the essence/ground which Marx understands as the *contradiction inherent in the essence/ground*. So the categorical "outfit" given in Ricardo's approach to the concept of value can be viewed as the *mediating link* between the categorical "outfit" of Newton's *Principia* and Marx's works in political economy.

But even if the "Ricardian" and "Newtonian" categorical outfits mutually differ, there exists a *fundamental similarity* between them that can be reconstructed by comparing Ricardo's idea of an invariable measure of value with Newton's claim, analyzed already in chapter 3, that "I am by no means affirming that gravity is essential to bodies. By inherent force I mean only the force of inertia. Gravity is diminished as bodies recede from earth" (1999, 796). According to Newton, only "those qualities of bodies that cannot be intended and remitted ... should be taken as qualities of all bodies universally" (1999, 795). So, for Newton, even if the gravity of each body is proportional to its own mass, because mutual attraction between bodies changes with the change of their mutual distance, it is not essential to bodies. Ricardo, similarly, when investigating into the nature of value, was in search of something that would not allow a more or less, and which he labeled as "the invariant measure of value."

In the case of Newton, as shown in chapter 3, one is dealing here with an inability, due to the movement of his thought in the framework characterized by the categories external measure and formal ground, to distinguish properly between the force of gravity as an absolute force and the accelerative effect of this force of gravity on other bodies. Newton's understanding, as shown also in chapter 3, of the search for essential properties of objects as a search for their quantitatively invariable properties is simply the (wrong) substitute for a lack of knowledge of the qualitative characteristics of the cause underlying the phenomena-effects. Similarly, Ricardo's antinomic construct labeled "invariant measure of value" is just the wrong expression of an unfinished, incomplete passage from the concept of money-price (external measure and invariable standard) to absolute value (the quantitative and quantitative determination of labor). So as he lacks a complete knowledge of the absolute value and has at his disposal only the knowledge of the quantitative determination of labor as the substance of value, he cannot properly distinguish between the external measure of the phenomena and the inherent measure of the essence/ground. For the former, invariability is a *must*, while for the latter, quantitative determinations can (or even have to, as in the case of labor with respect to value) *vary*. Stated otherwise: the meaning of the term "invariable measure of value" is just the messy product of an incomplete knowledge about what absolute value is before it generates its own phenomenon labeled by Ricardo as "proportional value."

4.2.2 Samuel Bailey versus Absolute Value

Bailey's *Critical Dissertation* of 1825 is – with respect to this chapter as well as chapter 3 – worth analyzing because it displays in its attitude to Ricardo's term "absolute value" a striking similarity with Ernst Mach's antimetaphysical attitude toward the Newtonian concept of force.

Bailey, contrary to Ricardo, refuses to accept any conceptual differentiation between relative (comparative, proportional) value and absolute (real, positive, natural) value. The determination of the value of a certain commodity A is, in his view, just the expression of its relation to another commodity B. He claims that (Bailey 1825, 3–5)

[t]his relation can be determined only by quantity. The value of A is expressed by the quantity of B for which it will exchange, and the value of B is in the same way expressed by the quantity of A. Hence the value of A may be termed the power which it possesses or confers of purchasing B, or commanding B in exchange ... it is essential to value that there should be two objects brought into comparison. It cannot be predicated of one thing considered alone, and without reference to another thing. If the value of an object is its power of purchasing, there must be something to purchase. Value denotes consequently nothing positive or intrinsic, but merely the relation in which two objects stand to each other as exchangeable commodities.

The fact that political economists introduce the term "absolute value" is, according to him, just the wrong conclusion from the obvious fact that (Bailey 1825, 8)

in speaking of the value of A being equal to the value of B, we are led to use the expression by the constant reference which we unavoidably make to the relations of these commodities to other commodities, particularly to money, and the import of our language, in its whole extent, is, that A and B bear an equal relation to a third commodity, or to commodities in general. It is from this circumstance of constant reference to other commodities, or to money, when we are speaking of the relation between any two commodities, that the notion of value, as something intrinsic and absolute, has arisen. When we compare objects with each other as exchangeable commodities, two relations necessary mix themselves in our comparison – the mutual relation of the objects, and their relations to other objects; and it is these latter which occasion the semblance of absolute value, because they seem independent of the former.

According to Bailey, the existence of money as the embodiment of value and as the third element entering into the relation between commodities A and B, generates the fiction of the existence of value as the third element with respect to the exchange value of A and the exchange value of B. So, in his view, the existence of a third element generates in us, with respect to the self-evident quantitative proportion between A and B, the false representation of a *metaphysical entity*. So, Bailey's epistemic/cognitive norm here is, as Marx expressed it, "to proceed from the surface to the depth is not allowed" Marx 1979, 1325). But a closer look at the following two claims by him also shows something else, which is again very similar to Ernst Mach's approach. He claims that for a third commodity, in which A and B express their mutual ratio, to exist, (Bailey 1825, 112)

the requisite condition ... is, that the commodities to be measured should be reduced to a common denomination, which may be done at all times with equal facility; or rather it is ready done to our hands, since it is the prices of commodities which are recorded, or their relations in value to money.

He states also that "estimating value is the same as expressing it" (Bailey 1825, 152). But because Bailey claims that C as a commodity and as a medium, expresses the value of A and B, then this means that he in fact bases the last of his claims mentioned earlier on the supposition that he explicitly refuses to accept, namely, that we already know that value is the common denomination of A, B, and C. What is worth noting is that he is well aware of this fact already in the very beginning of his *Critical Dissertation* where he refers to the exchange value of B as expressed by commodity A (Bailey 1825, 6–7):

It may be objected to this representation of the relative value of B, that when we say the value of A is equal to the value of B, the expression implies a quality intrinsic and absolute in each; for otherwise, how could we affirm that an equality existed between these values?

So, Bailey proceeds in a manner very similar to that of Mach. His view on the function of the money-commodity, as well as his antimetaphysical attempt to get rid of the term "absolute value" come only *post festum, after the previous development of political economy has already reduced the use-value and exchange value of commodities as appearances to the quantitative determination of their essence/ground, that is, to the quantitative determination of absolute value*. In his argument, he is already operating at the level of the *manifestation* of value, but he pretends and claims to be operating at the level of *appearance*. This is how we interpret Marx's commentary on Bailey's claim that estimating value is the same as expressing it; (Marx 1979, 1344):

As soon as the value of commodities is given as their common unity, does the measurement of their relative value and the expression of it coincide. But we cannot arrive at this expression as long as we do not arrive at a unity which is different from the immediate being of commodities.

There is also yet another similarity between Bailey's and Mach's antimetaphysical attitude. Bailey, like Mach, starts from the fact that entities (commodities) interact in certain quantitative ratios, but refuses to push further in order to discover the essence/ground of this interaction and of its quantitative determinations. This means that, with respect to previous development of the political economy, it represents an epistemic/cognitive drawback. This development had the intention and ambition of substantiating the exchange values of commodities in the process of their production, in their origin in labor, and succeeded even in discovering the quantitative determination of this essence/ground by discovering the quantitative determination of its substance, namely, labor.

4.3 Measurement

Let me now deal with the structure of measurement based on a scientific theory that grasps the essence/ground of the phenomena into which it investigates already at the level of the categories of the real ground and immanent measure

Reuten's claimed, as shown above in 4.1.1, that the concrete sizes of the coefficients a_i as well as of the coefficient m_i can be found out only in the market. So, as I discarded earlier the coefficients of the type a_i, Reuten's claim is relevant only for the coefficients of the type m_i and, thus, is equivalent to the problem *how one can find the concrete sizes* of the coefficient k expressed by equation /18/. In order to find this concrete size one has to find out the amount of labor necessary for the production of the chosen weight-unit of gold. But, and here Reuten is right, one *cannot* found out these amounts prior to the involvement of the gold-producers into exchange-relations with producers of other commodities. Reuten, however, does not explain *why* it is so. The *first* reason is that one is trying to compute the concrete size of a coefficient introduced in the process of theory-construction accomplished by the derivation of concepts and for this one has to turn, by means of data collection, to the objects referred to by these concepts and compute the concrete size of k on the basis of these data. And what are these objects? They are the properties and relations, given in a really existing economy that, however, does not fulfill *only* the essential labeled previously as C_E^A and C_E^B, presupposed as given in Chapters 1 through 3 of *Capital*, Volume I. It is a *capitalist* type of economy, that is, where, *in addition* to C_E^A and C_E^B, hold also the essential conditions jointly labeled previously as C_E^C, and for this type of economy, it holds that commodity production takes place on a *mass-scale* and is accomplished at the same time via *competition* for rare resources taking place both between companies in one production-branch and between companies from different production-branches. These last two characteristics we view as the *second* reason why the concrete size of the constant k cannot be found out prior to data-collection in the market. It is worth noting that Marx himself was well aware of the fact that, by shifting from the explanations based on concepts like work/labor, use-value/value, money, measure of value and standard of prices to that based on concepts like surplus-labor, surplus-value, and so on, the very issue of computation of the amount of expended labor undergoes a profound transformation. In the manuscript labeled "Sechstes Kapitel. Resultate des unmittelbaren Produktionsprozesses," he states (Marx 1988a, 33; 1976a, 954):

Not the individual commodities are given as the result of the process, but a *mass of commodities* ... The labor expended on the individual commodity – already ... because of the directly social and to the average-labor leveled out and estimated labor of many cooperating individuals – cannot be computed at all.

How can one, in an economy fulfilling simultaneously the essential conditions labeled in subchapter 4.1 as C_E^A, C_E^B and C_E^C, compute the amount of labor expended on the production of commodities, and what are the epistemic categories that characterize this computation? In order to answer these questions I start with a reconstruction of the relation of the sequence of explanations as given in Marx's *Manuscript 1861–63* and in Volume I, and the manuscript of Volume III of *Capital* to the very process of measurement based on based Marx's concepts of political economy. Then I deal with the procedure of the measurement of value and surplus-value based on the measurement of profits and production prices; here I follow an approach outlined for the first time, to the best of my knowledge, in (Tugan-Baranowsky 1905). Finally, I deal with the measurement of value and surplus-value based on purified new-product-data as given in (Moseley 1991).

4.3.1 Measurement based on Manuscript 1861–63, Capital, Volume I, and Manuscripts of Volume III

As mentioned already in subchapter 4.1, one of the central concepts of Marx in Volume I of *Capital*, with respect to commodities coming out of the production process are those of surplus-value, s; constant capital, c; and variable capital, v. As stated also in 4.1, what Marx deals with in Volume I of *Capital* is the explanation of the capitalist production process from the point of view of its essential conditions and, at the same time, as shown in subchapter 4.2, from the point of view of its *qualitative* characteristics. Everywhere where the essential conditions, reconstructed in subchapter 4.1, are given, surplus-value comes into being from surplus-labor. This means that at any location where a production process fulfilling those essential conditions takes place, surplus-value comes into being as well. At the same time the concepts value, surplus-value, and constant and variable capitals can be viewed as *magnitudes* because each of them unifies in itself a *qualitative* and a *quantitative* characteristic.

The specific feature of Marx's explanation procedure given in Volume I of *Capital* is that it takes the capitalist economy as characterized by *qualitatively identical* production (or industrial) capitals, while at the same time temporarily abstracting from any possible interactions between them, say, due to mutual competition, where their possible mutual *quantitative* differences – the mutual quantitative differences of these capitals expressed by different sizes of their characteristic magnitudes v, c and s – could bring in *new economic determinations of commodities*.

If v_i, c_i, and s_i stand for the variable capital, constant capital and surplus-value characteristic of the ith industrial capital, then I can, following Marx, bring in the following additional magnitudes characteristic of it:

$$s_i' = s_i/v_i \qquad /21/,$$

$$V_i = v_i + c_i + s_i \qquad /22/,$$

"s_i'" stands here for the rate of surplus-value and V_i for the value of the commodities produced by the ith industrial capital in a certain time-span.[173] If we take into account the constant k, introduced in subchapter 4.1 mediating between the magnitude of value and that of price, we obtain for the *price* of the commodities produced by the ith industrial capital:

$$P_i = k(v_i + c_i + s_i) \qquad /23/.$$

For the capitalist economy *here* understood as a simple *sum* of the determinants of singular industrial capitals, then, from the point of view of Volume I of *Capital*, it holds that

---

173 Here I presuppose that in the time-span in which the production process takes place, say, one year, the value of whole invested capital is crystallized in the value of the commodities, i.e., the whole capital circulates and thus fixed capital equals zero. Once I would abolish the last idealization I would have to take into account also Marx's conceptual sequence given in the manuscripts of *Capital*, Vol. II.

$$TV = \sum_i V_i \qquad /24/,$$

$$S = \sum_i s_i \qquad /25/,$$

$$V = \sum_i v_i \qquad /26/,$$

$$C = \sum_i c_i \qquad /27/,$$

$$P = \sum_i P_i \qquad /28/.$$

Here TV and S stand for the total value and surplus-value, respectively, produced in that economy; V and C stand for the total variable and constant capitals employed in it; P is the total price of all commodities produced in it in a certain time-period; and S' is the rate of surplus-value for the whole economy. From /23/, /25/, /26/, /27/, and /28/, one derives:

$$P = \sum_i k(v_i + c_i + s_i)$$
$$= k(\sum_i v_i + \sum_i c_i + \sum_i s_i)$$
$$= k(V + C + S) \qquad /29/.$$

And based on /21/, /25/, and /26/, one obtains for the rate of surplus-value in the economy as a whole:

$$S' = \sum_i (s_i/v_i)$$
$$= \sum_i s_i / \sum_i v_i$$
$$= \frac{S}{V} \qquad /30/.$$

The *causal nexus really* given between the characteristics of the singular industrial capitals and the aggregate characteristics of the economy as a whole, as well as the *direction of explanation* from the singular magnitudes $v_1, ..., v_n$; $c_1, ..., c_n$; $s_1, ..., s_n$; and so on to the aggregate magnitudes S and TV, defined by means of /24/ and /25/ as given in Volume I of *Capital* can then be represented by Figure 4.3 (Σ stands here for a sum over i):

251

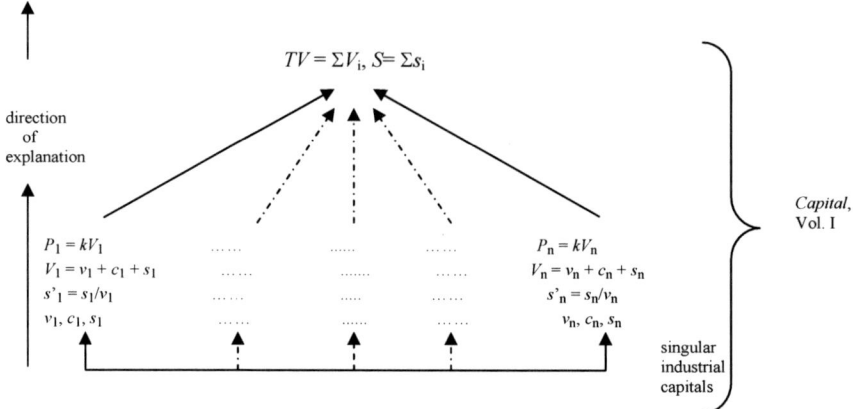

*Figure 4.3 The causal nexus and the direction of explanation in **Capital**, Volume I.*

As stated already above in subchapter 4.1, Marx does not deal in Volume I of *Capital* with the process of price-formation characteristic of capitalistically produced commodities. The latter is the subject matter of a part of the *Manuscript 1861–63*, as well as of the manuscripts that were later used by Engels for the publication of Volume III of *Capital*.

Marx employs in them the following strategy for the derivation of the concepts of political economy. He starts from the magnitude of value V_i of a commodity given in /22/ and introduces (Marx 1980d, 1611; 1992, 53; 1981b, 118) the magnitude of *costprice* (hereafter, k_i) so that it holds that

$$k_i = v_i + c_i \qquad /31/$$

/22/ is then transformed into

$$V_i = k_i + s_i \qquad /32/.$$

By bringing in the magnitude of costprice, Marx at the same time correlates the magnitude of the rate surplus-value with that of the *rate profit*. While the former relates the magnitude of surplus-value to that of the variable capital, the latter relates the magnitude of surplus-value to that of the total advanced capital, that is, to the costprice k_i, used by the *i*th capital in the production of commodities (Marx 1992, 52–53; Marx 1981b, 133–134); thus for the magnitude of the rate of profits it holds that

$p'_i = s_i/(v_i + c_i)$ /33/

Based on the magnitude of the rate of profit Marx starts to make reflections on the magnitude of *profit*, which is the excess of the price P_i for which the commodity is sold by the capitalist over its costprice k_i (Marx 1992, 55; 1981b, 127).

The introduction of the magnitude of the rate of profit as a function of three other magnitudes, namely, v_i, c_i, and s_i, enables Marx then to start considering the mutual quantitative differences of singular industrial capitals. He initially presupposes that different singular capitals operate with the same length of the working day, and with the same degree of exploitation and that the sizes of the singular capitals are mutually equal, then, the only possible source for the mutual difference of their respective rate of profit is the possible mutual differences in their composition, that is, in the ratio of v_i to c_i; the larger the percentage of the v_i-component, as compared to the percentage of the c_i-component, the larger the size of p'_i (Marx 1992, 222; 1981b, 248). Marx holds here rigorously to his understanding of value and surplus-value, where in both of them labor is viewed as the only substance in which value/surplus-value can originate: "The profit consists … in its essence from surplus-value" (Marx 1980d, 1626) and (Marx 1992, 222; 1981b, 248):

If a capital which for each 100 is composed of 9/10c or 90 constant components and 1/10v, or 10 variable components at the same degree of exploitation of labor would produce the same amount of surplus-value or profit as a capital which is composed of 1/10c and 9/10v or of 10c and 90v, it would be clear as daylight that surplus-value and hence value in general would have to have a completely different source than labor and that with this any rational basis of political economy would fall away.

From this Marx draws the conclusion (Marx 1992, 222; 1981b, 248–249):

Since capitals considered by percentage in different spheres of production – or capital of equal size – are unequally divided into constant and variable elements, set in motion unequal amounts of living labor and hence produce unequal amounts of surplus-value, i.e. profit, is the *rate of profit*, which consists precisely in the computation of the surplus-value as a percentage of the total capital, *different in them*.

Then yet another conclusion follows (Marx 1992, 222–223; 1981b, 249):

But if capitals of different spheres of production *computed by percentage*, thus *capitals of equal size* in different spheres of production increase with unequal *rates of profit*, as a consequence of different organic compositions, then it follows that the *profits* of unequal capitals in different spheres of production are *not proportional to the sizes of the capitals respectively employed in them*. For such an increase of the profit *pro rata* of the size of the employed capital would imply that from the point of view of percentage the profits are equal, hence *capitals of same size in different spheres of production increase in the same rates of profit* despite their different organic composition.

But the latter conclusion, as a consequence of Marx's theoretical integration of the concept of surplus-value into the law of value, contradicts the fact that "in reality ... the difference of the *average rate of profits* for the different branches of industry does not exist" (Marx 1992, 230; 1981b, 252). In order to deal with this contradiction of a consequence of his theory with an empirical fact, Marx introduces the terms "average rate of profit" and "average profit." The meaning of the former term should enable him to explain that "the rate of profit is the same for all capitals or, what is the same, that the masses of the profits behave directly and exactly as the sizes of the capitals" (Marx 1980d, 1625). The average rate of profit (hereafter, AP'), is given by the ratio of the aggregate surplus-value S to the total capital, K for short, where $K = k_i$, (Marx 1992, 231; 1981b, 255):

$$AP' = \frac{S}{K} \qquad /34/,$$

and then for the production price, or PP_i, of commodities produced by the *i*th singular industrial capital it holds that (Marx 1992, 234, 240; 1981b, 258, 265)

$$PP_i = k(k_i + k_i AP') \qquad /35/,$$

where k is the constant mediating between the magnitude of value and that of price, and k_i is the costprice. $k_i AP'$ is viewed by Marx as the "profit which, corresponding to this general rate of profit, falls to a capital of a given size, whatever its organic composition might be" (Marx 1992, 234, 240; 1981b, 257) and is labeled by him as the *average profit*.

From the point of view of the *sequence* of Marx's conceptual derivations as well as from the point of view of the *epistemic categories* employed in them by him, it is worth noting that he emphasizes the necessity to proceed via the conceptual sequence value & surplus-value → rate of profit &profit → average rate of profit & average profit; (Marx 1992, 234; 1981b, 257) (here m stands for surplus-value and C stands for the sum $v_i + c_i$):

Productionprices. Their presupposition is the existence of a *general rate of profit* which presupposes the average of the rates of profit of the particular spheres of production, which [are] particular rates of profit in each sphere of production = m/C and which can be developed only from the *value* of the commodity. Without this development the general rate of profit (and hence also the production price of the commodities) remains a meaningless representation devoid of any concepts (*eine sinn- und begriffslose Vorstellung*).

What is at the basis of the necessity to proceed in the explanation in such a sequence is that in Marx's economic theory not only, as shown in subchapter 4.1, is labor the substance of value and surplus-labor the substance of surplus-value, but also the *singular rates of profit* "serve as the substance, presupposition of *the general rate of profit*" (Marx 1980d, 1626). This then means that the concepts value/labor, surplus-labor/surplus-value, rate of profit, profit, and costprice are in the manuscripts of Volume III of *Capital* part of the *explanans*, while the concepts average rate of profit, average profit, and production price are part of the *explanandum*. Thus the former cluster of concepts has to be shifted from the *explanans* to the *explanandum* in the same manner as, as shown already in subchapter 4.1, the concepts of value and labor are shifted in the derivation of the concept of price in Chapter 1 of *Capital*, Volume I.

This in turn means that, from the point of view of the epistemic categories, that explanation has here the character of a movement form the *essence/ground to its manifestations* (Marx 1992, 7; 1981b, 117):

The configurations of capital, as we develop them in this book, thus approach step by step the form in which they appear (*auftreten*) on the surface of the society, in everyday consciousness of the very agents of production and, finally, in the action of the different capitals on one another, in competition.

The whole thought-movement given in Volume I and the manuscripts of Volume III of *Capital* from value and surplus-value up to the average rate of profit, average profit, and production price, considered in its unity, can then be represented by Figure 4.4 (Σ stands here again for a sum over i).[174]

How can the explanatory structure represented by this figure be used at all for the purpose of measurement? There are several passages in Marx's manuscripts that enable to give us an answer to this question. With respect to the measurement of the *total surplus-value* produced in an economy in a certain time-span, Marx gives the following hints:

1) "The sum of the profits in the different spheres of production = sum of surplus-values (Marx 1992, 249; 1981b, 273).
2) In the *Manuscript 1861–63* Marx states (Marx 1980d, 1627–1628):

In the same manner in which the surplus-value of the particular capital in each particular sphere of production is the measure of the absolute size of the profit – as far the latter is only the changed form of surplus-value – in the same manner the *total value* which the

---

[174] This figure reflects the idealization introduced in note 173 that the fixed capitals equal zero. If I would abolish this idealization, then I would have to take into account in this figure Marx's concept given in the manuscripts of *Capital*, Volume II.

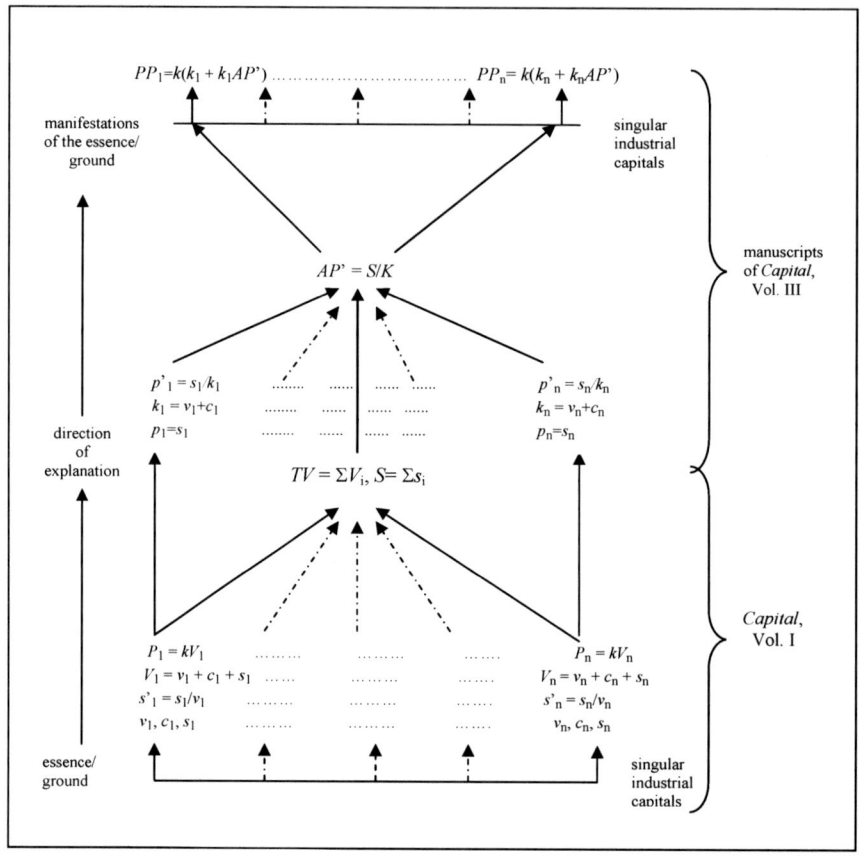

*Figure 4.4 Causal nexuses and directions of explanation in Volume I and the Manuscripts of Volume III of **Capital** considered in their unity.*

total capital, thus the whole class of capitalists produces, is the absolute *measure of the total profit of the total capital*, where under profit one should understand all forms of surplus-value like rent, interest, etc.

This means that while Marx starts his explanation-sequence from singular surplus-values s_i, so that for the capitalist economy as a whole it holds that $S = \sum_i s_i$, now it holds that

$$\sum_i AP_i = S \qquad /36/,$$

256

here AP_i stands for the average profit yielded by the *i*th industrial capital and where under "capital" Marx understands in the first quote above just *industrial capital*, that is, presupposing that no part of the surplus-value becomes rent, interest, and so on; and in the second quote, not only capital engaged in industry but also in agriculture, money-lending, and so on. But even if in Marx's approach the size of S in /36/ is *the immanent measure* of the size of $\sum_i AP_i$, and the size of $\sum_i AP_i$ is the *manifestation of the immanent measure* of the size of S, that is, *his thought-movement goes from the right side of this equation to the left side*, still this equation can be used as a starting point of a special type of "thought reversal" by means of which one obtains a recipe that be can employed for the purpose of the measurement of S. As we will see later: *by summing up the named numbers corresponding to the respective singular profits one obtains the total surplus-value produced in the economy*. I will show that to this "reversal" corresponds an important shift in the employed epistemic categories.

For the measurement of the *total value* of all commodities produced in an economy in a certain time-span one can employ several passages from the manuscripts of Volume III of *Capital*:

1. "... in the society itself – from the point of view of the totality of the societal branches of production – *the sum of the prices of production equals to the sum of their values* " (Marx 1992, 236; 1981b, 259).
2. "The sum of production-prices of the total product = sum of its values" (Marx 1992, 249; 1981b, 273).

So, one can state the following equation:

$$\sum_i PP_i = k \sum_i V_i, \qquad /37/,$$

and by taking into account the equation /24/ relating values of individual commodities to the total value, one obtains

$$\sum_i PP_i = kTV. \qquad /38/$$

This equation can also be used as a starting point of a special type of a "reversal" for the purpose of obtaining a recipe that can be employed for the purpose of measurement. As we will see soon: *the sum of production prices of all commodities produced in an economy yields the total value produced in it*.

What categories are correlated with the measurement procedures based on Marx's economic theory? Seemingly, the measurement of production prices of individual commodities and of singular profits – so as these magnitudes are conceptually derived by Marx as *manifestations* – starts from the manifestations as well. But, as shown earlier, once we view these magnitudes as manifestations given in the *explanandum*, then they are inseparably connected with concepts like labor, value, surplus-labor, and surplus-value shifted from the *explanans* to the *explanandum*. Thus, the measurement of singular profits and production prices pertaining to individual commodities as *manifestations* would require one to measure *in advance* and *prior* to them the size of the magnitudes v_i, c_i, and s_i, which are characteristic of the singular industrial capitals, that is, *straight at the level of the essence/ground. But such a measurement is not possible.* The data available at the level of singular industrial capitals and in fact of any capital (industrial, agricultural, bank-capital, etc.) are already data expressing the results of their mutual interactions, where their mutual quantitative differences are already at work; the sizes of the magnitudes characteristic of them are already quantitatively "distorted" by these interactions. It is exactly this inability to measure the "undistorted" or unmodified sizes of magnitudes v_i, c_i, and s_i pertaining to the singular capitals that forces one to perform the "reversal" of the equations /36/ and /38/. Hence, if one wants to measure surplus-value and value produced in an economy, one has to take as input data those "distorted" or modified data that have, not the status of *manifestations*, but that of *appearances* and on their bases compute the respective quantitative characteristics of the essence/ground of the underlying structures. *The measurement procedures based on Marx's economic theory stand for the thought-movement from the quantitative determinations of the appearances to the quantitative determinations of the essence/ground of their underlying structures.* So, *the former have with respect to the latter the epistemic status of external measures.*

By the reconstruction of the categories of appearance and external measure one can clarify the specific nature of the "reversal" of the equations /36/ and /38/ upon which the measurement of total surplus-value and total value is based. By this "reversal" of /36/ one does *not* obtain the equation

$$S = \sum_i AP_i. \qquad /39/$$

The AP_is still stand here for the singular average profits given at the level of *manifestations* and *not* at the level *appearances*. But the purpose of the "reversal" becomes readily apparent already here. By the "reversal" of the equation /36/, one obtains an equation which can serve as a recipe for the computation of the total surplus-value: the sum of the size of the singular average profits can serve as the *external measure* by means of which one can compute the size of the total surplus-value. The same holds for the "reversal" of the equation /38/. One does *not* obtain by means of it the equation

$$kTV = \sum_i PP_i. \qquad /40/.$$

The PP_is still stand here for the singular production prices given at the level of manifestations and not at the level of *appearances*. But the purpose of the "reversal" becomes readily seen also. By the "reversal" of the equation /38/, one obtains an equation that can serve as a recipe for the computation of the total value: the sum of sizes of the singular production prices is the *external measure* by means of which one can compute the size of the total value. Thus the "reversal" of the equations /36/ and /38/ is, from the point of view of the employed epistemic categories, in fact, an acknowledged thought-movement from the *manifestations* to the *appearances* and from the *manifestations of the immanent measure* of the essence/ground to its *external measure*. Later we will reconstruct the equations obtained by a "reversal" of the equations /36/ and /38/.

So as in equations obtained by the "reversal" of the equations /36/ and /38/, one moves from the phenomena as appearances to their ground/essence, the measurement procedures based on these "reversed" equations *partially* "copy" the path taken by political economy before Marx (Marx 1992, 52, 57; 1981, 134, 139):

In fact the rate of profit is that from which one historically departs. Surplus-value and rate of the surplus-value are, relatively, the unobservable essential to be investigated into, whereas the rate of profit and hence the form of surplus-value as profit show themselves on the surface of the manifestations. What is actually given as different from the rate of surplus-value and which obtrudes itself is the *rate of profit* … from it, not from the rate of surplus-value, one … actually departs; the rate of surplus-value, though its inner ground, thus its secret, hidden *prius* can be discovered from it only by analysis.

I have emphasized the word "partially"; *only after Marx has already accomplished the explanatory thought-movement from the essence/ground to its manifestations, can the epistemic move and strategy to use, via a reversal, the equations /36/ and /38/ as the basis of measurement be employed at all. Without the concepts of value and surplus-value, both at the level of singular industrial capitals and the total (aggregate) industrial capital, equations /35/ and /37/ could not be derived at all and thus reversed* in order to obtain the "reversed" equations. This means that Marx's explanatory thought-movement from the essence/ground to its manifestations as represented by Figure 4.4, *first*, gives the *impetus* for the collection of *relevant* data at the level of *appearances*. Here the relevant data to be collected are the size of production prices of individual commodities – $\mathfrak{P}_1$, $\mathfrak{P}_2$, …, $\mathfrak{P}_n$, for short, and the size of singular profits – p_1, p_2, …, p_n, for short. Thus by the "reversal" of equation /36/, one obtains the equation

$$S^* = \sum_i p_i, \qquad /41/$$

S^* stands here for the total surplus-value produced in an economy expressed in money terms. And by the "reversal" of equation /38/, one obtains the equation:

$$TV^* = \sum_i \mathfrak{P}_i, \qquad /42/$$

TV^* stands here for the total value produced in an economy expressed in money terms.

Second, Marx's explanatory thought-movement provides the framework and bases for the *ability to compute from these collected data the quantitative determinations* at the level of essence/ground of those appearances. So, measurements based on Marx's economic theory can labeled as *computational measurement*. Taking into account the above explicated "reversal" and, as we have seen, that one can compute the sizes of the aggregate magnitudes S and TV but *not* the size of the singular magnitude s_i and, thus, neither of s_i', V_i, P_i, p_i', nor of p_i, the relation of Marx's explanatory procedures to the measurement based on them can be expressed as shown in Figure 4.5.

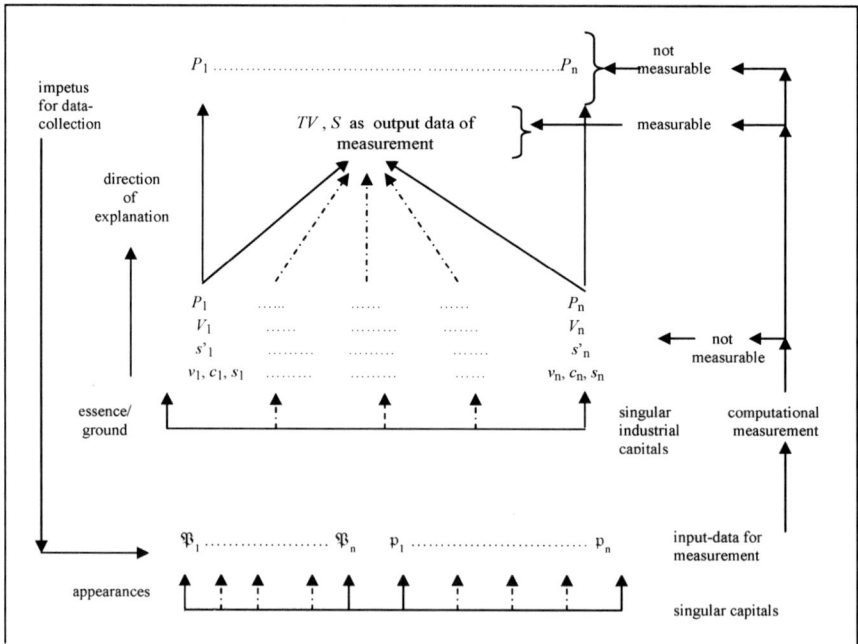

Figure 4.5 The relation of explanation and measurement in Marx's economic theory

Based on my analysis of the relation of the explanation and measurement in Marx's economic theory it is possible to state yet another criticism with respect to the views of Reuten. Even if the measurement of the amount of total value and, thus, of the total labor expended are guided by the knowledge of law of production price as scientific law in the sense that this knowledge, first, gives the impetus to collect the data pertaining to the sizes of production prices of individual commodities and, second, provides the instruction to perform a summation of these sizes, this law does not determine the very named numbers expressed in those data. The latter are given in the "accounting books" of the private companies and the knowledge used to collect and register them is the everyday knowledge or, with respect to the scientific theory given in Marx's economic works, *pre-theoretical* knowledge[175] of the actors participating in the very process of commodity production. So, while Reuten criticizes Marx's theory as not capable of providing the data relevant for the computation of the coefficients m_i, the *opposite* is true. It is the *positive feature* of Marx's economic theory that while providing, via the law of production prices as a scientific law, the possibility to derive a recipe for the measurement of the total labor expended in an economy, it does *not* determine the very data needed as inputs for that measurement. One can state this even in stronger terms. If Marx's economic theory would be able to provide not only that recipe, but would simultaneously determine the sizes of the magnitudes expressed in those data, then Marx's theory would turn into an all-embracing algorithm in the framework of which all future states of affairs in the world of capitalist economics could be computed without the use of any data independent of this algorithm; it would, thus, turn into a *completely analytic theory*. Not only is Marx's economic theory a nonanalytic, namely, *synthetic* theory but, in addition, it stands, as it is readily seen from Figure 4.5, in a *cyclical* relation to the level of data acquisition. On the one hand, Marx's economic theory gives the impetus to collect certain specific types of data and at the same time provides the recipes for the computational measurement based on this data. On the other hand, the data obtained and the results of the computations enable one to incorporate into that theory new information. Marx's economic theory is thus placed into a type of relation that, as shown in chapters 2 and 3, is characteristic also of Huygens' and Newton's theories.

Let me now reflect on the possible ways one could *really* compute the sizes of the magnitudes S and TV and obtain them as output data of measurement. Figure 4.3 suggests that one option could be to start the measurement by collecting data at the appearance-level for singular industrial capitals and then follow the paths along

---

175 Theoretical knowledge of other types can and in fact are involved, e.g., theories guiding the measurement of the weight amounts of gold.

which their respective profits are gradually split up into the profits of the industrial capitalists, the bankers, the land-owners, and so on. This procedure, by itself already very cumbersome, turns into an enterprise of tremendous proportions once we realize that in order to sum up all profits in all of their forms, one would have to perform this sum over the *whole world economy*, which is the stage where one can really speak about the total social capital and where singular profits are balanced out to an average profit. To the best of my knowledge, no actually performed measurements based on Marx's economic theory dared to compute the total surplus-value and total value produced in the world economy as a whole; instead these measurements opted for more-modest measurements restricted to national economies.[176] The possibility of such a restricted measurement was for the first time, to the best of my knowledge, at least *implicitly* indicated in (Tugan-Baranowsky 1905, 171–173).

4.3.2 From Average Profits and Production Prices to Surplus-Value and Value (M. Tugan-Baranowsky)

M. Tugan-Baranowsky bases his approach on the following four idealizations:

1. There exist three sectors of production. I: producing means of production; II: producing objects of consumption for the working class; III: producing objects of consumption for the capitalist class. Thus it is supposed that only these two classes exist.
2. For the time span of production of one year the turnover of the constant and variable capital is one year, that is, the fixed capital equals zero.
3. Simple reproduction takes place, that is, the total surplus-value produced in all three sectors taken together is consumed in the personal consumption of the capitalists.
4. The measure of profit is the same in all three sectors, namely, 25%.

Tugan-Baranowsky then supposes that p, r, and a in the table below stand for (in millions of German Marks) the *money-price (price of production) of the means of production employed in one year*, *the labor-wages* paid in one year, and the *money-profit* obtained in one year (*Rente* in his terminology).

	p	a	r	$p+a+r$
I	180	60	60	300
II	80	80	40	200
III	40	60	25	125

---

176 See, e.g., (Cockshott and Cottrell and Michelson 1995) and (Zachariah 2006).

Now the task is to express the money-prices in labor-vaues, that is, to transform prices into values. For this purpose Tugan-Baranowsky supposes that in sector I are, for the whole year, employed 150 000 workers. The workers produce, with the help of the means of production with the price of 180 millions of Mark (first cell in the first line in the table), an amount of commodities with a total price of 300 millions of Mark (fourth cell in the first line). If the *labor-value* of this amount of commodity is expressed in thousands of labor-years, X, then the value of the commodity consumed in branch I is given as $180X/300$. One then obtains the equation

$$\frac{180X}{300} + 150 = X$$

By its solution one finds out that $X = 375\ 000$ labor-years. So, the ratio of labor-value of the total product to production prices of the total product is equal to 375 labor-years/300 000 Marks.

Based on this ratio, one can then compute the labor-value of the means of production consumed in sector I: (375 labor years/300 000 marks) × 180 000 000 Marks = 225 000 labor-years; in sector II: (375 labor years/300 000 marks) × 80 000 000 Marks = 100 000 labor-years; and in sector III: (375 labor years/300 000 marks) × 40 000 000 Marks = 50 000 labor-years.

If we now take into account that in sector I we have 150 000 workers working for one year, then the total of 150 000 labor-years stands in sector I for the labor-value of the sum $(a + r)$. If we take into account also that on the wages of those 150 000 workers are spent (according to the second cell, first line in the table) 60 000 000 million marks, we find out that the one worker earns per year 400 Marks. Based on this per-year-income of a worker, one can compute the number of workers employed in sector II: 80 000 000 million Marks : 400 Marks per worker = 200 000 workers; and in sector III: 60 000 000 million Marks : 400 Marks per worker = 150 000 workers.

This in turn means that in sector II the labor-value of the sum $(a + r)$ is 200 000 labor-years, and in sector III it is 150 000 labor-years. In sector III we thus have the total labor-value of 200 000 labor-years, which equals the total surplus-value. In sector II we have a total labor-value of 300 000 labor years which equals the total variable capital expended. The rate of surplus-value for the whole society thus is 200 000 labor-years/300 000 labor-years $= \frac{2}{3}$. Based on this ratio we can split up the respective sums $(a + r)$ for the sectors I, II, and III and, thus, obtain the respective named numbers for a and r.

The results of Tugan-Baranowsky's computations are expressed in following table.

	p'	a'	r'	p' + a' + r'
I	225	90	60	375
II	100	120	80	300
III	50	90	60	200

Here the magnitudes p', a', and r', are the value-expressions corresponding to the price-expression p, a, and r.

What moral can be drawn from Tugan-Baranowsky's method of movement from p, a, and r to p', a', and r'? *First*, it shows that in order to perform the respective computations one has to know in advance the size of wages paid to the members of the working class. *Second*, this in turn requires knowing, in advance, who qualifies at all as the member of this class and how does the person *qualitatively* differ from the members of other social classes; a wrong qualitative framing of the term "member of the working class" would lead to the summing up of incomes that are qualitatively different – of wages with non-wages – and would thus lead to *distorted* quantitative results.

Third, contrary to the procedure employed by Tugan-Baranowsky, who starts from data pertaining solely to the industrial capitals, one has to – in order to escape the torturous path from singular industrial capitals to singular nonindustrial capitals – start with the collection of data pertaining from the very beginning to various types of capital: industrial, agricultural, commercial, banking, and so on. This means that – with respect to the restriction to a particular national economy – one has to draw from the very beginning on data pertaining to the GDP, to the national input and output (IO) accounts and to national income and product accounts (NIPA). This means also that from the very beginning one has to give up the idealization that the society is composed of just two social classes: the class of worker and the class of capitalist. This again requires having a *prior* understanding of the qualitative differences between all nonworking classes given in an economy for which the computational measurements should be performed.

Fourth, the knowledge about the class structure of a society then enables one to take into account the fact that in economies where the capitalist production dominates, there still exists (a) production of *exclusively use-values*, which find their way to the market and (b) production that results in commodities but in which no surplus-value is incorporated. But because in cases (a) and (b) the products are exchanged for money functioning as money-capital, this exchange still yields a *profit*. This then leads at the level of the national economy to a phenomenon not considered in detail in the manuscripts of Volume III of *Capital*, namely, that the *sum of profits diverges from the sum of surplus-values in that national economy*.[177]

Finally, *fifth*, if the data pertaining to a national economy are not from the very beginning collected on the bases of Marx's economic theory and his qualitative differentiations between memberships in different social classes, then those data can – *with respect to the recipes for the computational measurement of the total value and total surplus-value and the theoretical basis of these recipes* – be *ideologically distorted* in the sense indicated earlier; data pertaining to economic activities of different social classes can be mixed up and then lead – once taken as input-data for computations – to distorted results/outputs of these computations. The imperative thus is to *purify* or *split up* with respect to Marx's economic theory – before the computations start at all – the respective data so that they become relevant. Such a purification is at the basis of the approach given in (Moseley 1991).[178]

4.3.3 From the Net Product to Surplus-Value (F. Moseley 1991)

F. Moseley's 1991 book, targeting primarily the problem of changes of the rate of profit in the postwar (1947–1987) US economy, provides a highly interesting approach to how to derive measurement procedures from Marx's economic theory enabling to measure the total surplus-value produced in a national economy.

Based on Marx's concepts of constant capital, variable capital and surplus-value, Moseley introduces, first, the magnitude N^* – *the annual flow of new value* understood as the *money-equivalent* of the *sum* of the total surplus-value produced in an economy, S in my notation, and the total variable capital invested in this economy in one year, V in my notation. Thus, it holds (k is here, as earlier, the constant mediating between labor-values and prices):

$$N^* = k(S + V). \qquad /43/$$

Moseley introduces also the magnitude V^* as the money-equivalent of the magnitude V; it is "the annual flow of the wages of production workers ... measured in current prices" (Moseley 1991, 3). So it holds that:

---

177 On this see, e.g., (Shaikh 1984).
178 Such purification is accomplished also by A. M. Shaikh and E. A. Tonak in (Shaikh and Tonak 1994) who measure value and surplus-value by means of purified data from national input-output (IO) accounts and national income product accounts (NIPA).

$$V^* = kV. \qquad /44/$$

Based on the equation for the *rate of surplus-value* given an economy as a whole in one year, which, as shown earlier in /30/, is $S' = \frac{S}{V}$, one can – by using the equations /43/ and /44/ – derive the *rate of surplus-value expressed in money-terms* – S'^*, for short:

$$S'^* = \frac{N^* - V^*}{V^*} \qquad /45/.$$

The fact that Moseley operates in his approach with the Marx's magnitudes *expressed in terms of money* becomes understandable with respect to the following question he formulates: "Do the concepts of constant capital, variable capital, and surplus-value refer to observable quantities of money (prices) or to observable quantities of labor?" (1991, 26). By proving[179] that magnitudes pertaining to labor are not observable, he opts for the following answer: "Marx's concepts of constant capital, variable capital, and surplus-value are defined in terms of sums of money" (1991, 30).

Then, by means of the orientation to economic magnitudes expressed in money terms, and not in labor-time terms, Moseley can in his computations access the data given in NIPA and based on the concept of *net product of the Business sector* because the meaning of the term "net product of the Business sector" is quite close, but not identical in meaning to the term "new product." To obtain such an identity Moseley purifies the later, and then the data obtained on its basis, in three respects. *First*, the net product includes imputations not corresponding to commodities produced and then sold on the market by business enterprises; these imputations should account for the housing services of owner-occupied homes. But so as work these owners perform does not lead to the creation of value and surplus-value,[180] these imputations have to be eliminated for the data. *Second*, the net product does not take into account the nonproductive depreciations costs of the equipments and structures used in the *non-productive sectors*, namely, circulation and supervision; thus the respective named numbers have to be added to the data of the net product. *Third*, another distortion of the data of the net product has its origin in the fact that the results of the labor of self-employed proprietors and partners are taken into account. But since they are not workers employed in capitalist enterprises, the value of the results of their labor does not include any surplus-labor; thus the named numbers corresponding to them have to be removed from the NIPA. By these three purification procedures Moseley obtains

---

179 On this see (Moseley 1991, 27–31).
180 On this see (Moseley 1991, 38).

from the net product of the Business sector the data that can be viewed as the named-numerical expressions of the magnitude N^*.

In order to employ the equation /45/ F. Moseley has to purify also the NIPA-data pertaining to the paid wages because they misleadingly contain also wages of people performing nonproductive activities among which he lists *circulation activities* pertaining to the exchange of commodities and money as well as *supervisory activities*. Based on the data given in the *Census of Manufacturers* and the *Bureau of Labor Statistics Current Establishment Survey*, he purifies the original data given in NIPA, so that they involve only the wages paid to the *productive workers employed in capitalist enterprises*.

Once Moseley has the purified data both for N^* and V^*, he can, based on equation /45/, compute the rate of surplus-value and then, based on other equations, also the sizes of other magnitudes for the US economy for the years 1947–1987.[181]

The importance of Moseley's approach goes well beyond his aim of his 1991 book to compute the rate of profit for the US economy. It enables, as I will prove now, to compute the total surplus-value produced in an economy *terms, expressed in labor-time terms*, as well as the size of the constant k relating prices and values.

Based on the procedure proposed Moseley, one can compute the concrete size of S'^*; let it be p^*, thus $S'^* = p^*$ stands for the end-result of the computations procedure and, as can be readily seen from /45/, p^* stands for a *concrete pure (not named) number*. At the same time holds the equation /30/, which provides the equation $S' = \frac{S}{V}$. So as both S and V are expressed in labor-time terms, the values acquired by the magnitude S' are pure numbers. If the labor-time corresponding to V, the *necessary time*, is symbolized as NT, and the labor-time corresponding to S, the *surplus-time*, is symbolized as ST, the equation $S' = \frac{S}{V}$ turns into

$$S' = \frac{ST}{NT}. \qquad /46/$$

If, by means of computation, we find out $S'^* = p^*$, then from the last equation it follows that:

$$\frac{ST}{NT} = p^*. \qquad /47/$$

Thus, one obtains the equation $ST = NT \times p^*$ which is *one* equation with *two* variables. To solve it we need yet another equation with the same two variables. We obtain it by obtaining from the NIPA the datum providing the concrete value for the total time, H, for short, in which workers employed in capitalist produc-

---

181   For the results of these computations see (Moseley 1991, 48–102).

tion enterprises performed their labor in one year; let it be H* which stands for a *concrete named number*. The second equation is then follows:

$$H^* = ST + NT. \tag{48}$$

We have now a system of two equations /47/ and /48/ with variables *ST* and *NT*. By its solution we obtain:

$$NT = \frac{H^*}{1 + p^*}, \quad ST = \frac{H^* \times p^*}{1 + p^*}. \tag{49}$$

So, by knowing the concrete numbers H* and p* we can compute, based on /49/, the sizes of both *ST* and *NT*. But so as *ST* corresponds to *S* – the surplus value – and *NT* to *V* – the variable capital, /49/ can be written as follows:

$$V = \frac{H^*}{1 + p^*}, \quad S = \frac{H^* \times p^*}{1 + p^*}. \tag{50}$$

By combining $V = \frac{H^*}{1 + p^*}$ with the equation $k = \frac{V^*}{V}$, derived from the equation $V^* = kV$ given in /44/, we find out for the constant *k* mediating between prices and values the following equation:

$$k = \frac{V^* \times (1^* + p^*)}{H^*}. \tag{51}$$

By substituting the concrete numbers for *V** as well as the numbers p* and H* for a concrete economy, one computes the size of *k*.

As an empirical test of the correctness of the computation of the size of *k* on the basis of /51/, one can employ the second expression in /50/. So as according to equation /41/ for the sum of all profits in an economy, it holds that this sum equals the total surplus-value expressed in money terms produced in it, *S**, for short, one can compute *S** for which should hold at the same time *S** = *kS*. From the latter one obtains the equation $k = \frac{S^*}{S}$, which – once combined with $S = \frac{H^* \times p^*}{1 + p^*}$ from /50/, yields

$$k = \frac{S^* \times (1^* + p^*)}{H^* + p^*}. \tag{52}$$

Chapter 5: What is Grounded Theory? Methodological Reflections and Explications

5.1 Introduction

By Grounded Theory (hereafter, GT) one circumscribes in contemporary social science both a *method* and the *result* of application of this method, namely, a conceptually dense, social science theory anchored in (grounded on) data.

In the last 40 years GT gradually rose in prominence in social science as can be seen from Table 5.1 which comprises statistical information provided by the *Social Science Citation Index* on the frequency of the use of the term "grounded theory" in titles of scientific papers:[182]

Years	The number of occurrences of the term "grounded theory"
1973–1977	2
1978–1982	12
1983–1987	13
1988–1992	25
1993–1997	44
1998–2002	100
2003–2007	168
2008–2009	75

Table 5.1 Frequency of use of the term "grounded theory" in the years 1973–2009

Given the impact of GT on social science, it is surprising that the leading scholarly journals in the field of philosophy of science and philosophy of social science have not published in the last decades, to the best of my knowledge, a single article dealing with it. This chapter aims to remedy this oversight and will deal in detail with the methodological content and implications of GT procedures. As I will try to show, the importance of GT lies in its new approach to issues

---

[182] From 1985, computer based and narrowed down to areas of nursing, education, educational research, and behavioral sciences.

like induction, method of theory construction, and theory development. These issues are part of what I label here as the *general methodology of science*. GT, of course, cannot be limited just to this framework. It contains also issues relating to the socio-philosophical implications of American pragmatism and symbolic interactionism,[183] as well as to *social science epistemology*.[184]

Here I do not deal with those issues because, *first*, one can deal with issues related to general methodology of science prior and independently of those issues. *Second*, it is a fact that one finds in the works of social scientists working on GT a surprising *lack of reflection on the methods* they implicitly employ. This is acknowledged, for example, by Anselm L. Strauss, one of the founding fathers of the GT approach, who complained that "there is a remarkable absence of discussion about the nature of (grounded) theories, how to generate them, and how to go about verifying them" (Strauss 1995, 7) and that a detailed scrutiny of the subject index of the highly representative *Handbook of Qualitative Research* (Denzin and Lincoln 1994) with respect to the issue of scientific theory "is a greatly disappointing experience" (Strauss 1995, 8).

Third, even if Juliet Corbin promised that the third edition of *Basics of Qualitative Research* (authors Corbin and Strauss) will contain "philosophical notions or underpinnings of Strauss's approach to Grounded Theory" (Corbin and Cisneros-Puebla 2007, 86), one finds in that edition on this topic only the following truism (Strauss and Corbin 2008, 8):

> The methodological implications of the above can be summarized as follows. The world is very complex. There are no simple explanations for things ... Therefore any methodology that attempts to understand experience and explain situations will have to be complex. We believe that it is important to capture as much of this complexity in our research as possible.

Fourth, the general methodological orientation of the analyses of the GT method as presented here finds its justification also in a more recent claim of B. Glaser, another founding father of the GT approach, who states that "Grounded Theory is a general ... method which can use any type of data ... It is just a simple methodology ... [it is] possessed by no discipline" (Glaser 2005, 127, 141).

Finally, the need to remain at the level of a general methodology is prompted also by the fact that the methods employed in GT are often compared with and

---

183 On these implications see (Joas 1978), (Joas 1992) and (Shalin 1986).
184 Here I mean the discussions between the representatives of a constructivist approach to GT and those of a realistic approach to GT. On this see (Charmaz 1995), (Charmaz 2000), (Lomborg and Kirkevold 2003), (Mills and Bonner and Francis 2006a), (Mills and Bonner and Francis 2006b), and (Mills and Chapman and Bonner and Francis 2006).

evaluated against the hypothetico-deductivist methodologies – for example, in the works of Udo Kelle – a tendency which leads, as I will try to show below, to a misunderstanding of methods employed in GT and to a misevaluation of the hypothetico-deductivist methodologies. As I will try to prove, the importance of GT methodology is that through clarifying it the principal methodological deficiencies of the hypothetico-deductivist methodologies can be shown; *deficiencies that pertain not only to the methodology of social science but also, as shown in chapter 3, to that of physics.*

I would like to emphasize that here I do not deal with the possible methodological implications of the conflict between Glaser and Strauss which erupted in the nineties.[185] Neither do I deal here with possible methodological differences between substantive GT and formal GT.[186]

5.2 Grounded Theory – An Overview

From the point of view of a general methodology of science the main elements of procedures characteristic of the GT method are as follows:

- *Data Collection.*
- *theoretical coding* based on the *method of constant comparison,* on (substantive and theoretical) *codes,* the *coding paradigm*[187] and guided by the *concept-indicator model.*
- *theoretical sampling* and *theoretical saturation.*
- *concepts, categories, properties,* and *dimensions.*
- *Core category, the integration,* and *development of theory*

Let me deal with each of these elements.

Data collection. Given that the aim of the GT approach is "the discovery of theory from data" (Glaser and Strauss 1967, 1), the impact of preconceived ideas on the collection of data must be minimized because "initial decisions are not based on preconceived theoretical frameworks" (Glaser and Strauss 1967, 45). But since even initial data collection has to be guided by some antecedently given

---

[185] For the first shots of this conflict see (Glaser 1992); Strauss's reactions to them in memo see (Corbin 1998). For an analysis of this conflict see, e.g., (MacDonald 2001), (Strübing 2007a), and (Mey and Mruck 2007).
[186] On formal GT see, e.g., (Glaser 2007), (Kearney 2001), and (Kearney 2007).
[187] The coding paradigm is given prominence by Strauss. Glaser rejects this prominence, viewing it as just one of many theoretical codes that can be used to integrate a theory.

ideas, Glaser and Strauss state the following recipe (Glaser and Strauss 1967, 45):

> The initial decisions for theoretical collection of data are based only on a general sociological perspective and on a general subject or problem area. ... such as how ... policemen act toward Negroes or what happens to students in medical schools that turn them into doctors ... The sociologist may begin the research with a partial framework of "local" concepts designating a few principal or gross features of the structure and processes in the situations that he will study. For example, he knows before studying a hospital that there will be doctors, nurses, and aides, and wards and admission procedures. These concepts give him a beginning foothold on his research.

This means that the framework for the initial data collection is provided by a specific research problem or a set of general research problems or a general perspective of a (say sociological) discipline. Then, after certain concepts are derived from the data, as we will see below, these concepts start guiding the next stages of data collection. So, (1967, 47):

> beyond the decisions concerning initial collection of data, further collection cannot be planned in advance of the emerging theory ... The emerging theory points to the next steps – the sociologist does not know them until he is guided by emerging gaps in his theory and by research questions suggested by previous answers.

One aspect of this phase of GT construction is that data collection stands always for the choice of certain samples or groups (and of their subgroups) or cases involving a certain number of entities for which the data should hold; I will view these samples or groups (subgroups) of entities as subsets of *universes of entities* or of *universes of discourse*. The principal question one faces when dealing with the choice of a sample (or of several samples) is what will best serve the purpose of theory construction. This choice will, simultaneously, be driven by both antecedently given ideas and new derived concepts – in the terminology of Glaser and Strauss *theoretical sampling*. According to Glaser and Strauss is (1967, 47):

> The basic question in theoretical sampling ... is: *what* groups or subgroups does one turn to *next* in data collection? And for *what* theoretical purpose? In short, how does the sociologist select multiple comparison groups? The possibilities of multiple comparisons are infinite, and so groups must be chosen according to theoretical criteria.

And Glaser and Strauss provide the following, initially tentative answer (1967, 49):

> The basic criterion governing the selection of comparison groups for discovering theory is their *theoretical relevance* for furthering the development of emerging categories. ... Thus ... group comparisons are conceptual; they are made by comparing diverse or simi-

lar evidence indicating the same conceptual categories and properties, *not* by comparing the evidence for its own sake.

Theoretical coding. Under this term one understands the task of explicitly assigning codes and analyzing the already collected data or, stated otherwise, "coding means naming segments of data with a label that simultaneously categorizes, summarizes, and accounts for each piece of data" (Charmaz 2006, 43). This coding/analysis proceeds *via* the method of *constant comparison*[188] which stands for a comparison of "incident to incident with the purpose of establishing the underlying uniformity and its varying conditions" (Glaser 1978, 49). Based on their research on the loss of patients dying in hospital wards Glaser and Strauss consider the following incidents: a VIP is lying in the intensive care unit and several doctors are taking care of him/her. In another incident a lower class African American with a gaping knife wound is ignored on the city emergency ward. Based on the comparison of these discrete incidents the analyst can, according to them, generate the concepts *social loss* of a patient – *loss to family and occupation* – *medical attention* and then arrive at the tentative claim that "patients who have high social loss will receive better care than those who have low social loss" (Glaser and Strauss 1967, 24).

Here emerges the importance of the introduction of groups/subsets of universes of entities I mentioned above. According to Glaser and Strauss, one always frames groups/subsets of universes with respect to an incident appearing in these of groups/subsets of universes of entities and then they add the following defining rule for the method of constant comparison: "*while coding an incident for a category, compare it with the previous incidents in the same and different groups coded in the same category*" (1967, 106), and where "category" is understood as a higher order, more abstract concept that can initially – as we will see below – contrary to a property of category, stand for itself.

By this method of constant comparison one obtains concepts – "conceptual labels placed on discrete happenings, events, and instances of phenomena" (Strauss and Corbin 1990, 61), as well as categories for the respective groups/subsets of universes under investigations. Those concepts and categories are the basis for the – in the terminology of Glaser – *substantive codes* that "conceptualize the empirical substance of the area of research" (Glaser 1978, 55) and are, according to Glaser, different from the *theoretical codes* which "conceptualize how substantive codes may relate to each other as hypotheses to be integrated into the theory" (1978, 55). For example, once Glaser and Strauss obtained the substantive codes labeled "social loss" and "attention," they were theoretically coded into the

---

[188] This term was coined by Glaser in his 1965 article.

hypothesis that the higher the social loss of a dying patient on award, the more attention he or she receives from the nurses on this ward.[189]

Theoretical sampling stands for the employment of theoretical codes which (Glaser 1978, 72)

> conceptualize how the substantive codes may relate to each other as hypotheses to be integrated into a theory. ...They weave the fractured story back together again. ... Theoretical codes give integrative scope, broad pictures and a new perspective. This is why grounded theory is often "new" because of its *grounded integration*.

Glaser provides several sets of theoretical codes unified in what he calls *coding families*. He lists coding families that can be summarized as shown in Table 5.2 (Glaser 1978, 73–80):[190]

Family-name	Subject-matter	Elements
Six C's	Causal models	Causes, contexts, contingencies, consequences, covariance, and conditions
Process	Process-models	Stages, phases, progressions
Degree	Degree of attributes	Limit, range, intensity, etc.
Dimension	Patterns of connections	Elements, divisions, properties, etc.
Types	Types	Type, form, kinds, styles, etc.
Strategy	Strategies of action	Strategies, tactics, mechanisms, etc.
Interaction	Mutual interactions	Mutual effects, reciprocity, mutual trajectory, etc.
Identity-Self	Concepts of the self	Self-image, self-concept, self-worth, etc.
Cutting point	Cutting points	Boundary, critical juncture, turning point, etc.
Means, goal	Concepts of purposeful action	End, purpose, goal, etc.
Cultural	Cultural phenomena	Norms, values, beliefs, etc.
Consensus	Social consensus	Clusters, agreements, contracts, etc.
Mainline	Social integration	Social control, recruitment, socialization, etc.
Theoretical	Concepts for generating, criticizing, and judging theory	Parsimony, scope, integration, etc.

---

189 On this see (Glaser and Strauss 1964).
190 In this figure I draw also on (Mey and Mruck 2007) and (Dey 1999). For an enlargement of the list of coding families see (Glaser 1998, 170–175).

Ordering or elaborating	Ordering of data, concepts, and categories	Structural, temporal, conceptual
Unit	Units of social life	Collective, group, nation, etc
Reading	Way of Coding	Concepts, problems, and hypotheses
Models	Modeling of theory	Linear, spatial, etc.

*Table 5.2 Glaser's coding families in **Theoretical Sensitivity** of 1978*

Strauss, initially in his work 1987, and then in cooperation with Corbin in 1990 proposed, instead of Glaser's two-stage approach to substantive coding, a three-stage approach to it: *open*, *axial*, and *selective* coding. Open coding serves the breaking up, examining, comparing, conceptualizing, and categorizing data. In axial coding (Strauss and Corbin 1990, 97)

the focus is on specifying a category (*phenomenon*) in terms of the conditions that give rise to it; the *context* (its specific set of properties) in which it is embedded; the action/interactional *strategies* by which it is handled, managed, carried out; and the *consequences* of those strategies. These specific features of a category give it precision, thus we refer to them as *subcategories*.

In axial coding the subcategories are related to their respective category by means of the so-called paradigm model (Strauss and Corbin 1990, 99):

In grounded theory we link subcategories to a category in a set of relationships denoting causal conditions, phenomenon, context, intervening conditions, action/interactional strategies, and consequences. Highly simplified, the model looks something like this:

(A) CAUSAL CONDITIONS → (B) PHENOMENON →
(C) CONTEXT → (D) INTERVENING CONDITIONS →
(E) ACTION/INTERACTION STRATEGIES →
(F) CONSEQUENCES.

Finally, selective coding stands for the process of selecting the central (or core) category and systematically relating it to other categories, so that the core/central category should stand for the "central phenomenon around which all the remaining categories are integrated" (Strauss and Corbin 1990, 116).

A special place in the coding procedure is assigned, by both Glaser and Strauss, to the *concept-indicator model* going back to P. F. Lazarsfeld. This model should direct the conceptual coding of a set of empirical indicators. Glaser uses for this model in the framework of GT the diagram shown in Figure 5.1 (Glaser 1978, 62):[191]

---

191 Strauss uses the same diagram; (Strauss 1987, 25).

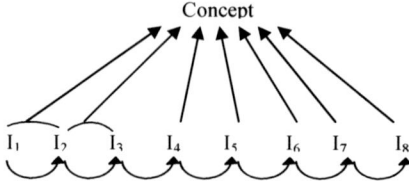

Fig. 5.1 Concept-indicator model of Glaser and Strauss

Here $I_1, I_2, ..., I_8$ stand for indicators which are initially mutually constantly compared so that from this comparison the "analyst is forced into confronting similarities, differences and degrees of consistency of meaning between indicators which generates an underlying uniformity which in turn results in a coded category and the beginning of properties of it" (Glaser 1978, 62). According to Glaser this concept-indicator model enables one to fulfill the condition, mentioned earlier, that concepts *earn* their way into theory instead of being taken *a priori* as relevant for a substantive area. It is in this respect that the approach of Glaser and Strauss to the concept-indicator model differs from that of P. F. Lazarsfeld. The latter used that model in order to establish empirical variables based on a concept *given already in advance*. He states the following (Lazarsfeld 1958, 100–101):[192]

There appears to be a typical process which reoccurs when we establish "variables" for measuring complex social objects. This process by which concepts are translated into empirical indices has four steps: an initial imagery of the concept, the specification of dimensions, the selection of observable indicators, and the combination of indicators into indices.

So the diagram in this case is as shown in Figure 5.2:

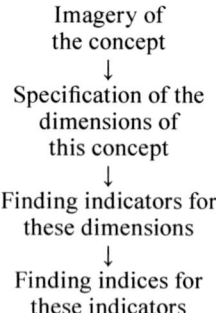

Figure. 5.2 Lazarsfeld's concept-indicator model

---

192   On this see also (Lazarsfeld 1966, 187–190).

Theoretical sampling and theoretical saturation. Theoretical sampling stands in the framework of the GT method for (Glaser and Strauss 1967, 45)

> the process of data collection for generating theory whereby the analyst jointly collects, codes, and analyzes his data and decides what data to collect next and where to find them, in order to develop his theory as it emerges. This process of data collection is controlled by the emerging theory.

Sampling, as stated already, stands for a choice of samples/groups/subsets of universes of discourse that are continuously mutually compared both for differences and similarities between them with the aim of obtaining concepts and categories. But so as the newly obtained concepts and categories in a *feedback manner* start to guide and direct the search for the relevant samples/groups/subsets of universes of entities for which one should obtain new data, (Glaser and Strauss 1967, 47, 49):

> further collection [of data] cannot be planned in advance of the emerging theory. ... The emerging theory points to the next steps – the sociologist does not know them until he is guided by emerging gaps in his theory and by research questions suggested by previous answers ... there can be no definite, prescribed, preplanned set of groups that are compared for all or even most categories.

From these characteristics it follows that in the framework of the GT approach the process of theory construction has with respect to data collection – a *cyclical* character. Each step in theory construction – by itself based on data collection and data analysis – drives and guides yet another data collection and data analysis. At the same time one can approach the already collected data from the point of view of newly developed concepts and categories in order to sift them through anew. Glaser therefore states: "The detailed, conceptual route from data collection to a finished writing is a process composed of a set of double-back steps. As one moves forward, one constantly goes back to previous steps ... going back and forth between data and concept as one generates theory" (1978, 16, 37).

The collection of data, with respect to a certain sample/group/subset of a universe of discourse, comes to an end at a point labeled by Glaser and Strauss as *theoretical saturation*, which means that with respect to that group no new data can be obtained any more that would lead to new concepts, categories and their properties. Here the concept-indicator model and specifically the technique of interchangeability of indices can be employed. Once the thought-movement from incident to incident for the same sample/group/subset of a universe of discourse does not yield new features of the category, it does not make sense to go on. According to Glaser "saturation has occurred because the incidents are interchangeable. Why keep collecting the same thing over and over with no yield" (1998, 26).

Concepts, categories, properties, and dimensions. In addition to concepts and categories, Glaser and Strauss also identify "properties of a category" and "dimensions of a property." As an example they consider the category name "nurse's perception of social loss" which names a nurse's view about what the degree of loss a patient's death will be for his or her family and occupation. As an illustration of a property of a category, Glaser and Strauss mention "loss rationales" naming the rationale a nurse uses to justify her own perception of the social loss of the dying patient. What unifies both categories and properties of categories is that once they are established *they are highly resistant to changes in the evidential basis on which they are based.* For example, if one introduces the name of the category "social loss" on the basis of incidents in one hospital and then discovers that the evidence is quite different in other hospitals then one has to find the difference in structural conditions between these hospitals: "the discovered theoretical category lives on until proven theoretically defunct for any class of data" (Glaser and Strauss 1967, 24). And they elaborate (Glaser and Strauss 1967, 36):

Once a category or property is conceived, a change in the evidence that introduced it will not necessarily alter, clarify or destroy it. It takes much more evidence – usually from different substantive areas – as well as the creation of a better category to achieve such changes in the original category.

Dimension should stand for a property located along a continuum (Strauss and Corbin 1990, 61). To make their differentiation between categories, properties and dimensions of properties more understandable, Strauss and Corbin give the following example which I quote in full length (Strauss and Corbin 1990, 70).

Let us look at the category of "color." Its properties include: shade, intensity, hue, and so forth. Each of these properties can be dimensionalized; that is, they vary along continua. Thus, color can vary in intensity from high to low; in hue from darker to lighter; and so forth. The properties listed above are *general properties* of color. They can pertain to color regardless of the situation in which color is found.

So, the diagram for categories, properties of categories and dimensions of properties for this example is as shown in Figure 5.3.[193]

Category	*Properties*	*Dimensional range*
color	shade	....
	intensity	low, ..., high
	hue	lighter, ..., darker

Figure 5.3 Strauss and Corbin on the category color with its properties and dimensions

---

[193] Here I draw on Strauss' and Corbin's representation of relations between categories, properties, and dimensions (Strauss and Corbin 1990, 72).

Core category and the integration and development of theory. The core category of a grounded theory is the center around which the generation of a unified theory takes place. According to Glaser it guides further data collection and the choice of samples/groups/subsets of universes of entities (i.e., theoretical sampling). The core category is usually related to a "core process" (Glaser 1978, 61) the conditions and consequences of which have to be identified by the analyst. This enables us to understand that (Glaser 1978, 93):

a core category accounts for most of the variation in a pattern of behavior, this has several important functions for generating theory ... Most of the categories and their properties are related to it, which makes it subject to much qualification and modification because it is so dependent on what is going on in the action. In addition, through these relations between categories and their properties it has the prime function of *integrating* the theory and rendering the theory *dense* and *saturated* as the relationships increase. These functions then lead to theoretical *completeness* – accounting for as much variation in a pattern of behavior with as few concepts as possible thereby maximizing parsimony and scope. Clearly integrating a theory around a core variable *delimits* it and thereby the research project.

According to Strauss and Corbin (1990, 116–118) identifying the core category – standing for the central phenomenon integrating all the other categories – is equivalent to the *explication of the story line*. As a next step the subsidiary categories should be related to the core category by means of the coding paradigm mentioned already above. Next steps then consist of

- relating the categories and the dimensional level;
- validation of those relationships against data; and
- filling in categories that may need further refinement and/or development.

Worth noting already here is that Glaser views the integration of theory around a core category as the delimitation of the theory, that is, the *identification of a set of processes which the theory is able, once unfolded, to cover conceptually.*

GT as a *result* of the application of the GT method should fulfill several criteria. Glaser lists three criteria. It must have *fit* in the sense that the categories of the theory must fit the data. It has to *work* in the sense that it should be able to explain what happened, predict what will happen, and provide an interpretation of what is going on in the area of inquiry. And it has to have *relevance* in the sense that "it allows core problems and core processes to emerge" (Glaser 1978, 5).

According to Strauss and Corbin (1990, 23) GT should, with respect to a certain phenomenon, fulfill the following four criteria. *Fit* in the sense that it should fit the substantive area under investigation. *Understanding* in the sense that it should make sense both to researchers and to those participating in the area. *Generality* in the sense that "the theory should be abstract enough and include sufficient variation to make it applicable to a variety of contexts related to that phenomenon" (Strauss and Corbin 1990, 23). And the theory should be able to provide *control* with respect to action oriented toward that phenomenon.

Finally, K. Charmaz lists the following criteria a complete GT should fulfill: *close fit with the data, durability over time, modifiability,* and *explanatory power* (Charmaz 2006, 6). With respect to durability and modifiability she states the following: "A grounded theory is durable because it accounts for variation; it is flexible because researchers can modify their emerging or established analyses as conditions change or further data are gathered" (Charmaz 2000, 511).

5. 3 "Categories" and categories, "properties" and properties – A semantical clarification

Let me now turn to a clarification – in the framework of semantics – of what categories and properties are *about*. I will draw on this clarification below when I will deal with the specific nature of inferences applied in the GT approach.

In works on GT, one can find two approaches to that *about-ness*. One, by and large correct; and another one that is highly confusing. Let me start with the first one. Already in *Discovery of Grounded Theory* one can find the following claim: "Concepts should be analytic – sufficiently generalized to designate characteristics of concrete entities, not the entities themselves" (Glaser and Strauss 1967, 38). In a similar vein Strauss and Corbin claim the following (Strauss and Corbin 1990, 61, 62, 65, 67, 99):

[1.] *Concepts*: Conceptual labels placed on discrete happenings, events, and other instances of phenomena.
[2.] *Open coding*: ... is part of analysis that pertains specifically to the naming and categorizing of phenomena.
[3.] The phenomenon represented by a category is given a conceptual name.
[4.] How do categories get named? ... The name you choose is usually the one that seems most logically related to the data it represents, and should be graphic enough to remind you quickly of its referent ... The important thing is to name a category.
[5.] In grounded theory we link subcategories to a category in a set of relationships denoting causal conditions, phenomenon, context, intervening conditions, action/interactional strategies, and consequences.

So, the semantics that can be obtained from those quotes and to which Glaser, Strauss, and Corbin implicitly hold is as shown in Figure 5.4.

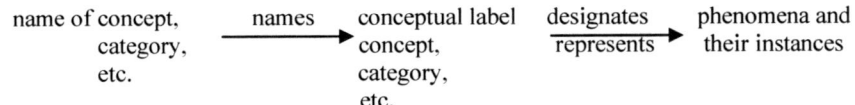

Fig. 5.4 Glaser's, Strauss', and Corbin's semantics for categories, properties, etc.

Worth noting here is the distinction between three entities: *the name*; *named entity* (conceptual label, concept, category etc.), and what is *designated/represented* by the latter (phenomena and their instances). This, by and large, corresponds to the three entities considered by modern semantical theories.[194]

Let me now turn to the confusing approaches. So, for example, K. Charmaz states the following (Charmaz 2006, 43):

Coding means naming segments of data with a label that simultaneously categorizes, summarizes and accounts for each piece of data. Coding is the first step in moving beyond concrete statements in the data to making analytic relationships. We aim to make an interpretative rendering that begins with coding and illuminates studied life.

Where Charmaz clearly deviates from the approach of Glaser, Strauss and of Corbin is in the use of the term "statements" which stands for "declarative sentences," that is, for certain *names*, but at the same "statements" should stand for *interpretations*, that is, for certain *operations with meanings*. Thus, the earlier quote from Charmaz' 2006 work comes very close to the confusion between a *language expression (name)* and *its meaning*.

The confusion surfaces in the following quote from a article by Janice M. Morse (Morse 2004, 1389):

Principles of induction demand that concepts first be derived by identifying common segments of data, accruing these data to form a category and applying a label, or *emic tag*. An emic tag is a label derived from the category that actually occurs in the data and that best describes the category as a whole. The next step is to develop a detailed description of the category. If this description fits the description of a concept in the lay or published scientific literature, then the concept label already in use replaces the emic tag.

Here it is claimed, on the one hand, that the emic tag is derived from a *category*, and so by its very nature must be a *meaning-entity*. But, on the other hand, once the descriptive equivalent is already present in the literature, a concept-label – and not a concept – should replace the emic tag; thus the latter *is not*

---

194 Here I mean, e.g., G. Frege's differentiation between the name, its meaning (*Sinn*), and its reference (*Bedeutung*), or R. Carnap's differentiation between the designator, its extension, and its intension dealt with in chapter 1.

a meaning-entity but a name-entity. That we face a confusion between a name and its meaning is readily seen also in Morse's claim that her aim "is to prevent the proliferation of the same or similar concepts with different names, basically referring to the same sets of behaviors, from cluttering literature" (Morse 2004, 1389). What she has in mind, once one clearly differentiates between a name and its meaning, is as follows: *to prevent the proliferation of different names expressing the same or similar concepts, basically referring to the same sets of behavior, from cluttering literature.*

That confusion then befalls her understanding of the very relation between concept and concept label; (Morse 2004, 1389–1390):

Furthermore, due care must be taken to avoid developing a single concept with many slightly different meanings applied to the same concept label ... If a description of the concept cannot be located in the literature, then the researcher has the prerogative of replacing the emic tag with a new name and introducing it as a new concept.

Thus Morse abolishes the difference between concept and concept-label/concept-name; suddenly we have one *concept/concept-label* with (i) slightly different *meanings*, and (ii) the *name* – replacing the emic tag – serves us as a new *concept*.

To escape such confusions I will from now on clearly differentiate between, on the one hand, a *linguistic expression* (*name*) and, on the other hand, its *meaning*, that is, the *extralinguistic-entity*. Based on subchapter 1.4, I will bring in additional semantical entities of Transparent Intensional Logic in order to analyze statements of GT.

5.4 Grounded Theory's Black Eyes: Deduction, Enumerative Incomplete Induction, Abduction and the Enumerative Thought-Universal

Let me now turn to GT's self-reflection on its own method. Already in *Discovery of Grounded Theory* the authors view their own method of constant comparison leading to concepts and categories as an *inductive* method (Glaser and Strauss 1967, 114). More than a decade later Glaser states that GT "is, of course, a theory induced or emerged after data collection starts" (Glaser 1978, 37). *But neither Glaser nor Strauss provides any general characterization of this method which would correspond to the method of induction in the framework of the GT method.* Close to a decade later Strauss still views the method by means of which a scientific theory is conceived as that of induction which he describes as follows: "Induction refers to the actions that lead to discovery of a hypothesis – that is having a hunch or an idea, then converting it into a hypothesis" (Strauss 1987, 11).

Together with Corbin he later states that GT "is one that is inductively derived from the study of the phenomena it represents. That is, it is discovered, developed ... through systematic data collection and analysis of data pertaining to that phenomenon" (Strauss and Corbin 1990, 23).[195] The term induction is used also by K. Charmaz. Initially she introduced it with respect to theoretical sampling as follows: "Theoretical sampling means sampling aimed toward the development of the emerging theory. ... As an inductive technique, theoretical sampling exemplifies the inductive logic of the grounded theory approach" (Charmaz 1983, 124–125). Later, she generalizes her understanding of induction with respect to GT stating: "Essentially, grounded theory methods consist of systematic inductive guidelines for collecting and analyzing data to build middle-range theoretical frameworks that explain the collected data" (Charmaz 2000, 509).

In addition to the concept of induction the previously-mentioned authors bring in also the term "deduction." Already in *Discovery of Grounded Theory*, one can find the term "logico-deductive" process of arriving at theory (Glaser and Strauss 1967, 31), where deduction should start from a "grand" theory (e.g., of T. Parsons) and which they view as opposed to the method of theory-generation by induction from data. This opposition is stated by Charmaz as follows: "The grounded theory method stresses discovery and theory development rather than logical deductive reasoning which relies on prior theoretical frameworks" (Charmaz 1983, 110).

Logico-deductive reasoning was, as shown in chapter 1, viewed in the framework of the hypothetico-deductivistically oriented philosophy of science as the central method of construction of empirical theories. Glaser goes even further than this philosophy of science in his view on the logico-deductive method and differentiates between two of its varieties as follows (Glaser 1978, 40):

Deductive elaborating is vital to the theoretical sampling phase of grounded theory. This we call conceptual elaboration as opposed to the logical elaboration founded in deductive hypothesis testing research. Conceptual elaboration during theoretical sampling is the systematic deduction from the emerging theory of theoretical possibilities and probabilities for elaborating the theory as to explanations and interpretations. These become hypotheses which guide the researcher back to locations and comparative groups in the field to discover more ideas and connections from data. In contrast, logical deductions are a re-entry of the primarily deductive approach after a bit of grounded theory making.

---

195 But at the same time they emphasize that a "too rigid a conception of induction can lead to boring or sterile studies. Alas, grounded theory has been used as a justification for such studies. This has occurred as a result of the initial presentation of grounded theory in *Discovery* that has led to a persistent and unfortunate misunderstanding about what was being advocated. Because of the partly rhetorical purpose of that book and the authors' emphasis on the need for *grounded* theory, Glaser and Strauss overplayed the inductive aspects" (Strauss and Corbin 1994, 277).

This means that *deduction*₁ (= conceptual elaboration) should stand for theory elaboration by means of explanation and interpretation, while *deduction*₂ (= logical elaboration) should stand for research leading to hypotheses-testing, leaving thus the differences between those two types of deduction in the dark.

Finally, in addition to induction and deduction one can identify in the discussions of the GT method also the term "abduction" introduced by C. S. Peirce.[196] While Strauss relegates it to a footnote which runs as follows: "See the writings of Charles Peirce ... whose concept of abduction strongly emphasized the crucial role of experience in ... [the] first phase of research" (Strauss 1987, 12), K. Charmaz uses it in the following way (Charmaz 2006, 103–104):

The particular form of reasoning invoked in grounded theory makes it an abductive method, because grounded theory includes *reasoning* about experience for making theoretical conjectures and then checking them through further experience. Abductive reasoning about the data starts with the data and subsequently moves toward hypothesis formation. In brief, abductive inference entails considering all possible theoretical explanations for the data, forming hypotheses for each possible explanation, checking them empirically by examining data, and pursuing the most plausible explanation.

And in a note Charmaz adds that abductive reasoning "underlies the pragmatist tradition of problem solving" (Charmaz 2006, 122).

With respect to this chapter worth are mentioning Peirce's early views on abduction.[197] He claims: "We conclude the existence of a fact quite different from anything observed, from which, according to known laws, something observed would necessarily result" (Peirce 1960–1966, 2.636). By using his terminology of *rule* (law), (observed) *result*, and *case* (*fact*) together with his well known "bean" example one obtains the following reconstruction of the structure of abductive inference (Peirce 1960–1966, 2. 623):

Rule. – All the beans from this bag are white.
Result. – These beans are white
∴**Case**. – These beans are from this bag.

If one views, as does Peirce in his early period, the *result* as a *singular effect* of the *case* as its *singular cause*, then abduction is a *thought-movement from a singular effect to its singular cause* (Peirce 1960–1966, 2.636), *given a certain general theory*.

This central feature of Peirce's early understanding of abduction was reconstructed by Hilary Putnam as a case when a general theory is employed to solve

---

196   On Peirce's theory of abduction see, e.g., (Fann 1970) and (Anderson 1986).
197   At that time he labeled it "hypothesis."

the following scientific problem.[198] One has at his/her disposal both a *theory* and a *fact* (*result* in Peirce's terminology) to be explained, and the problem is to find some *auxiliary statements* (hereafter, AS_1, ..., AS_k, ..., AS_s) (*case* in Peirce's terminology), so that the latter, once conjoined with a theory, would yield an explanation of this fact. This situation can be expressed as shown in Figure 5.5 (Putnam 1974, 428):

Theory
??AS_1??, ...??AS_k??..., ??AS_s??
Fact to be explained

Figure 5.5 Putnam on abductive inference

Later I will partially draw on this figure in order to highlight the positive features of the GT method.

In order to understand what is the basis of the understanding of induction and deduction as given here, one has to turn to those social scientists who contrary to Glaser, Strauss, Corbin and Charmaz, make attempts at explicating the methods employed in the GT approach; here I mean especially J. Strübing and U. Kelle. Strübing reflects on the mutual relations of induction, deduction and abduction in the framework of a pragmatist logic of research and speaks about an "iterative-cyclic process of experimental testing, in which from qualitative inductions as well as from abductions are obtained *ad hoc* hypotheses, which are in the next step of the process related in a deductive movement again to data" (Strübing 2004, 46). The unity of these three types of reasoning Strübing represents as shown in Figure 5.6 (Strübing 2007b, 595).

Kelle pushes even further and makes an attempt to compare the GT method with the methodologies developed in the framework of the 20th century philosophies of science. In that comparison he complains that *the qualitative research methods developed in sociology lack a philosophy of science which would be as explicit, coherent and persuasive as the hypothetico-deductivistically oriented philosophies of science* (Kelle 1994, 13). In a next step, when dealing with the problem of the employment of the inductive method in the GT approach, he draws on K. R. Popper's critique of that method (Kelle 1994, 120–122) and evaluates on its basis inductive arguments as follows: "Inductive arguments can be used only for the generalization of singular phenomena, but they can produce no substantially new hypotheses" (Kelle 1994, 143).

---

198 Putnam does not, however, refer here to Peirce.

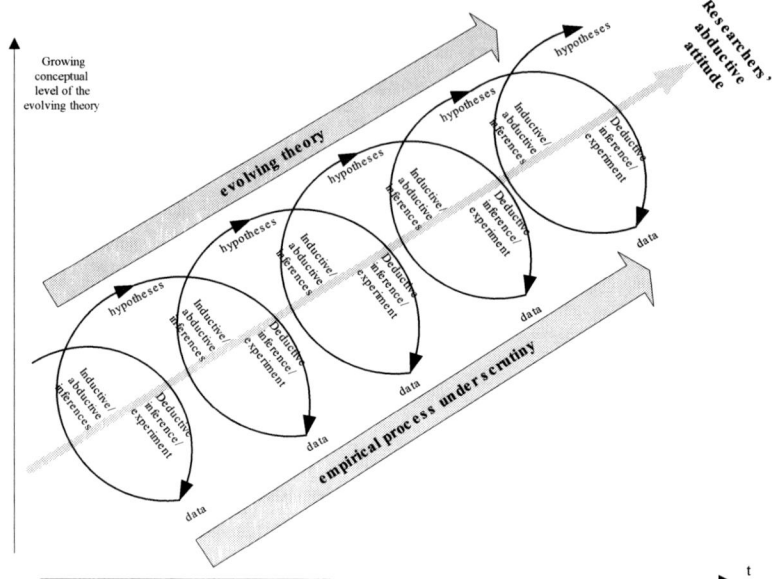

Figure 5.6 J. Strübing on the unity of induction, deduction and abduction in the pragmatist logic of research

But even if, based on Popper's critique, Kelle speaks about induction in general, it is readily seen that what he has in mind is *enumerative induction*. The latter is a type of inference that can be characterized as shown in Figure 5.7.

$P(a_1) \land Q(a_1)$

$P(a_2) \land Q(a_2)$

$\vdots \qquad \vdots$

$\underline{P(a_k) \land Q(a_k)}$

$\forall x\, [P(x) \to Q(x)]$

Figure 5.7 Scheme of enumerative induction

Here "a_1," ..., "a_k" stand for the names of individuals, "P" a "Q" for names of properties, "$\land$" for the sentential connective "and," "$\forall$" for the universal quantifier, "x" for the individual variable, and "$\to$" for the sentential connective "if ..., then ...". If x ranges over individuals a_1, ..., a_n and it holds that $k < n$, then one deals with *enumerative incomplete induction* and if $k = n$, then, with *enumerative complete induction*. The former is more relevant than the latter with respect to inferences employed in science, and it can be characterized by the following features:

- The conclusion, as compared to the premises, contains new information.
- Truth is not always "transferred" from the premises to the conclusion in the sense that there can be a case when the premises are true and the conclusion false.
- The conclusion pertains to all entities from a certain universe of discourse while all the premises together pertain only to a sample/subset of entities from the same universe of discourse, that is, the universe of discourse of the premises and conclusion is the same.
- Any expression appearing both in the premises and the conclusion has a stable, fixed meaning.
- In the conclusion, as compared to the premises, no new language expressions are given.

Deduction, on the other hand, can be viewed as a thought-sequence where from certain sentences we derive other sentences, so that between them holds the relation of *entailment*. A sentence W is entailed by a set M of sentences if and only if there cannot be the situation that all sentences from M are true and W would be false. Deduction can be characterized by the following features:

- The conclusion, as compared to the premises, does not contain new information.
- Truth is always "transferred" from the premises to the conclusion.
- Both the premises and the conclusion pertain to the same universe of discourse.
- Any expression appearing both in the premises and the conclusion has the same, fixed meaning.
- In the conclusion, as compared to the premises, no new language expressions are given.

As stated already, Kelle in his critique of the method of induction employed in the GT approach, draws on Popper's critique of enumerative incomplete induction but, surprisingly, without taking into account the *basis* of Popper's critique. This basis can be characterized as *nominalism*, that is, the view that the aim of scientific knowledge *is not the discovery of some pattern/mechanism underlying certain observable phenomena, but only to state thought-universals just as regularities in those* phenomena and where these *regularities are just our creations lacking, contrary to singular terms and statements, any counterpart in the external world.* So, for example in (Popper 1948), when dealing with the issue of causal explanation based on a *thought-universal* (universal scientific law) *u* and the singular statement *i*, together yielding the singular statement *e*, Popper unambiguously states that *e* and *i* have counterparts *E* and *I*, respectively, in the external world, but he does not at all address the issue if *u* has an *U* as a counterpart in the real world (Popper 1948, 145–146).

Popper also indicates that nominalism has certain *methodological* implications, namely, that it blocks all attempts to move in cognition to a thought-reconstruction of the functioning of the mechanisms/patterns underlying the observed data; those attempts he labels as *methodological essentialism* to which he opposes *methodological nominalism* which he holds to, claiming that "most people will admit that methodological nominalism has been victorious in natural sciences" (Popper 1964, 29). Even if he does not spell out in all details the methodological implications of his nominalism for his own understanding of induction as well as for the creation of a thought-universal, still the following can be stated. A thought-universal is, *according to Popper*, always just our conjecture based on particular instances, for example, in Figure 5.7 "$\forall x[P(x) \rightarrow Q(x)]$" is, for $k < n$, based on but *not induced* from "$P(a_1) \wedge Q(a_1)$," ..., "$P(a_k) \wedge Q(a_k)$,"; I label such thought-universals as the *enumerative type of thought-universal*.[199] At the same time one is not allowed to move from "$P(a_1) \wedge Q(a_1)$," ..., "$P(a_k) \wedge Q(a_k)$" to new terms with new meanings that would correspond to patterns/mechanisms underlying the observed instances expressed by those sentences and to a new universe that would differ from that to which individuals $a_1, ..., a_k$ belong.

Thus Popper's philosophical nominalism and methodological nominalism are the basis of the fact that Popper's philosophy of science[200] reflects on induction as the method of creation of a thought-universal only as enumerative incomplete induction and on thought-universals only as enumerative thought-universals.

Nominalism as a certain philosophical position, together with its methodological implications in the form of methodological nominalism are just "isms", that is, certain views and positions held inside philosophy and philosophy of science which can be confronted with other philosophical and philosophy-of-science "isms." But the inner-philosophical and inner-philosophy-of-science "isms" acquire a completely new basis once they are confronted with approaches to

---

199 What I label here as the "enumerative type of thought-universal" corresponds to what Popper labels in *Logic of Scientific Discovery* as "strictly universal statement" (Popper 1968, 62) or as "all-statement" (Popper 1968, 63). The thought-universal that is the result of enumerative complete induction is labeled by Popper as "numerically universal statement" (Popper 1968, 62).
200 Here I mean primarily his *Logic of Scientific Discovery*.

thought-universals as practiced not by philosophy and/or philosophy of science but by empirical science. Stated otherwise, *if one tries, as does Kelle, to evaluate and understand the inductive procedures practiced in a concrete empirical science by means of a certain understanding of induction as given in a certain (say, Popper's) philosophy of science, then the question one faces is whether the "isms" of that philosophy of science match at all the approach to thought-universals in that empirical science.* As I will show now, Popper's "isms" do not match GT's approach to thought-universals and, thus, one should not evaluate the methods of the GT approach by means of Popper's philosophy of science.

5.5 Grounded Theory's Feathers in the Cap: Cyclical Method, Unit Busting and the Nonenumerative Thought-Universal

The specific feature of the GT method is that it views the very process of theory construction not as a *one-directional* but, as shown already in the case of data collection and theoretical sampling, as a *cyclical process*; thus it clearly deviates from the views of Hempel and Carnap as presented in chapter 1. Therefore Jane Hood claims that theoretical sampling "really makes grounded theory special and it is the major strength of grounded theory because theoretical sampling allows you to tighten what I call the corkscrew or the hermeneutic spiral so that you end up with a theory that perfectly matches your data" (Charmaz 2006, 102).[201] Glaser can thus give an overall characterization of GT as follows: "The stages of grounded theory research may very well cycle in circles" (Glaser 1998, 15).

Now what are the distinctive features of the GT method from the point of view of *employed methods*? The distinctive feature of the GT method is, *first*, the choice of certain groups/subsets of universes of entities. Initially that choice is driven only by a general sociological perspective,[202] the concepts of which enable one to focus on the respective groups. Then the elements of the same group, as well elements from different groups, are mutually compared in order to find differences and similarities between them and in order to formulate categories and properties together with the structural conditions under which they vary. This enables one to develop both the *scope of the theory under construction and its generality*. In addition, according to Glaser and Strauss, (Glaser and Strauss 1967, 55–56)

---

201 This passage was edited by J. Hood and then quoted in (Charmaz 2006).
202 U. Kelle characterizes that general sociological perspective as that of social interactionism going back to the works of Thomas, Mead and Blumer (Kelle 1994, 308–309) and (Kelle 1996, 30–31).

comparison groups also provide simultaneous maximization or minimization of both the differences and the similarities of data that bear on the categories being studied. *This control over similarities and differences is vital* for discovering categories, and for developing and relating their theoretical properties, all necessary for the further development of an emergent theory. By maximizing or minimizing differences between concrete groups, the sociologist can control the theoretical relevance of his data collection. Comparing as many differences and similarities in data as possible ... tends to force the analyst to generate categories, their properties and the interrelations as he tries to understand his data.

However, Glaser and Strauss emphasize that the aim of theoretical sampling is not to fix some regularity universally given in the available data (1967, 30); this is blocked off because this they view as the point, as shown previously, when *theoretical saturation* is already achieved. Instead, approaching the data from various points of view[203] enables us, by grasping the various structural conditions of groups, to find out how the changes of these conditions change the respective properties, and, by means of this, find out the *mechanisms underlying these changes* (Glaser and Strauss 1967, 66), or as stated by Judith Holton, it should lead us to the discovery of a "latent pattern of behavior ... in the social setting under study" (Holton 2007, 244).

Drawing partially on figure 5.5 as well as on (Sintonen 2005), the GT type of thought-movement from slices of data to the underlying pattern/mechanisms delineated by concepts, categories, their properties core categories and theory, can be represented as shown in Figure 5.8.

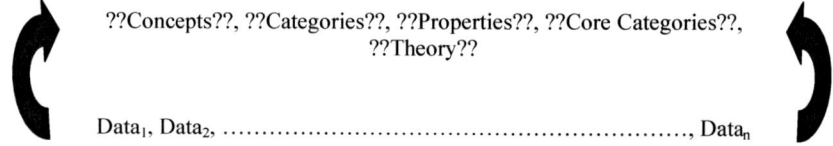

Figure 5.8 GT type of movement from slices of data to concepts, categories and so forth.

What has to be emphasized here is that the thought-movement represented in Figure 5.8 stands for an attempt to discover previously unknown mechanisms/patterns to which the respective data fit *qualitatively*. If the data really qualitatively fit the discovered mechanisms/patterns, then in the following type of thought-movement one tries, after stating the respective concepts, categories, properties and core categories, to *move back to the initially given data*, which now should be derived as fitting the respective underlying mechanisms/patterns. At the same

---

203 Glaser and Strauss label data approached from various points of views as "slices of data" (Glaser and Strauss 1967, 65).

time this type of thought-movement guides the research to the discovery of new, relevant data. This can be represented by Figure 5.9 below. Here "$??AS_1??$, ..., $??AS_k??$, ..., $??AS_s??$" stands for the as yet undiscovered missing links that should mediate between the concepts, categories, properties, and core categories on the one hand, and the data on the other. These conditions have the specific feature that they acquire their status – as missing links – only against the background of both the concepts, categories, and so on, *and* the data; they should *hook up*, so to speak, *both to the explanans and the explanandum*. Below I will explicate how the *derived* data differ from those that were the *point of departure* of the thought-movement *to* the patterns/mechanisms; here I symbolize this difference simply by adding the symbol "*."

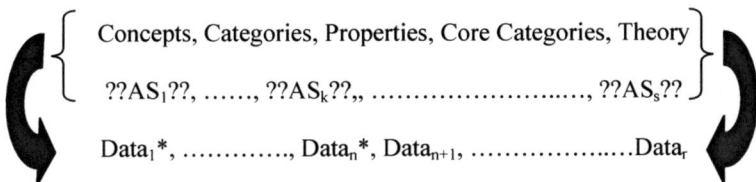

Figure 5.9 Derivation of data from concepts, categories, etc.

Let me first deal with the thought-movement to $Data_1^*$, ..., $Data_n^*$. To understand the nature of the missing link in the case of the derivation of these data let me consider the case mentioned already: A VIP is lying in the intensive care unit and several doctors are taking care of him/her. In another incident a lower class African American with a gaping knife wound is ignored on the city emergency ward. Based on the comparison of these discrete incidents the analyst can, according to Glaser and Strauss, generate the concepts *social loss* of a patient and *medical attention* and then state the hypothesis "Patients who have high social loss will receive better care than those who have low social loss" (Glaser and Strauss 1967, 24). From this last statement one should then be able to explain why VIPs will receive a higher degree of medical attention as compared to lower-class African Americans; the mediating link is here an auxiliary statement such as: "VIP patients have a higher social loss than lower-class African American patients." The data* derived then has the form : "VIPs receive a higher degree of medical attention as compared to lower class African Americans because VIP patients have higher social loss than lower-class African American patients." Thus in this derivation of the initially given data in the form of data* the crucial difference between *data* and *data** is that the term "high social loss" (and its meaning) is *shifted from the explanans to the explanandum*. Data* differs from data due to the fact that the latter, unlike the former, contains concepts, categories, and so on, which were initially derived from the data or, stated otherwise: *data* are the initial data but reinterpreted in the light of the discovered concepts, properties, categories, and so on*. Figure 5.9 has to be, with respect to the initially given data, modified as shown in Figure 5.10.

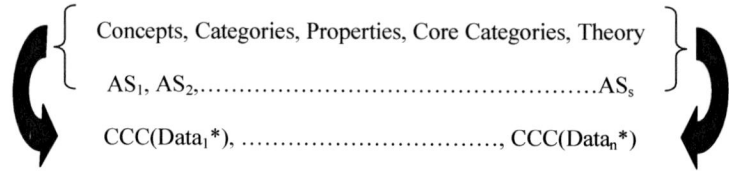

Figure 5.10 Derivation of data* from concepts, categories, etc.

Here "AS_1," "AS_2," ..., "AS_s" stand for auxiliary statements and "CCC(Data*)" for the reinterpretation of data by means of concepts, categories, core categories, and so forth.

Let me now deal with the problem of how it is possible to predict completely new data, that is, data not known before. A solution to this problem is given in A. L. Strauss's 1970 article,[204] and it comes very close to the method of gradual concretization presented in chapter 3. Strauss, initially, as an example, draws on F. Davis' 1961 article dealing with stages of management by person a visible disability of his/her strained interaction with a "normal" person (one who has no visible signs of disability). From the point of view of this chapter one has to mention that Davis's paper deals with:

(i) people with visible signs of disability
(ii) interaction between, at most, two persons with disabilities and, at most, two "normal" persons
(iii) interaction takes place in a "sociable" situation
(iv) the person with a disability tries to minimize his/her own disability
(v) the control of the interaction is vested completely in the hands of the person with the disability while the "normal" person either does not resist to this "game" or even does not recognize it
(vi) the person with the disability is quite trained in managing strained interactions with a "normal" person who is relatively inexperienced in this type of "game."

---

204 For a development of Strauss' views from this paper see (Gerson 1991).

This means that Davis's theory in fact involves the following *idealizations*:

(i*) there is no immediately nonvisible handicap
(ii*) the number of interacting persons never exceeds two on each side of the interaction
(iii*) there are no impersonal or intimate situations of interaction
(iv*) there are no attempts by the person with the disability to maximize his/her disability
(v*) there are no attempts by the "normal" person to seize control of the interaction
(vi*) the experience of the "normal" person in dealing with a person with a disability is, as compared to the experience of the person with a disability to deal with the "normal" person, very small.

Strauss makes now an interesting thought-move; he regards Davis's theory as a core of a theory that can be developed further by abolishing some of these idealizations. He considers, for example, the case of a person with a relatively invisible (but potentially visible) handicap interacting with a "normal" person, that is, he abolishes the first idealization and at the same time holds to the idealization (vi*): the person with the disability is still much more experienced than his/her "normal" counterpart. Yet another thought-case arises if the person with the disability is not experienced with his/her disability, for example, in the case of a woman only recently have ungone mastectomy. The lack of experience will have certain effects on his/her behavior, for example, if the woman wishes to keep her "disability" secret, she may put special emphasis on her choice of clothing. Strauss's considerations thus mean that the idealizations given earlier can be gradually abolished, which means that the *following conditions can be at work*:

(i**) a person with a nonvisible disability is involved in interaction
(ii**) the number of persons involved in interaction exceeds two on both sides
(iii**) interaction takes place in an impersonal or nonintimate situation[205]
(iv**) the visibly handicapped tries to maximize his/her disability
(v**) the "normal" person tries to seize control over the interaction
(vi**) the person with the disability and the "normal" person have approximately the same level of experience in strained interactions.

By taking into account these conditions it is possible to develop Davis's original theory so that (Strauss 1970, 49–50)

---

205 Here the sentential connective "or" stands for "either ..., or ..., but only one of them."

[t]his filling in what has been left out of the extant theory is a useful first step toward extending its scope. We have supplemented the original theory. ... Supplementation has led to the generation of additional categories. By ... [this] we have begun (in imagination) to sample theoretically; we could, in fact now either interview or seek existing data. ... A moment's reflection about those comparison groups of handicapped – visibly or invisibly – tells us that we have generated additional categories.

Now what happens with the structure of the original theory, understood now as containing the previously given idealizations, once the idealizations are gradually abolished? Let me represent the internal structure of the original theory with the idealizations (i*) through (vi*) as shown in Figure 5.11:

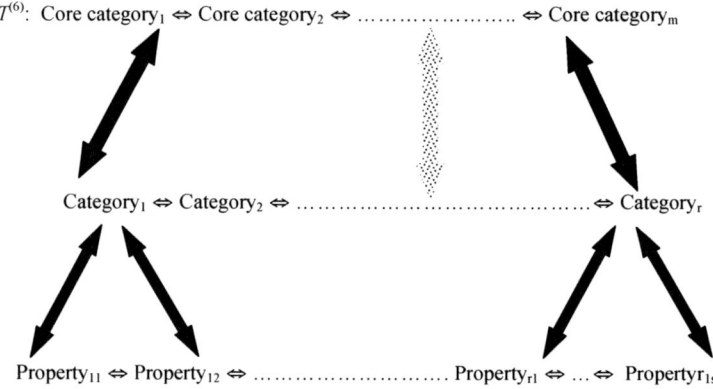

Figure 5.11 The internal structure of theory $T^{(6)}$

Here "$T^{(6)}$" stands for a theory involving six idealizations while all the arrows stand for mutual relations between core categories, categories and properties.

If one gradually abolishes the above-given idealizations (i*) through (vi*), that is, takes gradually into account the conditions (i**) through (vi**), then according to Strauss "we have generated additional categories. Possibly some may become core categories. ... If we pursue this analysis ... we can eventually develop testable hypothesis. ... The hypothesis are designed ... to add density of conceptual detail to our evolving theory" (Strauss 1970, 50). This means that for a theory that contains less than those six idealizations, say, five ($T^{(5)}$), one obtains the diagram for its internal structure shown in Figure 5.12:

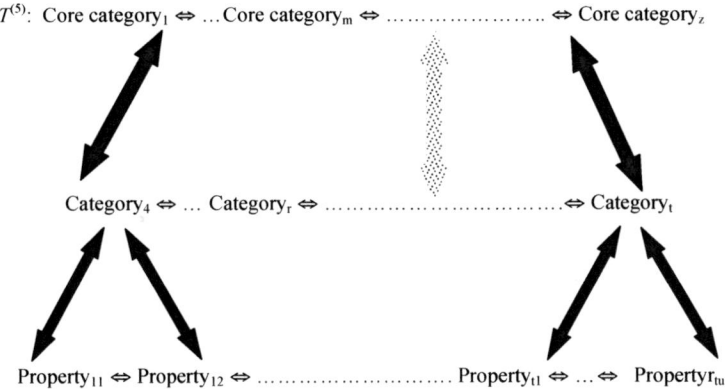

Figure 5.12 The internal structure of theory $T^{(5)}$

Here I have already taken into account the conclusion given earlier that the *introduction of new concepts and categories into an already given network of concepts and categories changes the internal structure of the latter*; this is my first explication of Strauss' phrase "to add density of conceptual detail to our evolving theory" (Strauss 1970, 50).

Based on what is represented in Figures 5.11 and 5.12, I can represent the mutual relations between theories symbolized there in a manner that corresponds to Figure 3.12 in chapter 3, as in Figure 5.13.[206]

Here "⊥" stands for an abolishment of an idealization and a simultaneous change of concepts, categories, properties and core categories; the lower indices indicate which idealizations have already been abolished, for example, $T_1^{(5)}$ indicates that the theory involves five idealizations and that in it the first idealization from $T^{(6)}$ has already been abolished. That change of concepts, categories, properties, and core categories has profound implications for the very meanings of the employed terms. For example the term "stages of management of strained interaction" introduced by F. Davis in his theory, here represented as $T^{(6)}$, changes its meaning inside theories $T_1^{(5)}$, ..., $T^{(0)}$ because in each of the theories this meaning is

---

[206] For the sake of simplicity I presuppose here that all idealizations can be abolished *independently of each other*. I presuppose also that the abolishment of the second idealization in $T_1^{(5)}$ and of the first idealization in $T_2^{(5)}$ yields the same theory, i.e., that $T_{1,2}^{(4)}$ and $T_{2,1}^{(4)}$ are equivalent (and the same should hold for $T_{5,6}^{(4)}$ and $T_{6,5}^{(4)}$, etc.). With respect to Figures 5.11 and 5.12, it is readily seen that such a "combinatorics" need not work in the case of interactions between humans in social groups; see (Abbott 1988).

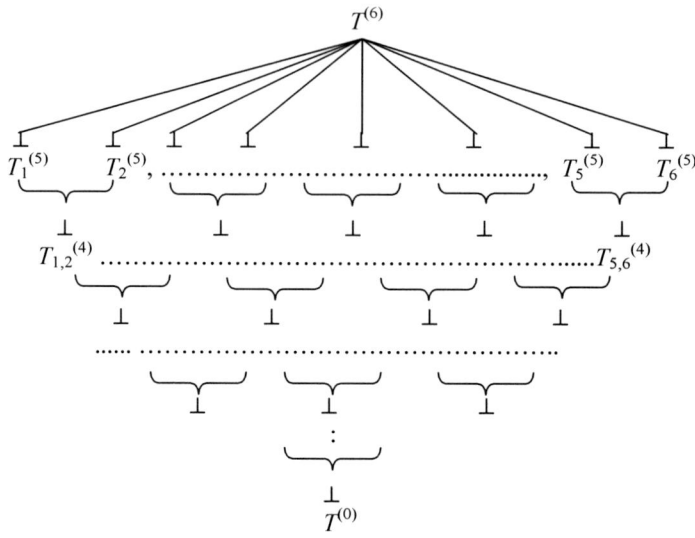

Figure 5.13 Network of theories obtained by abolishing idealizations

constructed by means of a different network of concepts, categories, properties, and core categories.

What conclusions can be drawn from the fact that a sequence of changes of meaning occur in the process of theory construction? First, it shows again that Carnap's and Hempel's view of theory construction has it wrong. So as *deduction* should have a central place in it, meaning should not change in that process. Second, it shows also that Strübing's representation of the logic of research in Figure 5.6 is wrong in an important aspect; *the road from hypothesis to data is not centered around deduction*. Finally, it shows that *Discovery of Grounded Theory* was written in a time when a philosophy of science was dominating in the field of social science which was attempting to base its views on methods of science primarily on a logical analysis of propositions of scientific theories while presupposing that these propositions do not change. This is readily seen in the following statement of Glaser and Strauss (Glaser and Strauss 1967, 32):

Our strategy of comparative analysis for generating theory puts a high emphasis on *theory as process*; that is, theory as an ever developing entity. ... The discussional form of formulating theory gives a feeling of "ever developing" to theory, allows it to become quite rich, complex, and dense. ... On the other hand, to state a theory in propositional form would make it less complex, dense and rich. ... It would also tend by implications to "freeze" the theory instead of giving the feeling of a need for continued development.

From this analysis I draw the conclusion that both in the process of theory construction, as well as in an already constructed theory, *meanings shift even if both stand for derivations of propositions from other propositions, that is, even if both have a propositional structure.* By means of this, I explicate B. Glaser's claim that "concept generation is a meaning making activity" (Glaser 1998, 140), as well as Strauss's and Corbin's claim that the GT approach enables "developing theory … with considerable meaning variation" (Strauss and Corbin 1994, 274). If one unifies this claim with my analysis given in chapter 3, then it becomes readily apparent that a change of meaning is not limited only to theories of natural science, for example, as shown earlier, to classical mechanics, but takes place also in GT as a social-science-type of theory. Thus, the *hypothetico-deductivist view of the method of theory construction holds neither for physical nor for social theories.*

Based on theories $T_1^{(5)}, \ldots, T^{(0)}$, one can start different theoretical samplings, different because they are driven by different theories and then, based on these samplings, predict the existence of previously unknown data that can then be looked for in the areas under investigation. I can thus add Figure 5.14, standing for the next stage of construction of a GT:[207]

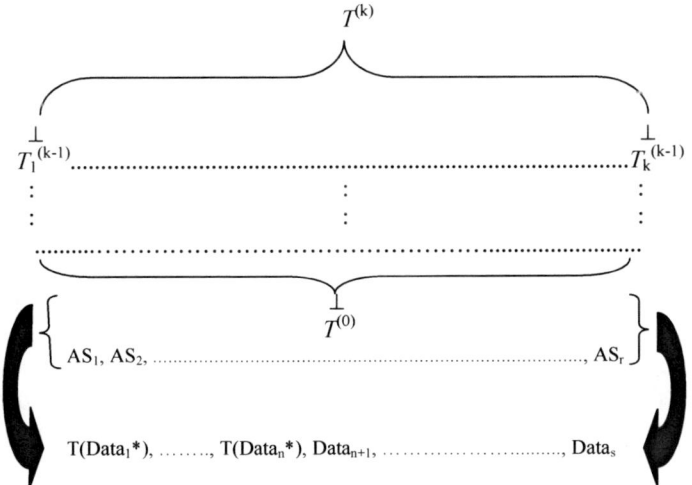

Figure 5.14 Derivation of data from a network of theories

---

[207] In order not to overburden the figure I presuppose that the auxiliary statements are conjoined only with the theory of type $T^{(0)}$. In respect to footnote 117 it holds that the respective auxiliary statements can be conjoined with any $T^{(j)}$ from that network, where $j > 0$.

Here each theory stands for a specific infrastructure of concepts, categories and core categories as indicated by Figure 5.11 for $T^{(6)}$ and Figure 5.12 for $T^{(5)}$; "T(Data*)" stand for the data from which the initial theory was derived (now understood as $T^{(k)}$), but now reinterpreted by means of concepts, categories and core categories of theories $T^{(k)}$, $T_1^{(k-1)}$, ..., $T^{(0)}$, while "Data$_{n+1}$," ..., "Data$_s$" stand for data predicted on the basis of $T^{(k)}$, $T_1^{(k-1)}$, ..., $T^{(0)}$ and unknown at the time of thought-movement from data to the initial theory.

Let me now return to the already above-mentioned example, used by Strauss and Corbin, of a network of terms for the name of the category color and let me compare it with the reflections on GT in this part of chapter 5. The network integrated around the category color as considered by them (Strauss and Corbin 1990, 70) can be represented as shown in Figure 5.15.

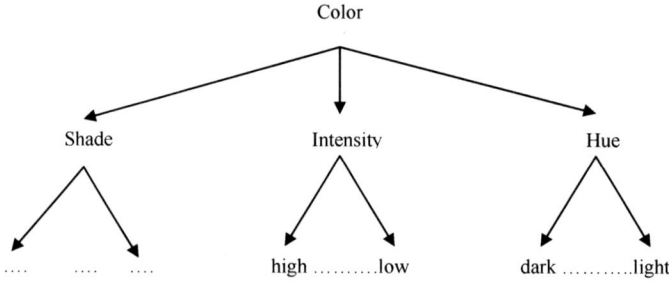

Figure 5.15 Network of concepts and properties for the category color

Strauss and Corbin use this network as an example of how one can classify certain entities (e.g., flowers). Whatever the merits of that network are, still this example can be, once compared with the GT method, quite misleading with respect to this method. What I have in mind is that one deals here with an already given set of terms and their meanings to the effect that the very nature of the GT approach to concepts, categories, and so on, gets here lost, namely, the movement and growth of knowledge from data to concepts, categories, and so on. If I take into account what was said previously and illustrated in diagrams above, then I have to "split up" Figure 5.15 as follows. I can, first, take into account, as an *imagined case*, the situation in which one has at his/her disposal only the data terms like "high," ..., "low"; "dark," ..., "light"; and so on, and on their bases one discovers the concepts corresponding to their meanings: *shade, intensity,* and *hue*. Thus, the thought-movement and growth of knowledge goes initially in the direction indicated by the arrows in Figure 5.16.

298

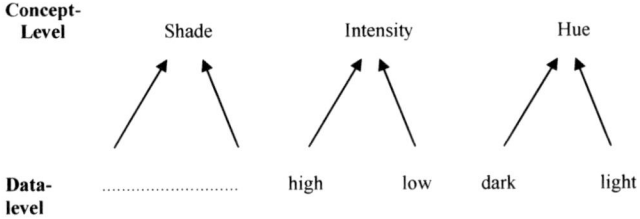

Figure 5.16 Thought-movement from data to concepts

What could happen, according to the GT approach, in a next step? According to Figure 5.11 I then could find out, say, with the help of abolishment of idealizations and auxiliary statements, that high and low are high and low *intensity*, that dark and light are dark and light *hue*, and so on. Thus we have a thought-movement and growth of knowledge going into the direction indicated by the arrows in Figure 5.17.

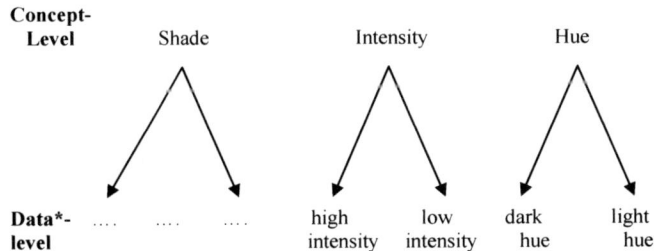

Figure 5.17 Thought-movement from concepts to data*

If now one compares Figure 5.17 with Figure 5.16, it becomes evident that once the antecedently given data are derived from their respective concepts, the names of the latter are shifted into names of the former, for example, instead of saying "*y* is dark," one has to say in the imagined case "*y* is of dark hue." Thus, due to a shift of concepts from the explanans to the explanandum, the *original meanings of the names referring to the initial data shift*.

What could follow next in this imagined case? Once the concepts of shade, intensity and hue are given, one could discover the category color. Thus we have thought-movement and growth of knowledge going as shown in Figure 5.18.

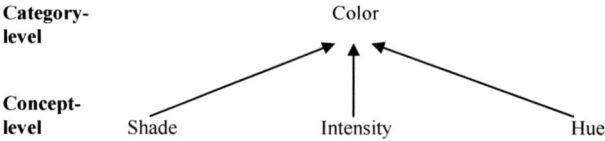

Figure 5.18 Thought-movement from concepts to category

One would then understand *shade, intensity,* and *hue* as *shade of color, intensity of color,* and *hue of color,* that is, one would face here a change of the *initial meanings of names of concepts.* Finally, due to these changes of meanings, one would obtain *yet another change in the meanings of names of data.*

That one faces here really a change of meaning of language expression can be shown by applying Transparent Intensional Logic to the expressions "*x* is dark," "*x* is of dark hue," and "*x* is of dark hue of color." Sentences of the form "X is F" are analyzed in Transparent Intensional Logic as $\lambda w \lambda t\ [^0F_{wt}\ ^0X]$. The first expression is then reconstructed as follows (here the expression "dark" names a property of individuals):

$\lambda w \lambda t\ [\lambda x\ [^0Dark_{wt}\ x]\ ]$.

The second expression is reconstructed as follows (in this expression "dark" names a function from properties to properties, that is, an object of the type $(\varphi\varphi)$, while "hue" names here a property of individuals):

$\lambda w \lambda t\ [\lambda x\ [^0Dark^{(\varphi\varphi)}\ ^0Hue]_{wt}\ x]$.

Finally, the third expression is reconstructed as follows (here "dark" and "hue" name functions from properties to properties, that is, objects of the type $(\varphi\varphi)$, while "color" names a property of individuals):

$\lambda w \lambda t\ [\lambda x\ [^0Dark^{(\varphi\varphi)}\ [^0Hue^{(\varphi\varphi)}\ ^0Color]\ ]_{wt}\ x]$.

Thus one faces in my reconstruction of color-example of Strauss and Corbin a sequence of meaning changes.

The "final" diagram in my imagined case is, therefore, as shown in Figure 5.19 (the number of stars indicates the number of reinterpretations of meaning the respective level underwent).

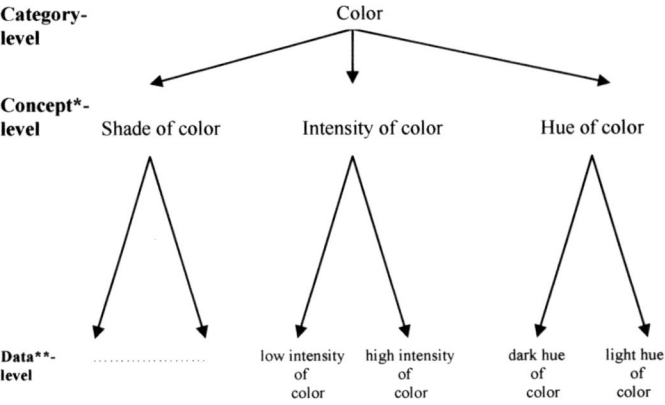

Figure 5.19 Thought-movement from categories to concepts* and data**

A *second* important feature of the GT method is a *continuous change in the universes of discourse*. By means of (i) method of theoretical sampling, driving the choice of comparison groups/subsets of universes, then (ii) via thought-operations varying the respective conditions for the elements of these groups and (iii) by going back to data collection and picking new relevant data, a gradual shift in the universes of discourse takes place. Glaser labels this shift as *unit busting*, stating (Glaser 1998, 160):

> Theoretical sampling is unit busting in two ways. Grounded theory generalizes to a conceptual unit which is a core category. It readily takes sampling outside the boundaries of the unit it may have begun in. Researchers typically like to start a study in a unit where a clear instant of their interest is located and with a tacit feeling of generalizing to the unit. By theoretical sampling this breaks this vision down relatively soon, as the researcher discovers his core variable. ... By constantly coding and analyzing the original unit is soon left behind as too descriptive as the researcher goes conceptual. ... Breaking the bound of a unit is part of the fracturing nature of grounded theory. One breaks up the unit conceptually as an area for a grounded theory. Later the theory can be applied to the original unit, as well as to others. By theoretical sampling the theory has left time and place, and expanded well beyond the limited properties of the original unit. Obviously, not going beyond the boundaries of a unit severely constrains the grounded theory and its completeness. The theory will be too thin and specific.

For a better understanding of the unit-busting nature of the GT method, let me turn again to (Strauss 1970). As shown already, from the initial universe set up by, at most two persons with nonvisible disabilities and, at most, two "normal" persons with their characteristics stated in conditions (i) through (vi), one shifts, by abolishing idealizations (i**) through (vi**), to other universes with persons with other characteristics, that is, theories $T^{(6)}$, $T_1^{(5)}$, ..., $T^{(0)}$ *differ mutually by their respective universes of discourse to which they pertain.*

The *third* important feature of the GT method is the feature where by the thought-movement from data to the underlying patterns/mechanisms, by the changes of universes of discourses, and by the collection of new data, *new terms are introduced referring to those patterns/mechanisms, universes and data.* This feature of the GT method I label *transdiction*,[208] which I characterize as follows. Transdiction is an operation by means of which one introduces names whose meanings enable, initially, the thought-movement from data for certain phenomena to the patterns/mechanisms underlying that phenomena, then, the application of the knowledge about those patterns/mechanisms in a reinterpretation of the knowledge about already known phenomena and, at the same time, its application to previously unknown phenomena which one fixes by turning to new data collection.

The *fourth* important feature of the GT method is that it is based on a type of thought-universal that differs from the enumerative type of thought-universal mentioned already above. From the point of view of *philosophy* and *philosophy of science* their difference can be spelled out as follows. The enumerative thought-universal, as it is just the conclusion of inference – where this conclusion pertains to all elements of a universe of discourse known already from the premises of that inference – is, paraphrasing Glaser (1998, 160), *conceptually thin and concrete.* The nonenumerative thought-universal aimed at and used in the GT approach is *conceptually dense and at the same time general.* This was expressed by Strauss and Corbin, when reflecting on the criteria of generality that a GT – pertaining to a phenomenon – should meet (Strauss and Corbin 1990, 23):

If the data upon which it is based are comprehensive and the interpretation conceptual and broad, then the theory should be abstract enough and include sufficient variation to make it applicable to a variety of contexts related to that phenomenon.

What are the *explanatory potentialities* given in the enumerative and nonenumerative types of thought-universal, respectively? The former, so as it is conceptually thin and concrete, can serve only as a base of explanations with the trivial structure shown in Figure 5.20.

---

208   This term was used in (Mandelbaum 1964); there, however, it is restricted to the movement from data to the underlying patterns/mechanisms where the latter are understood as "in principle unobservable" (Mandelbaum 1964, 63).

$$\forall x \, [P(x) \rightarrow Q(x)]$$
$$\underline{P(a)}$$
$$Q(a)$$

Figure 5.20 Structure of explanation based on an enumerative type of thought-universal

It is trivial, because it enables to derive *just one explanandum*. The nonenumerative type of thought-universal, on the other hand, can serve as the basis of explanations with a much more differentiated structure as shown previously in my reconstructions of the explanations of theories $T_1^{(5)}, \ldots, T^{(0)}$; here one derives a *whole variety of different explananda theories* each of which can, in turn, serve as an *explanans* for the derivation of *explananda data** and *explananda data*. Now one can readily see that the explanations of theories $T_1^{(5)}, \ldots, T^{(0)}$ stand for a *thought-movement from a conceptually dense and general thought-universal to conceptually denser and more general thought-universals*. This is my second explication of Strauss's phrase "to add density of conceptual detail to our evolving theory" (1970, 50).

Based on what has been stated up to this point, I can, finally, as an overview and summary of this chapter, compare deduction, enumerative incomplete induction and the GT method by means of table Table 5.3.[209]

Characteristics of	New information in conclusion as compared to premises	Change of meaning	"Transfer" of truth	Transdiction	Type of universal involved
Deductive argument	No	No	Always	No	?
Enumerative incomplete induction	Yes	No	Not always	No	Enumerative (concrete and thin)
GT-method	Yes	Yes	Not always	Yes	Nonenumerative (general and dense)

Table 5.3 Comparison of characteristics of deductive argument, enumerative incomplete induction, and GT method

---

209 I leave it open what type of thought-universal is involved in deductive arguments.

Chapter 6: Beyond the Qualitative-Quantitative Divide in the Social Sciences

The aim of this chapter is to deal with views that approach qualitative and qualitative research methods employed in social sciences as either *in principle separable* or, even, as *irreconcilable* methods of social sciences. I start with a characterization of these views and show how they deal with the various aspects or dimensions of the qualitative-quantitative divide, namely, the *technical, methodological*, and *epistemological*. Next I try to go to the roots of that divide by dealing with certain aspects of the works of Herbert Blumer, as well as with the views of Thomas P. Wilson from the 1970s and 1980s. Finally, I show, by means of an analysis of the *categories* of quantity, quality, and measure as well as of scientific law and scientific explanation, that the qualitative-quantitative divide is based on an incorrect, one-sided understanding of the qualitative and quantitative approaches which in fact can neither be separated nor be put into opposition.

6.1 The qualitative-quantitative divide

Thirty years ago Reichardt and Cook presented a summary, in the form of table (see Table 6.1), of the qualitative-quantitative divide, understood as a clash of paradigms, each characterized by the attributes shown in the table (Reichardt and Cook 1979, 10).[210]

Qualitative Paradigm	Quantitative Paradigm
Advocates the use of qualitative methods	Advocates the use of quantitative methods
Phenomenologism and verstehen; concerned with *understanding* human behavior from the actor's own frame of reference	Logical-positivism; seeks the *facts* or *causes* of social phenomena with little regard for the subjective states of individuals
Naturalistic and uncontrolled observation	Obtrusive and controlled measurement
Subjective	Objective

---

210 For another characterization of the qualitative-quantitative divide see, e.g., (Kusá 2005, 235–236).

Grounded, discovery-oriented, exploratory, expansionist, descriptive, and inductive	Ungrounded, verification-oriented, confirmatory, reductionist, inferential, and hypothetico-deductive
Process-oriented	Outcome-oriented.
Valid; "real", "rich" and "deep" data	Reliable; "hard" and replicable data
Ungeneralizable; single case studies	Generalizable; multiple case studies
Holistic	Particularistic
Assumes a dynamic reality	Assumes a stable reality

Table 6.1 Reichardt and Cook on attributes of the qualitative and quantitative paradigms

Here the divide is approached by means of a possible link between the respective method and the attributes of a paradigm. At the same time these authors provide the division of social-science *techniques/methods* into the qualitative and quantitative groups, as shown in Table 6.2 (Reichardt and Cook 1979, 7).

Qualitative methods/techniques	Quantitative methods/techniques
Ethnography, case studies, in-depth interviews, participant interviews	Randomized experiments, quasi-experiments, paper and pencil "objective" tests, multivariate statistical analyses, sample surveys

Table 6.2 Reichardt and Cook on methods/techniques of qualitative and quantitative research

A more detailed listing of the research methods/techniques is given by D. Silverman as shown in Table 6.3 (1998, 82, 84).

Qualitative Research		Quantitative Research	
Method	Features	Method	Features
Observation	Extended periods of contact	Social survey	Random samples Measured variables
Text and documents	Attention to organization and use of such material	Experiment	Experimental stimulus Control group not exposed to stimulus
Interviews	Relatively unstructured and open-ended	Official statistics	Analyses of previously collected data

| Audio and video-recording | Precise transcripts of naturally occurring interactions | Structured observation | Observation recorded on predetermined schedule |
| | | Content analysis | Predetermined categories used to count content of mass media products |

Table 6.3 Silverman on qualitative and quantitative research methods and their features

Another characterization of the nature of the qualitative-quantitative divide is presented by A. Bryman as shown in Table 6.4 (Bryman 1988, 94).

	Quantitative research	**Qualitative research**
Role of qualitative research	preparatory	means to exploration of actors' interpretations
Relationship between researcher and subject	distant	close
Researcher's stance in relation to subject	outsider	insider
Relationship between theory/concept and research	confirmation	emergent
Research strategy	structured	unstructured
Scope of findings	nomothetic	ideographic
Image of social reality	static and external to actor	processual and socially constructed by actor
Nature of data	hard, reliable	rich, deep

Table 6.4 Bryman on differences between quantitative and qualitative research

Based on such a characterization of the differences between qualitative and quantitative research, one can approach the very issue of the qualitative-quantitative divide in social sciences from two points of view. The first, *moderate*, according to which one deals only with two *different sets of techniques* that can be, if required, *mutually combined*,[211] and the second, *radical*, according to which the divide and the respective techniques/methods are rooted in *paradigmatically opposed epistemologies* and, thus, the respective methods/techniques cannot be combined at will.

---

211 A critical view on the possibility of a simple combination of qualitative and quantitative methods, while at the same time refusing to embrace the paradigmatic view, was stated by M. Hammersley in (Hammersley 1992; 1996). For his critique of "pure" qualitative and quantitative methods see (Hammersley 2007; 2008).

According to the moderate view (Henwood 1996, 28):

> the choice of numerical and nonnumerical methods is based primarily on pragmatic considerations. These include, for example, the scope for and constraints upon operationalizing particular "variables"; the availability of time and resources (for example, to conduct and analyze extensive interviews rather than using the more "pre-programmed" methods provided by questionnaires); and the compromises involved in making decisions about sampling.

According to the radical view, the quantitative methods/techniques have their intellectual underpinning in the positivist and realist paradigms, while the qualitative methods/techniques have their intellectual underpinning in interpretativist, constructivist, and naturalist[212] paradigms. The opposition between these paradigms was succinctly characterized by E. G. Guba as follows: "The one precludes the other just as surely as belief in a round world precludes believing in a flat one" (Guba 1987, 31). The opposition between these paradigms is then expressed as shown in Table 6.5 (Lincoln and Guba 1985, 37).

Axioms About	Positivist Paradigm	Naturalist Paradigm
The nature of reality	Reality is single, tangible, and fragmentable	Realities are multiple, constructed, and holistic
The relationship of the knower to the known	Knower and known are independent, a dualism	Knower and known are interactive, inseparable
The possibility of generalization	Time- and context-free generalizations (nomothetic statements) are possible	Only time- and context-bound working hypotheses (idiographic statements) are possible
The possibility of causal linkages	There are real causes, temporally precedent or simultaneous with their effects	All entities are in a state of mutual simultaneous shaping, so that it is impossible to distinguish causes from effects
The role of values	Inquiry is value-free	Inquiry is value-bound

Table 6.5 Contrasting of positivist's and naturalist's paradigms (Lincoln and Guba 1985, 37)

---

212 For the explication of the meaning of the term "naturalist" in this context see (Guba 1987), (Guba and Lincoln 1982) and (Lincoln and Guba 1985). For the various labels assigned to quantitative and qualitative approaches see (Bryman 1992).

In a similar vain, Jana Plichtová claims that the differences between the quantitative and qualitative research strategies are based on a paradigmatically different understanding of the subject matter of and the sense of cognition in the *very* social sciences, that is (Plichtová 2002, 9):

The quantitative approach starts from the premise that we can arrive at trustworthy knowledge only if the human being is reduced to a set of measurable variables between which we can presuppose the relations of causation. It sees the sense of cognition in prediction and control of human behavior.	The qualitative approach does not agree with that reduction because it degrades the human being to a reacting mechanism. It proposes such research strategies which respect the fact that the human being is an acting being pursuing certain intentions, that it is a semiotic being creating and understanding meanings, that it is a socio-cultural being whose adaptation has a mediated, social character.

In my view, as I will try to show, later, the premise stated by Plichtová is *not* an integral part of the *very* quantitative approach *inside* the social sciences but a part of *metareflections about the methods and categories employed in empirical science*. Such metareflections can, of course, have a negative feedback effect on the very aims and methods applied in the praxis of social sciences – *negative* in the sense that they can block off certain types of metareflections on those aims and methods, as well as the development and application of other, more suitable methods.

The standard characterizations of qualitative and quantitative approaches in social sciences I sum up in table 6.6.[213]

Approach	Techniques & methods	Intellectual underpinning	Methodology	Central category
Qualitative	Observation, use of text and documents, unstructured and open-ended interviews, transcripts of naturally occurring interactions by means of audio- and video-recording	Symbolic interactionism, phenomenology, naturalism, ethogenics, etc.	Idiographic based on understanding	Quality

---

213 The term "ethogenics" was coined by Harré and Secord in (Harré and Secord 1972).

Quantitative	Random sampling and introduction of measured variables in social survey, generation of stimuli in experiments, analyses of data collected in official statistics, structured observation	Positivism Realism	Explanation based on nomothetic statements	Quantity

Table 6.6 Standard characterization of qualitative and quantitative approaches

6.2 The qualitative-quantitative divide: Back to the roots

6.2.1 Herbert Blumer – symbolic interactionism and the logical premises of variable analysis

Keeping in mind the aims of this chapter, one can trace the roots of the qualitative-quantitative divide, at least partially, back to certain features of Blumer's work.[214] The following four features of that work are worth mentioning.

First, in his paper of 1940, he postulates the following difference between the natural and social sciences, namely, that different types of observations are applied to their phenomena (Blumer 1940, 714–715):

---

214 For a detailed analysis of the works of Blumer with respect to the qualitative approaches see (Hammersley 1989).

In the observation of human conduct one kind of item that the observer can detect and identify readily is what can be called the physical action – such as moving an arm, clenching the hand, running, cutting with a knife, and carrying some object. Such kinds of activities can be directly perceived and easily identified; designations or descriptive accounts of them can be readily verified. For, in the last analysis ... they can be translated into a space-time framework or brought inside what George Mead has called the touch-sight field.

However, there is another kind of item disclosed in observation of human behavior which is of a markedly different nature, as when we observe that a person is acting aggressively, or belligerently, or respectfully, or hatefully, or jealously, or kindly. This kind of activity cannot be reduced to a physical act or translated into a space-time framework and still retain the character suggested by the adverbs employed. It is such a kind of act which is genuinely social. ... The observation that detects such a kind of act is different from that which reveals the physical act, and, incidentally, is of a complicated nature. It is complicated in that it comes in the form of a judgment based on sensing the social relations of the situation in which the behavior occurs and on applying some social norm present in the experience of the observer.

Second, already in 1940, while dealing with the problem of the imprecise and vague nature of most concepts in social psychology, Blumer states, with respect to the method of operational definition of concepts, his first critique of the quantitative approach, this critique is as follows (Blumer 1940, 710–711):

This method, apparently, would confine the meaning of a concept to quantitative and mensurative data secured with reference to it. Prevailing concepts – or at least some of them – would be accepted; counting and measuring devices would be used in the case of each concept; the resulting information would constitute the content and meaning of the concept. ... However, critical consideration of this method should convince one that it does not offer a solution to the problem. ... The operational procedure ... could be successful in meeting the problem of vague concepts in social psychology only if the problems out of which the concepts arose and the items to which they refer were themselves essentially quantitative in nature ... unless it be shown that their nonquantitative aspects are spurious, the "operational" method is not a means of meeting the problem considered in this paper.

Third, when returning to the issue of vagueness of concepts, now in social science in general, Blumer differentiates between two types of concepts, namely, *definitive* and *sensitizing* ones (Blumer 1954, 7).[215]

---

215 In this paper I deal only with Blumer's approach to sensitizing concepts; for an analysis of his approach to definitive concepts see (Bulmer 1979).

A definitive concept refers precisely to what is common to a class of objects, by the aid of a clear definition in terms of attributes or fixed bench marks. This definition, or the bench marks, serve as a means of clearly identifying the individual instance of the class and the make-up of that instance that is covered by the concept.	A sensitizing concept lacks such specification of attributes or bench marks and consequently it does not enable the user to move directly to the instance and its relevant content. Instead, it gives the user a general sense of reference and guidance in approaching empirical instances.

At the same time he claims that while "definitive concepts provide prescriptions of what we see" (Blumer 1954, 7), sensitizing concepts "merely suggest directions along which to look. ... They lack precise reference and have no bench marks which allow a clean-cut identification of a specific instance and of its content. Instead, they rest on a general sense of what is relevant" (Blumer 1954, 7). And, as a continuation of his paper of 1940, he now clearly states that the *concepts employed in social sciences can be only of the sensitizing type* (Blumer 1954, 7–8).

I take it that the empirical world of our discipline is the natural social world of every-day experience. In this natural world every object of our consideration ... has a distinctive, particular or unique character and lies in a context of a similar distinctive character. I think that it is this distinctive character of the empirical instance and of its setting which explains why our concepts are sensitizing and not definitive. In handling an empirical instance of a concept for purposes of study or analysis we do not, and apparently cannot meaningfully, confine our consideration of it strictly to what is covered by the abstract reference of the concept. We do not cleave aside what gives each instance its peculiar character and restrict ourselves to what it has in common with the other instances in the class covered by the concept.

If one take also into account what Blumer states about definitive concepts, namely, that they enable us to describe what we see, and that such seeing is part of the observation of phenomena belonging to the framework of natural sciences (Blumer 1940, 714), then definitive concepts find their application exactly in that framework. Below, I will return to Blumer's characterization of sensitizing concepts and provide a critique of it.

Fourth, Blumer returns to the issue of applying quantification procedures in social science, now under the heading of "variable analysis," which he subjects to the following critique. He starts from one of the premises of symbolic interactionism to which he himself holds, namely, that human action and social life in human groups is based on a *continuous process of interpretation*,[216] and the latter "gives a character to human group life that seems to be at variance with the logical premises of variable analysis" (Blumer 1956, 685). That variance he then characterizes as follows (Blumer 1956, 687):

---

216   On this see (Blumer 1962; 1966; 1969).

Now the question arises, how can variable analysis include the process of interpretation? ... Interpretation is a formative process in its own right. It constructs meanings which ... are not predetermined or determined by the independent variable. If one accepts this fact and proposes to treat the act of interpretation as a formative process, then the question arises, how one is to characterize it as a variable. What quality is one to assign to it, what property or set of properties?

As an answer to that question he states that the process of interpretation cannot be characterized either by its result or by the presuppositions entering it as constitutive elements; the latter "vary from one instance of interpretation to another and, further, shift from point to point in the development of the act" (Blumer 1956, 687).

Finally, Blumer confronts the application of variables with the very fact that meanings shift in the process of interpretation. He claims that this (Blumer 1956, 687–688)

varying and shifting content offers no basis for making the act of interpretation into a variable. ... The question of how the act of interpretation can be given the qualitative constancy that is logically required in a variable has so far not been answered ... the need is to catch it as a variable, or set of variables, in a manner that reflects its functioning in transforming experience into activity. This is the problem, indeed dilemma, which confronts variable analysis in our field. I see no answer to it inside the logical framework of variable analysis.

Below I will return to this issue and show how one can unify meaning changes and quality-constancy with the introduction of variables.

6.2.2 Wilson on context, historicity, laws, and explanations

Thomas. P. Wilson presented in his 1982 article, his views on the qualitative-quantitative divide which are, in a certain aspect, path breaking with respect to that divide, but due to the fact that its English original (Wilson 1981) never appeared in print, those views are not widely known. My analysis here tries to remedy this.

Wilson characterizes the qualitative-quantitative as a clash between two opposing paradigms: *scientism* and *historism*. According to scientism, social sciences should be modeled so as to correspond as closely as possible to the natural sciences which extensively use quantitatively formulated scientific laws in explanations, according to historism, the principal aim is, not explanation on the basis of scientific laws, but the grasping of the historico-cultural contexts of social phenomena. Even if, according to Wilson, attempts are made to unify the qualitative and quantitative techniques/methods, still such unifications lack a "systematic theoretical foundation" (Wilson 192, 488; 1981, 3). Wilson thus sets for himself as the aim, "the development of a perspective in which ... the integrative basis (*Ansätze*) is developed systematically and not only from the points of view of utility" (1982, 490).

As a point of departure for such a development, he chooses the fact that social science deals with a world constituted by means of *situated actions* of members of society (Wilson 1982, 490; 1981, 5). In a next step Wilson states the following three features of situated actions (Wilson 1981, 7; 1982, 491):

1. *The objectivity of social structure.* Members of society tend to treat social categories, customs, norms, recurrent patterns of events and the like as existing "out there" and independently of any particular individual's action
2. *The transparency of displays.* Within a particular social group, it is in most instances plainly evident to the members what others are doing. ... Gestural and verbal displays ... are transparent in the sense that members can usually apprehend directly the concrete, situated actions being performed.
3. *The context-dependency of meaning.* The meaning of a gestural or verbal display depends on the context of its occurrence.

Based on such a characterization of the three features of situated action, Wilson views the extreme positions with respect to qualitative and quantitative approaches as the result of selective emphasis and at the same time neglect of certain features of situated actions (Wilson 1981, 7; 1982, 491):

The radical quantitative view focuses entirely on the experienced objectivity of social structure and transparency of displays while treating the context-dependency of meaning as merely a technical nuisance ... without theoretical or methodological importance.	The radical qualitative position emphasizes the context dependency of meaning but neglects the objectivity of social structure and the transparency of displays.

What scientism and historicism have in common is their lack of understanding for the nature of the context dependency of meaning, so that those who speak out in favor of purely quantitative methods completely underestimate the importance of that context, while those who speak out in favor of purely qualitative methods elevate the context dependency of meaning to the single over-riding principle (Wilson 1982, 492; 1981, 7–8).

Based on such an analysis, Wilson then derives a crucial methodological implication for the very qualitative-quantitative divide, namely, that one has to *abandon completely the utility of the nomothetic-ideographic distinction* for the research practice in social sciences because it is untenable with respect to the latter. He views, to be more specific, that distinction as a completely misleading opposition between the following two attempts: "the attempt to explain social phenomena as the result of transhistorical universally valid laws on the model of the natural sciences and the attempt to understand them in their concrete individuality and complexity" (Wilson 1982, 499; 1981, 20). And why is that opposition misleading? Because, according to Wilson, each of those opposites fails in the framework of social sciences on its own ambitions (Wilson 1981, 20–21; 1982, 499–500):

On the one hand, nomothetic explanation consists of showing that the facts to be explained can be deduced logically from a conjunction of universal laws and further facts about particular situations that are taken to be given for the purpose of explanation. For this kind of explanation to make sense, the terms appearing in the laws and descriptions of phenomena must have the same meaning no matter where or when they are applied, since otherwise one cannot claim to have used the same law to explain facts in different situations. However, we have seen that situated actions are reflectively tied to the social structural contexts within which they occur, and these contexts vary across cultural traditions and over historical periods. Consequently, the only possibility for genuine nomothetic explanation of social phenomena is to describe them in terms that are entirely independent of the meanings of situated actions, which would require abandoning most of the topics of interest to social science. Thus, the search for non-trivial, non-metaphorical trans-historical laws of social phenomena will be as barren in the future as it has been in the past.

On the other hand, the idea that one can ignore regularities in patterns of situated action is equally misleading. For, we have also seen that social structural categories reflecting trans-situational regularities enter into the constitution of the meanings of situated actions, so that one cannot make sense of what is going on here and now without reference to regularities in the social environment. Thus, understanding social phenomena cannot be limited to grasping complexes of meaning, even as the question of meaning cannot be dismissed as "metaphysical" or "subjective."

Below I will analyze Wilson's approach to scientific laws and explanations and provide a critique of it.

6.3 To new shores

I will now try to give a critique of the views of Blumer and Wilson, discussed earlier, by going beyond the standard approaches in at least three aspects of the qualitative-quantitative divide. First, I regard what is usually labeled as "theory" as a unity of *applied categories, applied methods*, and *employed theoretical concepts*. This unity enables one to deal with and solve empirical issues/problems stated in empirical concepts which then, in turn, lead to new empirical issues/problems.

Second, I try to approach the qualitative-quantitative divide by viewing quality and quantity as *categories* of thinking that are at work both in the natural and the social sciences. What categories are was explicated by me already in the introduction of this book.

It is surprising that even if, from all the dichotomies tormenting contemporary sociology,[217] only the qualitative-quantitative dichotomy *directly* suggests itself for a category analysis, the whole discussion on that dichotomy has until now completely bypassed such an analysis. I will approach those categories from the point of view of a *metacategory* level and will draw again primarily on Hegel's explications. This metacategory level, I view as an integral part of the paradigmatico-epistemological aspect of the qualitative-quantitative divide.

Third, as another integral part of the paradigmatico-epistemological aspect of the qualitative-quantitative divide, I regard the analysis of methods of explanation based on scientific laws; this analysis I view as part of what we label as *methodology*; here I will draw primarily on the explications given in subchapter 3.2.

Fourth, I unify the analysis at the level of methodology with that at the level of metacategories, so that together they set up what can be tentatively labeled as "metatheory." I use here quotations marks because the analysis at the levels of metacategories and methodology pertains not only to theories but also to empirical issues/problems.

My approach here can thus be represented by means of Figure 6.1.[218]

---

217 For in-depth analyses of these dichotomies see the articles in (Jenks 1998).
218 The line starting from **E** and leading, via theory, to **E'** stands for the theoretical solution of empirical issues/problems that leads to new empirical issues/problems.

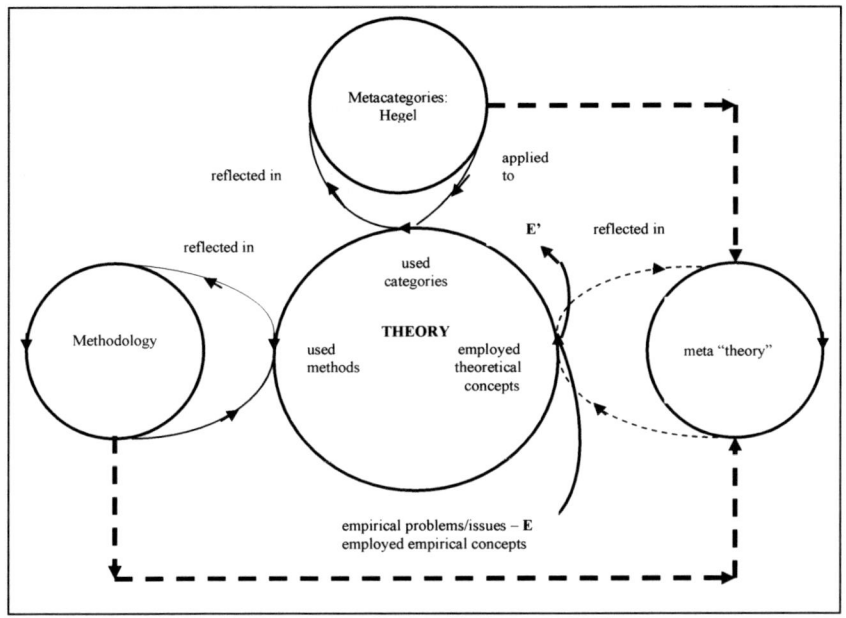

Figure 6.1 The relation of categories, applied methods, and empirical/theoretical concepts to metacategories, methodology and "metatheory"

6.3.1 Categories of quantity, quality, and measure: Howe and Hegel Versus Ellis

The characteristic of categories given in the introduction of this book, namely, that they unify in themselves an epistemic and ontological dimension, was used, even if in an unacknowledged way, by K. R. Howe in his argument against the qualitative-quantitative divide with respect to the data of social sciences. He claims that once data are viewed *via* the prism of the qualitative-quantitative divide, this divide acquires an *ontological* sense and a *measurement* sense, so that one obtains the fourfold division of data represented in Table 6.7 (Howe 1988, 11).

Data are \ Sense	Ontological	Measurement
Qualitative	if they are intentionalist, i.e., incorporate values, beliefs, and intentions	if they fit a categorical measurement scale
Quantitative	if they are nonintentionalist, i.e., they exclude values, beliefs, and intentions	if they fit an ordinal, interval, or ratio scale

Table 6.7 Howe on the ontological and measurement characteristics of data

This in turn means that one can deal in social sciences with four types of data as illustrated in Table 6.8 (Howe 1988, 11):

		Ontological	
M e a s u r e m.		Qualitative	Quantitative
	Quantitative	(I) e.g., cooperative/uncooperative	(III) e.g., greater/less than 12 years of school
	Qualitative	(II) e.g., critical thinking (on the Cornell)	(IV) e.g., income (in $)

Table 6.8 Howe on possible types of data

Based on Table 6.8, the view that qualitative and quantitative methods are paradigmatically incompatible means that the *locus* of this incompatibility has to be located in the mutual incompatibility of its rows, in the mutual incompatibility of its columns, and in its cells in the sense that the incompatibility is located in certain combinations of measurement and ontological types of data, that is, that certain cells cannot exist at all. Howe is, however, able to show that the incompatibility cannot be located in incompatibility of the rows because one can *combine variables to which different scales are assigned.* Neither can it be located in the mutual incompatibility of the columns because one can combine demographic variables like years of schooling with action variables like cooperativeness and critical thinking skills. Finally, it cannot be located in the cells because any of the four cells cannot be bared. This holds, for example, for cell (I) because one can move from speaking, for example, about the presence/absence of cooperativeness to speaking of cooperativeness, say, in terms of 0–100.

Howe's ability to unify the ontological and the measurement aspect with respect to the categories of quantity and quality brings us now to Hegel's reflections on the very *categories* of quantity and quality as given in his *Science of Logic*

(1969) and his *Lectures on Logic* of 1831 (2001). These reflections, even if dating back nearly two hundred years, still have a profound methodological importance for the qualitative-quantitative divide in the social sciences and especially for what Blumer has labeled as the "logical premises of variable analysis." This importance is given by the fact that Hegel, while analyzing the categories of quality and quantity, shows that in empirical (i.e., nonmathematical) sciences one deals always with their unity to which he assigns the category *measure*. Worth mentioning here is especially Hegel's reconstruction of the levels/stages through which the unity of quality and quantity in measure develops in empirical sciences.

For Hegel the point of departure is the immediate unity of quality and quantity in a certain entity to which one can assign what Hegel labels as "Größe" and what we translate as "*magnitude*"; the latter expresses the knowledge of that unity at the level of the category of measure. We thus represent a magnitude q assigned to an entity A by means of the ordered pair $<\{q\}, [q]>_A$. Here $\{q\}$ stands for the quantitative aspect of the magnitude q, while $[q]$, for its qualitative aspect.

A specific terminological difficulty one encounters when dealing with the categories of quantity, quality, and measure is the ambiguous meaning of the term "magnitude." Hegel removed this ambiguity at the level of metareflections on categories by distinguishing between magnitude as a "determined quantity" (*bestimmte Quantität*) (Hegel 2001, 125), which he regards as belonging to the category *quantity*, and magnitude as a unity of quantity and quality, which he regards as belonging to the category *measure*; "in measure magnitude is the quality itself" (Hegel 2001, 95). One is confronted with those two meanings of the term "magnitude" when dealing with the problem of how to grasp terminologically a concrete value/size of a magnitude as, for example, in the expression: "The length of the rod A is 20 meters." Here the numeral "20" expresses knowledge about a determined quantity; and that knowledge is part of the knowledge, expressed by the name "20 meters," about the unity of quality and quantity. In order to escape the meaning-ambiguity of the term "magnitude" we use the term "size/value of a magnitude," instead of the confusing English "magnitude of a magnitude" or German "Größe einer Größe."

In a second step, Hegel considers the situation in which the same magnitude is assigned to several entities, say A, B, C, and so on, and one tries to find out what happens – due to the interactions between them – with the size/value of this magnitude on the entities B, C, so on, if its size/value varies on entity A. This level of knowledge about measure we express as follows (here the sign "Δ" stands for variation and "$\rightarrow$" for a certain type of relation, say, direct or inverse proportionality):

$$<\Delta\{q\}, [q]>_A \rightarrow <\Delta\{q\}, [q]>_B. \qquad /1/$$

In the third step, Hegel considers a more general and complicated situation when one has a *prior* knowledge about several different magnitudes $q_1, q_2, ..., q_n$, that is, has also knowledge about different qualities and, *then*, one reflects on their possible mutual relations. This I express as follows:

$$<\{q_1\}, [q_1]> \rightarrow <\{q_2\}, [q_2]>. \qquad /2/$$

Finally, Hegel reflects on the situation when, based on *prior* knowledge about magnitudes $q_1, q_2, ..., q_n$, one introduces a *new*, that is, *previously not known* magnitude q_{n+1} and, thus, *discovers* at the same time a new *quality*. This situation I represent as follows (here " $=_{Df}$ " stands for definition):

$$<\{q_{n+1}\}, [q_{n+1}]> =_{Df} <\{q_1\}, [q_1]> \rightarrow <\{q_2\}, [q_2]> \rightarrow ... \rightarrow <\{q_n\}, [q_n]>. \qquad /3/$$

With respect to Hegel's views on quality, quantity, and measure, the following facts are worth mentioning. *First*, Hegel in his reflections on category of magnitude brings in the category of scientific law (Hegel 1928, 202–203; 1977, 163); a scientific law stands for the knowledge of the unity of quality and quantity, that is, he assigns the category of scientific law to that of the category of measure. Below, when dealing with Wilson's views, we will provide a more detailed reconstruction of the structure of scientific laws in which appear expressions with the structure of /1/ through /3/.

Second, for Hegel, magnitudes of empirical sciences stand for a type of knowledge which inseparably unifies the knowledge about quality and quantity. This aspect of Hegel's reflections on magnitudes inside framework of the category of measure can be made more understandable by comparing it with the views presented in Brian Ellis's 1960 article. Ellis sets at the very beginning of his article the *categorial framework* for his reflections on the issue of measurement (1960, 37–38):

Measurement must yield the measure of some quantity. But under what conditions can a thing be said to have or possess a quantity? It can be said ... that if some things have a quantity in common, whether it be length, area, mass, pressure, charge or temperature, then it must make sense to compare these things in respect of this quantity, and say whether one is greater than, equal to, or less than another in this respect ... Sometimes it makes sense to say that two things are equal to one another in some respect when it makes no sense to say that one is greater or less than another in this respect. ... An equality of this type may be described as a *qualitative equality*. ... if two things can be said to be equal in respect of some quantity, it must also make sense to say that one is greater or less than another in this respect.

This means that Ellis, contrary to Hegel, within the framework of the issue of measurement, puts the category of quality in irreconcilable opposition to the category of quantity so that they cannot be unified in some type of knowledge. From the point of view of Hegel, what Ellis labels as "quantity in common," is in fact a quality shared by several entities, that is, it is *one and the same quality common to all of them.*[219] And if that quality allows for what can be tentatively labeled as "more or less," then one can speak about a unity of quality and quantity in what Hegel labels, in the framework of the category of measure, as "magnitude." *Measurement is thus the determination of the value/size of a magnitude.*

The moral drawn from the analysis of Hegel's reflections and their comparison with the views of Ellis in (Ellis 1960) is that *all attempts to dismiss measurement in empirical science as a one-sided, purely quantitative enterprise are untenable.* Measurement is always based on the explicit knowledge of the qualities[220] of objects under investigation. Based on this we can now return to Blumer's understanding of what he labels as the "logic of variable analysis."

6.3.2 H. Blumer, symbol, meaning, and shifts of meaning

My aim here is, not to criticize Blumer's symbolic interactionism, but to provide an *alternative* to his understanding of the logical premises of variable analyses, so that this alternative and symbolic interactionism would fit each other.

Based on my analyses of the structure of magnitude it apparent that we have here always a *name* of a magnitude and the *meaning* expressed by this name, the latter being constituted by the *knowledge of quality and quantity*. This, with respect to this knowledge, means that Blumer's requirement that variables, in order to grasp the very nature of human interactions, would have to shift their meaning, can be taken into account by assigning to one and the same name of a magnitude ("variable" in Blumer's terminology) various meanings. Stated otherwise: *in the description of human interaction one can use in its various sections and subsections the same name of a magnitude (variable) but with different meanings, so that each of these meanings stands for knowledge about a certain fixed, constant quality.*

Based on Hegel's reflections on the category of measure it becomes apparent also that one can come to terms with the issue of meaning-shifts – with respect to magnitudes (variables) – by introducing in the course of descriptions *new*

---

219 On this see (Koslow 1982; 1992).
220 This knowledge of quality can pass, of course, as shown in chapter 3 and 4, through various levels.

symbols of magnitudes with new meanings. This, finally, enables one to come to terms also with Blumer's preference for sensitizing concepts in social science and his claim about the inapplicability of definitive concepts in it. Once we view the *description* of human interactions as going on in the way of a sequence of definitive expressions with shifting meanings, then one can apply such sequences to grasp particular happenings in the social world.

6.3.3 Laws and explanations: Universality, historicity, and shifts of meaning

Let me now deal with Wilson's claims about scientific laws and explanations. His criticism of attempts to employ in social sciences explanations based on scientific laws is correct if and only if one accepts his understanding of the nature of scientific explanations and of scientific laws employed in them. This understanding, with one exemption dealt with later, goes back to Hempel's understanding of explanation and of scientific laws and is *wrong in some crucial aspects, not only with respect to social science, but also, as shown in chapter 3, to natural science.*

First, Wilson, even if he speaks out correctly against the possibility of stating "transhistorical laws of social phenomena" (Wilson 1981, 21; 1982, 499), still wrongly *identifies transhistoricity of a scientific law with its universal validity* (Wilson 1981, 20; 1982, 499). So, for example, he claims that (Wilson 1982, 504):

> a physical law like Ohm's law is thought ... as universally valid, but only under certain restrictions with regard to temperature, voltage, etc. The claim about universal validity thus means that Ohm's law is valid independently of place and time only if these conditions are fulfilled,

and, as we will see, *falsely* assigns that identification to the very natural sciences, claiming that "rather than seeking to emulate the natural sciences for universal transhistorical concepts and laws, the task of sociology is the analysis of specific social structures and how they work within the context of their historical development" (Maynard and Wilson 1980, 311). Such identification seems to be quite common among social scientists.[221]

That wrong identification goes back, as shown in chapter 3, to Hempel's superficial reflections on the structure of scientific laws due to which he left out in his reconstruction one crucial aspect, namely, that *scientific laws are always*

---

[221] So, for example, K. Danziger claims, when dealing with the transformation of psychology into a science: "Such sacred and unquestioned emblems of scientific status included features like ... the search for universal (i.e., ahistorical) laws" (Danziger 1990, 120).

stated for entities of a certain kind. And it is not important whether those entities exist everywhere or only in certain regions of space and/or intervals of time or how many of those entities exist. Thus, *one should not make the law-like status of a universally valid statement dependent on the number and distribution of its instance in space and/or time*. For Ohm's law, for example, this means that its central aspect is that it can be applied for the purpose of explanation only and only where there are entities of a certain kind, namely, of the electric-conductor kind. At the very end of this subchapter, I provide a reconstruction of the structure of Ohm's law, which takes this already into account.

Second, Wilson presents us with a view of explanation that is Hempelian in its nature, that is, that the *explanandum* is logically deduced from the *explanans* (Wilson 1981, 20; 1982, 499), and that in the process of explanation the *very statement of law and its terms do not change their meaning*. At the very end of this chapter, I provide as an example also a reconstruction of the explanation procedures, based on Ohm's law, in the course of which the statement of this law and one of its terms, namely, "voltage" changes its meaning.

There is, however, one characteristic of scientific explanation mentioned by Wilson that goes well beyond what Hempel's D-N model of scientific explanation has achieved to reconstruct, namely, that the process of explanation stands in fact for the explanation of *different* situations (Wilson 1981, 20; 1982, 499). Hempel's D-N model is not capable of reconstructing this obvious characteristic of the explanation process in empirical science because Hempel, in his explication of the structure of scientific explanation in Part III of (Hempel and Oppenheim 1948), has chosen a simple model language without numerical expression. As a result of such a choice he was not able to reconstruct the structure of laws containing magnitudes in the sense explicated earlier. This forces him to collapse the structure of laws like such as, "Whenever a body falls freely from rest in a vacuum near the surface of the earth, the distance it covers in t seconds is $16t^2$ feet" (Hempel 1966, 54), into that of the conditional $(x)(Fx \rightarrow Gx)$. And conditionals with that structure force him to stick to the following reconstruction of the structure of explanation:

$$(x)(Fx \rightarrow Gx)$$
$$\underline{Fa}$$
$$Ga$$

that is, where only an explanation of just *one* situation is possible, for example, as in the case:

All ravens are black
This is a raven
This is black

6.4 Explanation Based on Ohm's Law

Drawing on subchapter 3.2, I reconstruct the structure of Ohm's law as follows:

$L^{(4)} : (x)\{Conx \:\&\: \Delta Tx = 0 \:\&\: ex = 0 \:\&\: Lx = 0 \:\&\: \frac{1}{Cx} = 0 \rightarrow Ex = RxIx\}.$

Here "*Con*" denotes an electric conductor; "*T*" denotes temperature; "*e*" denotes electromotive force; "*L*," self-inductance; "*C*," capacitance of a capacitor; "*E*," voltage; "*R*," resistance; and "*I*," current. Ohm's law as stated here contains four idealizations (therefore L with the superscript (4)): (i) deviation of the temperature of the electric conductor from a certain referential temperature – say, room temperature – is equal to zero, $\Delta T = 0$; (ii) electromotive force equals zero, $e = 0$; (iii) self-inductance equals zero, $L = 0$; and (iv) capacitance of the capacitor is infinite, $\frac{1}{C} 0$. Based on $L^{(4)}$, one can derive various laws by gradually abolishing the respective idealizations. For example, by abolishing the first idealization, so that now it holds that $\Delta T \neq 0$, one obtains:

$L^{(3)} : (x)\{Conx \:\&\: \Delta Tx \neq 0 \:\&\: ex = 0 \:\&\: Lx = 0 \:\&\: \frac{1}{Cx} = 0 \rightarrow Ex = Rx[k(Tx - T_ox) + 1]Ix\}.$

"*T*" stands here for the temperature of the conductor, "T_o" stands for the referential temperature, and "*k*" stands for a material constant. By abolishing the second idealizations one obtains:

$L^{(2)} : (x)\{Conx \:\&\: \Delta Tx \neq 0 \:\&\: ex \neq 0 \:\&\: Lx = 0 \:\&\: \frac{1}{Cx} = 0 \rightarrow Ex = Rx[k(Tx - T_ox) + 1]Ix + ex\}.$

And by abolishing the third and fourth idealizations, one obtains ("*Q*" stands here for the charge of the capacitor):

$L^{(0)} : (x)\{Conx \:\&\:\&\: \Delta Tx \neq 0 \:\&\: ex \neq 0 \:\&\: Lx \neq 0 \:\&\: \frac{Qx}{Cx} \neq 0 \rightarrow Ex = Rx[k(Tx - T_ox) + 1]xIx + ex + Lx + \frac{1}{Cx}\}.$

Based on such a reconstruction of the structure of Ohm's law and of explanations based on it, the following methodological conclusions can be stated. *First*, the crucial aspect of Ohm's law is that it can be applied *only and only where entities of a certain type exist*, namely, electric conductors, and, as stated earlier, the number and distribution of those entities in space-time are irrelevant for the law-like nature of statements with the structure $L^{(4)}$, $L^{(3)}$, $L^{(2)}$ and $L^{(0)}$. The latter hold universally for all entities of that type but they are *not* transhistorical; once there are worlds where are no electric conductors, laws $L^{(4)}$, $L^{(3)}$, $L^{(2)}$ and $L^{(0)}$ cannot be applied to them.

Second, by drawing on Transparent Intensional Logic, it could be proven, in the same manner as in chapter 3, that in the process of explanation by gradual concretization leading from $L^{(4)}$ through $L^{(3)}$ to $L^{(2)}$ and $L^{(0)}$, the meaning of the term "voltage," that is, the meaning of symbol "E" changes due to changes in the construction through which the meaning of the symbol "E" is given in the respective statements of Ohm's law. Here again, as in the case of explanation by gradual concretization dealt with in chapters 3 and 5, it holds that *in explanation by gradual concretization, the meaning of the expressions given in the statements of the scientific law undergoes a sequence of gradual shifts*. Hempel's D-N model cannot grasp these shifts because it reconstructs scientific explanation as a simple procedure of substituting into the *explanans*-law the respective *singular* (initial and/or boundary) conditions. By such substitution the scientific law "disappears," so to speak, because it is transformed into the *explanandum*-phenomenon. It, thus, *seems* that in the process of explanation, the statement of the *explanans*-law and the terms given in it have fixed meanings with respect to the explanations of different situations given in the *explanandum*. Contrary to this, the earlier given reconstruction of explanation procedure by gradual concretization shows that it brings about, in an irreducible manner, an *approach of the law-statement given in the explanans to what is stated in the explanandum*.

Third, explanation procedure by gradual concretization enable, as shown in chapters 3 and 5, the derivation of different *explananda*. So, for example, on the basis of $L^{(4)}$ one can derive not only laws $L^{(3)}$, $L^{(2)}$ and $L^{(0)}$, but – by gradually abolishing in a different order, as compared to that given earlier – the four idealizations in $L^{(4)}$, one obtains even more different *explananda*-laws.

Chapter 7: Historical Sociology versus Rational Choice Theory: The Adventures of Nominalism

7.1 Introduction

The aim of this chapter is to provide – from the point of view of methodology – a comparative analysis of Historical Sociology (hereafter, HS) and of Rational Choice Theory (hereafter, RCT) as two important approaches in contemporary sociology. This analysis can be regarded at the same time as an attempt to bring to a conclusion the discussion between the proponents of HS (Quadagno and Knapp 1992; Skocpol 1994a; Somers 1998) and of RCT (Kiser and Hechter 1991; 1998).[222] The standard differentiation between HS and RTC can be represented as shown in Figure 7.1.[223]

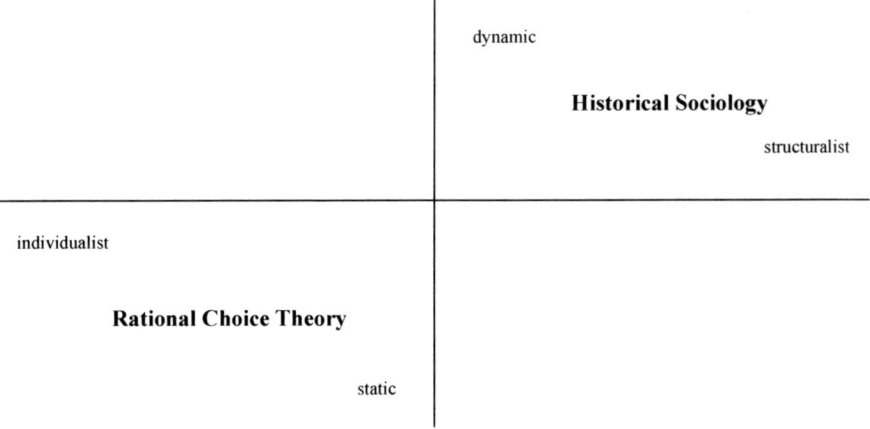

Figure 7.1 Two dimensions by which RCT and HS are delineated

RCT is delineated by an individualist strategy combined with a static approach, while HS takes a dynamic and structuralist perspective.

I start with a brief characterization of HS, next I analyze RCT, and then I bring in the main concepts of HS such as path-dependence, narrative, and conjuncture.

---

222 See also the articles (Goldstone 1998) and (Calhoun 1998), which contributed to that discussion.
223 Here I draw on (Swaan 1996).

Finally I make an attempt at an integration of these concepts into a unified system enabling us to bring to the surface the more deeply rooted differences between RCT and HS.

7.2 Historical Sociology

Historical Sociology as a direction in the craft of sociology can be characterized by the following features (Skocpol 1984a, 1–2):

1. It asks questions about social structures or processes understood to be concretely situated in time and space.
2. It addresses processes over time, and takes temporal sequences seriously in accounting for outcomes.
3. It attends to the interplay of meaningful actions and structural contexts, in order to make sense of the unfolding of unintended as well as intended outcomes in individual lives and social transformations.
4. It highlights the particular and varying features of specific kinds of social structures and patterns of change. Intrinsically of interest are, along with temporal processes and contexts, social and cultural differences. World's past is not seen as a unified developmental story or as a set of standardized sequences. Instead, it is understood that groups or organizations have chosen, or stumbled into, varying paths in the past. Earlier "choices," in turn, both limit and open up, alternative possibilities for further change, leading toward no predetermined end.

The six main ideas of HS are then as follows (Spohn 1996, 363):

(i) the unity, based on logic science, of sociology and history;
(ii) the methodological mediation of past and present;
(iii) the linking, based on social theory, of process, structure, and action;
(iv) the analytical connection of the sociohistoric micro- and macro-levels;
(v) the methodological importance of comparison;
(vi) the theoretical appropriateness of methodical procedures.

Theda Skocpol in her path breaking book *States and Social Revolutions* characterizes social revolutions as (Skocpol 1979, 4):

rapid, basic transformations of a society's state and class structures. ... Social revolutions are set apart from other sorts of conflicts and transformative processes above all by the combination of two coincidences: the coincidence of societal structural change with class upheaval; and the coincidence of political with social transformation.

Based on this understanding of the term "social revolution" Skocpol can approach the French revolution (1789), the Chinese revolution (1911–1949), and the Russian revolution (1917–1921) as "three comparable instances of a single coherent social-revolutionary pattern" (Skocpol 1979, xi). Based on this, she then views the task of a comparative historian "in establishing the interest and prima facie validity of an overall argument about causal regularities across the various historical cases" (Skocpol 1979, xiv).

Her own conceptual approach to social revolutions she relates to a (Skocpol 1979, 5–6)[224]

structural perspective, with a special attention devoted to international contexts and to developments at home and abroad that effect the breakdown of the state organizations of old regimes and the buildup of new, revolutionary state organizations. ... Comparative historical analysis is the most appropriate way to develop explanations of revolutions that are at once historically grounded and generalizable beyond unique cases.

The logic of comparative history, Skocpol employs in her 1979 work is characterized by her as "macrosocial analysis" and is based on manipulations "of groups of cases to control sources of variation in order to make causal inferences when ... data are available about a ... number of cases" (Skocpol and Somers 1980, 182).

An inquiry in the framework of comparative history is characterized by Skocpol et al. as embodying the following three features (Mahoney and Rueschemeyer 2003, 11–13):

(a) it focuses on the identification and explanation of causes producing "major outcomes of interest ... the causal argument is central to the analysis";
(b) it analyzes historical sequences and the unfolding of processes over time; and
(c) it stands for a contextualized comparison of cases both similar and contrasting, small in number.

James Mahoney distinguishes in comparative HS two basic strategies of *causal analysis*, across-cases comparison and within-case comparison while defining them along the dimensions of level (type) of measurement and level of aggregation; they can be represented as shown in Table 7.1 (Mahoney 2003, 338).

---

[224] This structural perspective she labels also a "nonvoluntaristic" one (Skocpol 1979, 14).

		Level of Measurement		
		Nominal	Ordinal	Interval
Level of aggregation	Aggregated	Nominal strategy	Ordinal strategy	Not typically used
	Disaggregated		Within-case analysis	

Table 7.1 Mahoney on strategies of causal inference

Nominal and ordinal strategies differ in the employment of the type of measurement;[225] the former makes comparisons based on nominal measurement (i.e., the use of mutually exclusive categories), and the latter employs ordinal measurement (rank ordering of cases using variables with at least three values based on the degree to which the phenomenon under investigation is present). What still unifies these two strategies is that they make comparisons across highly aggregated units.

In addition to cross-case comparison, HS engages also in *within-case* analyses of processes drawn from within particular cases. These analyses can be characterized by the following three "techniques" (Mahoney 2003, 360–365):

1. *pattern matching*: the technique of testing hypotheses initially derived from cross-case comparison by confronting them with observations from within specific cases;
2. *process tracing*: the technique of locating the causal mechanism linking a hypothesized explanatory variable to an outcome;
3. *causal narrative*: the technique of "aggregated cross-case associations by 'braking apart' variables into constituent sequences of disaggregated events and comparing these disaggregated sequences across cases."

The concept of *causal narrative* appears also as one of the concepts introduced by M. Somers in order to give "an answer to epistemology's central methodological question: How does a set of hypotheses, if true, explain why something happened" (Somers 1998, 736). The answer to this question is, according to Somers, it should be a *causal* type of explanation that is *temporally* and *narratively* constructed and depicts what happened as a *contingent, path-dependent process, as driven by a spatio-temporal mechanism* but *not as a result of the operation of a universal law* (Somers 1998, 737).

---

225 On these strategies see also (Mahoney 1999).

Skocpol explicitly links her macrosocial comparative approach to J. S. Mill's method of agreement and method of difference (Skocpol 1979, 33–40; Skocpol and Somers 1980, 183).[226] Because of this linking and because J. S. Mill viewed these methods as *inductive* methods, L. J. Griffin claims that "Skocpol's generalizations are procedurally based on inductive logical comparisons" (Griffin 1992, 425) and that her generalizations about the causes of both the presence of social revolutions in France, Russia, and China, and their absence in England in 1640–1688, in Germany in 1807–1814 and 1848–1849, and Japan in 1868–1873, are "inductively derived" (Griffin 1992, 409).

Does such a characterization of the methods employed in Skocpol's work 1979 really hold? To answer this question let me take a closer look at Griffin's explication of the research-explanation methods employed in (Skocpol 1979). He shows that from a set of countries that underwent important social transformations, only three – France, Russia, and China – underwent a change corresponding to the concept of social revolution delineated earlier. The data pertaining to those three countries are then subject to the method of agreement and method of difference, and the cause of revolution they underwent is located in "the conjuncture of international crisis and the adaptive constraints induced by their agrarian class structures and political institutions" (Griffin 1992, 409). In the next and final step, the particular cases of social revolution are explained on the basis of the generalization about the causes of revolutions and where this generalization covers these three cases.

Even with this brief overview of how Griffin characterizes the procedures employed in Skocpol's *State and Social Revolutions*, it becomes readily apparent that they contain at least two components that have to be given in advance in order to provide the explanation and are thus not the result of the inductive procedures employed here. *First*, in order to choose France, Russia, and China as the "correct" cases, one has to have in advance the definition of the term "social revolution" as well as the set of theories/hypotheses enabling one to provide this definition. Only if one accepts in advance Skocpol's definition of this term (and the hypotheses/theories behind it), one can choose those three countries as objects of comparative analyses; that is, this definition guides the selection of the "correct" cases.

Second, in order to perform the research procedures in (Skocpol 1979), one has to have also a prior knowledge of the methods of agreement and difference. The sequence of procedures employed in (Skocpol 1979) I thus represent as shown in Figure 7.2.

---

226 For a critique of this link see (Burawoy 1989, 765–768), (Paige 1999, 790–791), (Lieberson 1991; 1994), and (Goldstone 1997). For its defense see (Salvolainen 1994).

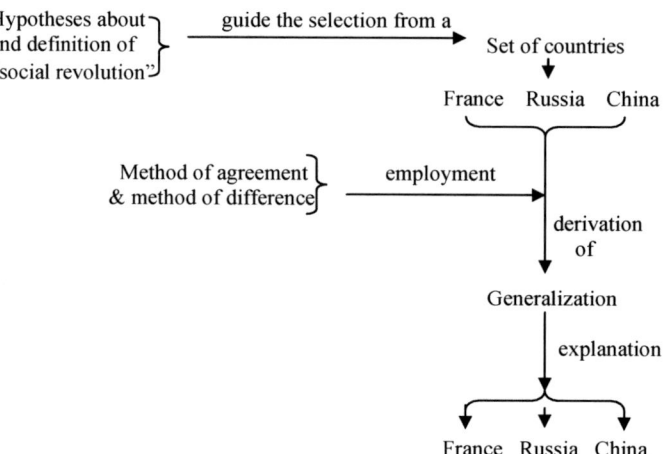

Figure 7.2 Sequence of procedures employed in T. Skocpol's **States and Social Revolutions**

Finally, let me show that Skocpol's approach to HS displays, from the point of view of methodology of social sciences, a highly problematic feature due to which HS, as given in (Skocpol 1979), as we will show later, comes in a certain aspect very close to RCT. This feature comes to the surface when one deals with what Skocpol labels as "comparative historical analysis" (hereafter, CHA). The latter should denote "the most appropriate way to develop explanations of revolutions that are at once historically grounded and generalizable beyond unique cases" (Skocpol 1979, 5–6). CHA, she claims, "is distinctively appropriate for developing explanations of macro-historical phenomena of which there are inherently only a few cases. … [It] is, in fact, the mode of multivariate analysis to which one resorts when there are too many variables and not enough cases" (Skocpol 1979, 36). And it "works best when applied to a set of a few cases that share certain basic features" (Skocpol 1979, 39). So, with respect to attempts to explain social revolutions, it holds that one has "to find important regularities across given historical instances – including similar institutional and historical patterns in the situations where revolutions have occurred and similar patterns of conflict in the process by which they have developed" (Skocpol 1979, 18). Thus, she identifies a pattern with certain features holding for each case in the specified set of cases, and, with respect to the application of CHA to social revolutions in France, Russia, and China, this means that she (at least implicitly) assumes that they "are in fact a uniform class of objects governed by identical causal laws" (Sewell 1996b,

258). But then she faces the following crucial question: *Why is that pattern uniformly present in that set?* – or to put it more specifically – *Why was the pattern of social revolution at work in France of the 18th century also at work in Russia and China in the 20th century? And by what mechanisms was this pattern reproduced in three different space-time locations, namely France, Russia, and China?* One possible answer to these questions is that there is no such mechanism at work, and it is just the historical sociologist's endeavour that creates this uniformity; there is no real mechanism at work, and only due to historical sociologist's endeavour that creates this uniformity; there is no real mechanism at work and only due to his or her conceptual definitions does one regard the France's, Russia's and China's societal transformations as elements of the same class of objects. To this alternative, which I label here *nominalism*, holds Skocpol in her *States and Social Revolutions*, because she provides no explanation why that pattern is reproduced in those three locations. Then, of course, she rightly becomes subject to the charge of "*freezing history*" (Burawoy 1989, 769) and of "fracturing the congealed block of historical time into artificially interchangeable units" (Sewell 1996b, 258). Stated in the terminology of HS, this means that once she does not explain why the same social causal law is permanently at work, she can be criticized for not taking into account that "new classes and new class relations arise over time" (Sewell 1996b, 258) and, thus, not taking into account also that this "might well alter the conditions necessary and sufficient for social revolution" (Sewell 1996b, 258).

The other option is that once we have unified these countries into one set, we start looking for the mechanism of (re)production due to which they are objects of one class. This alternative I label *realism*; I will explicate this option later in subchapter 7.5.

7.3 Rational Choice Theory

RCT is, according to its proponents, based on the following five assumptions:[227]

1. Rational[228] action involves utility maximization, i.e., the actor performing it picks out, once confronted with several options, the one he/she believes best serves his/her objectives.
2. Two consistency requirements have to be part of the definition of rationality: (i) all options available to an agent can be rank-ordered, and (ii) preference orderings are transitive.

---

227 Here I draw on (Green and Shapiro 1994, 14–17).
228 On the assumption of rationality see, e.g., (MacDonald 2003, 552).

3. Each individual maximizes the expected utility of his/her payoff, measured on a utility scale.
4. The maximizing agents are always individuals.
5. RCT models are applicable to all persons under study.

These assumptions figured prominently in the discussion that ensued after the publication of the article (Kiser and Hechter 1991)[229] which targeted the practice of comparative HS. That practice, according to Kiser and Hechter, attacks the role of general theory and instead of explaining events on the basis of causal laws that are, in turn, based on the descriptions of mechanisms imputed from a general theory, relies on the method of induction and Mill's methods of agreement and difference.

Worth mentioning, from the point of view of this chapter, are the views of Kiser and Hechter on causal laws and mechanisms. Causal laws, they claim, are either *universal* or at least *conditionally universal*[230] and at the same time are *omnitemporal* statements expressing a certain regularity. The description of a mechanism enables one to relate conceptually the cause and its effects (expressed by means of variables) and can be derived, according to Kiser and Hechter, by two different ways. One stands for a *direct* derivation as given by the derivation of micro-properties from macro-properties, for example, in physics;[231] the second, practiced in the social sciences (due to their inexactness and because social mechanisms are unobservable), stands for an imputation from a general theory of the description of the mechanism (Kiser and Hechter 1991, 19; 1998, 786).

Based on those views, Somers states the following profound criticism of Kiser and Hechter (Somers 1998, 751–752):

[The] journey ... began with a call for laws (constant conjunction), argued it necessary also to find deep causes (causal mechanisms), and, having faced the problem of the unobservability of mechanisms, retreated back to the laws and from there elevated to "general theory" ... what we have now learned is that they are also the very same mechanisms that Kiser and Hechter tell us are the central components of all adequate explanation. ... This is remarkable: The very causal mechanisms required for an adequate explanation of a problem at hand are to be inferred from a general theory composed about the assumptions about *those same causal mechanisms*. ... Kiser and Hechter have taken us on a dizzying circular trip: they have told us that something called general theory can give us the causal essence of a phenomenon, and from this essence they are handily able to explain a social problem by imputing it to the same causal property that already lies at the definitional core of the general theory.

---

229 For reactions to this paper see (Skocpol 1994a) and (Quadagno and Knapp 1992).
230 This means that the scope conditions under which it holds are specified in the law (Kiser and Hechter 1991, 6).
231 Here Kiser and Hechter refer to (Salmon 1984).

This, thus, means that between HS and RCT there exists a profound difference: the former but not the latter derives the description of the mechanism at work, not only by drawing on a general theory, but also by investigating into the fact, data and empirical material pertaining to the events and processes under investigation.

Even with that difference there exists, at least in my view, one feature of RTC that brings it quite close to CHA in Skocpol's *States and Social Revolutions*, namely, nominalism, even if it is, contrary to (Skocpol 1979), located in the RTC at the level of general theory. If one focuses on the axioms of RCT spelled out earlier, one finds out that according to its proponents they can be applied everywhere where there are human beings. In (Hechter and Kanazawa 1997) one finds an extensive list of applications of RCT: family and demography, gender, organizations, political sociology, stratification, as well as race and ethnic relations. And, RTC – it is claimed – need not be limited in its application to modern societies. As shown in (Little 1991, 40), it should be applicable also to such distant phenomena as European feudalism. But RCT, and this brings it quite close to HCA as performed in (Skocpol 1979), never addresses the crucial issue: *What makes RCT applicable to the past, present, and (one hopes) the future*? That this is so is readily seen, for example, in (Coleman 1990), where the model of purposive action and the use of the concept of maximization of utility are viewed as basic at the level of metatheory, but the question of why they can be applied for explanations of all specific events and processes is not addressed at all.

The following additional critical point can also be made against RCT. This theory, according to Ferejohn and Satz, is (Ferejohn and Satz 1996, 81)

> most credible under conditions of scarcity, where human choice is severely constrained. ... All of the cases where rational-choice theory is a strong predictive theory are ones in which the theory is relying on interests that are determined by features of the agent's environment.

This means that RCT bases its explanations, not on a description of the mental outfit of individual actors, but on "the environmental constraints they face" (Ferejohn and Satz 1996, 74) and, thus, on "illuminating *structures* of social interaction" (Ferejohn and Satz 1996, 74). But under such an interpretation of RCT, the latter faces the following questions: Where do the structural constraints come from? Are they, with respect to mental outfit and actions of actors, just "externalities"? To these questions one does not find any answer even in a depsychologized RCT as presented by Ferejohn and Satz. So, a conceptualization of the relation between structure and action has to be integrated into RCT; this would enable one to understand *why both rational action* and the *structures constraining it are reproduced and thus endure*, and thereby explain why the concepts of RTC, because their referents endure, can be applied everywhere.

The tools – concepts – for an overcoming of nominalism of HS as given in (Skocpol 1979) are already given in the very works on HS. I will now deal with these concepts and then integrate them into one conceptual network.

7.4 Mechanism, Path Dependence, Conjuncture and All That

There are the following ten general concepts that have to be integrated into one coherent network.

1. *Mechanism*. Mechanisms play a central role in the explanation of phenomena; "Accounts of empirical phenomena are *explanatory* ... insofar as they show how those phenomena can be seen as instances of causal laws or *mechanisms*. Explanations ... identify mechanisms that are held to apply across all relevantly similar contexts" (Ferejohn and Satz 1996, 74). Mechanisms can be viewed also as abstract properties of basic units of analysis that produce outcomes and associations (Mahoney 2004, 460–461).

2. *Path-dependency*. If *path* stands for a sequential order of events, then *path-dependency* stands for a property of a system in which a particular outcome in any given "run" of the system depends on the choices or outcomes of events intermediate between the initial conditions and this outcome (Goldstone 1998, 834). From the point of view of social sciences, path-dependency "refers to the notion that for any given trajectory past choices and temporally remote events can help to explain subsequent paths of development and contemporary outcomes. It suggests that decisions made at a particular point in time delimit future options" (Aminzade 1992, 462–463). Once one reflects on path-dependency, then one has to (a) take into account that intermediate events/choices can stand for the forks/bifurcations in the paths of the system; (b) investigate into the mechanisms as determinants of those choices; and (c) pinpoint the mechanisms sustaining the movement along the given path (Aminzade 1992, 463).

A certain class of paths displays the property of *self-reinforcement* in the sense that certain types of events are continuously renewed/reinforced which in turn leads to a preservation of patterns of certain events (Mahoney 2000, 508, 512–516). There are also paths in which the events involved stand in relations that can be labeled "reactive sequences," that is, they are chains of events temporarily ordered and causally related in the sense that an event is both a reaction to an antecedent event and a cause of a subsequent one. In these paths all events are transformed but *not reinforced* (Mahoney 2000, 526).

3. *Conjuncture*. It stands for "a particular combination of structural causes and events in a particular time and place [and] may create unique outcomes that will not necessarily be repeated in other contexts" (Paige 1999, 782).

4. *Contingency*. A particular event is a contingent one if it originates in an intersection/coming together of two or more already existing, separately determined sequences of events and, thus, "cannot be explained on the basis of prior historical conditions" (Mahoney 2000, 507).

5. *Time/Temporality*. As a general concept, time enables one capture the processual unfolding of social action (Isaac 1997, 6). Three concepts, in addition to that of trajectory given earlier have to be stated here: (i) *duration* "refers to the amount of time elapsed for a given event or sequence of events" (Aminzade 1992, 459); (ii) *pace/tempo* stating the number of similar, that is, repetitive events given in a certain time (Aminzade 1992, 461); and (iii) *cycle* referring to the time interval between two recurrences of the same phenomenon (Aminzade 1992, 468).[232]

6. *Causality*. For causality as applied to social processes it holds that (i) the pacing of causes can change; (ii) the relevance of causes change with respect to different events; and (iii) causes can appear and disappear.

7. *Structure and Action, Mesostructure*. Structures have a dual character (Giddens) in the sense that they are "composed simultaneously of schemas which are virtual, and of resources which are actual" (Sewell 1992, 13), where "schemas" stand for procedures applied and applicable in the reproduction/enactment of social life (Sewell 1992, 8), while "resources" stands for human and nonhuman resources (medias) of power; and it holds that schemas and resources stand in the mutual relation of cause and effect (Sewell 1992, 13).

Based on the understanding of structures, both made up of both schemas and resources, it is possible to understand how structures are *transformed into different structures*. At the basis of such transformations are, according to Sewell, (i) *the multiplicity of structure*; "social actors are capable of applying a wide range of different and even incompatible schemas and have access to heterogeneous arrays of resources" (1992, 17); (ii) *the transposability of schemas*: "the schemas to which actors have access can be applied across a wide range of circumstances" (1992, 17); and (iii) *the unpredictability of resource accumulation*; by the "very fact that schemas are ... capable of being transposed or extended means that the resources consequences of the enactment of cultural schemas is never entirely predictable" (1992, 18). So, (Sewell 1992, 19):

structures, then, are sets of mutually sustaining schemas and resources that empower and constrain social action and that tend to be reproduced by that social action. But their reproduction is never automatic. Structures are at risk ... in all of the social encounters they

---

232 On temporality see also (Sewell 1996b).

shape – because structures are multiple and intersecting, because schemas are transposable, and because resources are polysemic and accumulate unpredictably.

The term "mesostructure" ("mesodomain") stands for the *intersection of process, history, structure, and social action* (Hall 1991, 397).[233] Based on this, one can understand that (i) structures set certain limits to social actions and processes (Pestello and Voydanoff 1991, 107), (ii) structures are enacted and at the same time condition this enactment (Hall 1987, 10), and (iii) structures can be changed in action.

8. *Narrative*. It is "the portrayal of social phenomena as temporally ordered, sequential, unfolding, and open-ended 'stories' fraught with conjunctures and contingency" (Griffin 1992, 405), and it is also "the organization of material in chronologically sequential order and the focusing of the content into a single coherent story, albeit with subplots" (Griffin 1993, 1097), while Somers characterizes it as *"causal emplotment"* (Somers 1992, 601). It is viewed as a processually oriented action approach to social reality that assumes that "sociohistorical processes happen in sequences of actions embedded in social relations endowed with constraining and enabling capacities" (Isaac 1997, 9).

9. *Comparison and Comparative Method*. The central issue of comparison and comparative method is the choice of an analytic unit so that particular units can be understood as configurational "cases" (McMichael 1992, 355). There are four strategies employed in comparison (Tilly 1984, 82, 125):

a. *individualization* contrasting particular instances of a given phenomenon in order to understand the peculiarities of each of these cases
b. *universalization* enabling one to understand that every instance of a phenomenon follows essentially the same rule
c. *variation-finding* by which investigating into the differences of instances establishes a variation in the nature and intensity of the phenomenon
d. *encompassing comparison* explaining variation among cases due to their relationship to the whole

In order to escape the predicament of *either* postulating in advance a whole with respect to which the particular cases are just instances or postulating in advance the particular instances as wholes in themselves which then can be mutually compared and, thus, to subject the very choice of units to theoretical investigation, one should employ a method ensuring that (i) the units of analysis are conceptu-

---

[233] On this term see also (Maines 1979; 1982), (Hall 1987), and (Mahoney and Snyder 1999).

alized as fluid and (ii) the whole is conceptualized as an *emergent* whole (McMichael 1992, 359). Once this is achieved, the method of comparison can acquire a *diachronic* form and a *synchronic* form. The former "involves comparison across time of multiple instances of a single historical process" (McMichael 1992, 359). The latter stands for (McMichael 1992, 360)

a comparison across space within a single world-historical conjuncture. ... It is essentially a 'cross-sectional' comparison of segments of a contradictory whole, in which segments ... embody distinct and overlapping 'social times.' The segments are comparable because they are brought into relation through some competitive, or contentious, common process. ... That is, the conjuncture is defined as a juxtaposition of historically distinct segments.

10. *Scientific Laws and Historically Conditional Theory*. With respect to the concept of scientific law the dispute between representatives of RCT and HS displays multiple "rifts." On the one hand, Kiser and Hechter as representatives of RCT speak out in favor of using *omnitemporal* scientific causal laws (Kiser and Hechter 1991, 6; 1998, 793) serving as the basis of causal explanations. On the other hand, Somers proposes to use a "notion of causality based on narrativity, and the centrality of meaning, sequence, and contingency, rather than universality and predictive law" (Somers 1996, 63). She proposes also the use of description of *historic specific conditions* and of *specific mechanisms* as elements of narrative, relational explanations that would allow one to bypass the employment of general laws (Somers 1998, 736–738). Quadagno and Knapp state the following (Quadagno and Knapp 1992, 495–496, 499):

Historical sociologists find general laws troublesome because they fail to recognize the social world as fundamentally heterogeneous in character and constitution. ... Historical sociologists have abandoned general laws because such invariant explanations deny the fundamentally historical and temporal nature of historical events. ... Explaining fundamentally historical events through atemporal causal models violates the historical character of the event and calls into question the intelligibility and usefulness of the model's explanatory features. ... What historical sociologists repudiate are the characteristics of universality and omnitemporality.

From this W. H. Sewell then draws the conclusion that "sociology's epic quest for social laws is illusory" (1996b, 272).

A more differentiated position is taken by J. Paige who conceptually distinguishes between two types of social science theories: a *universal* type of theory and a *historically conditional* type of theory, so that while the former stands for theories "without time space specification" (Paige 1999, 785), the latter stands for theories that explicitly specify "the range of historical conditions under which the theory is thought to apply" (Paige 1999, 785). This latter type of social science theory, "unlike conjunctural or contingent explanation, does involve generalizations beyond the individual case ... and contributes to the development of conditional theoretical frameworks, not simply to the understanding of particular historic conjuncture" (Paige 1999, 785). Theories of this type "are resolving in practice the artificial antinomy between historical specificity and theoretical generalization that underlies current epistemological debates over narrative, comparison and general theory" (Paige 1999, 785–786), and they "undercut attempts to create universal generalizations valid in all times and places" (Paige 1999, 788). So, according to Paige, this "indicates not that sociology's epic quest for causal laws is illusionary, as Sewell contends, but that it has been looking for the wrong kinds of laws – universal rather than conditional" (Paige 1999, 789).

7.5 Attempt at Synthesis: Beyond the Nominalism of RCT and Skocpol's *States and Social Revolutions*

Let me now try to unify the above discussed general concepts into one conceptual network. The framework should enable one to reach the aim of HS which was stated by G. Steinmetz as follows (Steinmetz 2007, 5):

history and sociology are both concerned with human social practice in its capacity for willed or unintentional change – its capacity for producing events, writ large, revolutionary ruptures in the existing socio-symbolic order. Sociology and history are both interested in the equally paradoxical *reproduction* of social structures, that is, in the ways social structures are perpetuated such that they appear to be natural and unhistorical.

How can such an aim be reached at all? There are, at least in my view, two different, but still closely interrelated components enabling one to reach such an aim: *general social theory* and what J. Paige labels as *historically conditional social theory*.[234] General social theory stands here for a set of concepts that describe general characteristics of *any society* under investigation, for example, the relation of agency and structure, action and order, the character of processes going on it, and so on. Those concepts stand for what Jeffrey C. Alexander labels "presuppositions in social scientific argument" (Alexander 1982, 37), which are *truly general* in the sense that they cannot be subsumed by any more empirically oriented level of sociological research, and truly decisive in the sense that they "must have significant repercussions at every more specific level of sociological

---

[234] This differentiation comes quite close to the differentiation between *metatheoretical conception* and specific substantive theory as given in (Berger and Wagner and Zelditch 1992).

analysis" (Alexander 1982, 37). For general social theory it holds that it *guides and frames research going in the direction of formulating historical conditional social theory (theories)*. The entities to which a general social theory is applicable are societies viewed as units unrelated in space/time that do not reproduce some universal process; thus general social theory is *thoroughly nominalistic*.

Such a nominalist feature is displayed, for example, by Marx's well-known claim that "the history of all hitherto existing society is the history of class struggles." Such a claim is not based on any investigation into some deeply rooted mechanism by means of which class struggle is reproduced in all societies; but still it guides investigations into the specific mechanisms (re)producing class struggle in the framework of each respective society. How can such a mechanism be discovered and theoretically explained?

For social causal mechanisms it should hold, according to Kiser and Hechter, that – contrary to a mechanism at work in the realm of nature – they are not established in a direct way from certain empirically given social (macro) characteristics (Kiser and Hechter 1991, 6). Already here one finds a profound misunderstanding on the part of Kiser and Hechter. When they introduce the concept of causal mechanism, they mention, as an example,[235] the computation of Avogadro's number N stating the number of elementary entities (usually molecules or atoms) in one mole of any substance. With respect to that computation, they claim that it is performed in a rather direct way without, however, as I will show in chapter 8, taking into account that it stands in fact for an indirect way of measurement/quantification, namely, the computation of a quantitative characteristic of a micro-property by means of the quantitative characteristics of macro-properties. This computation takes place outside the framework of what Kiser and Hechter label "general theory"; general theory stands here for the contention going back to ancient Greek philosophy that matter is composed of atoms and frames and drives the above mentioned indirect measurement/quantification.

It is my contention that social causal mechanisms, like causal mechanisms that are at work in the realm of nature, are described, explained, and measured/quantified in the framework of a theory that is *different* from a general (and nominalistic) theory driving and guiding that measurement/quantifications and explanations.

What are the features of such descriptions and explanations? In my view concepts like causal mechanism and general (scientific) law have here a central role that can be explicated by drawing at least partially on the article (Goldstone 1998). Goldstone claims, on the one hand, that "mechanisms are not the functional equivalent of 'general laws' and in no way substitute for them in explanation" (1998, 832). On the other hand, however, when dealing with the concept of

---

235 Here they refer to (Salmon 1984) and (Nye 1972).

general (scientific) law he does not relate it to the concept of causal mechanism and instead claims that "discovering or positing the 'mechanism' ... is quite a different matter than stating the general law" (1998, 833), while the concept of general law he understands as (Goldstone 1998, 832):

> a statement of necessary or probable connection between two kinds of events. ... What makes a general law "general" is that it applies to a range of initial conditions and asserts a necessary or probable connection between particular initial conditions and a subsequent event or events.

A scientific law can, as shown in chapter 4, stand for the relation/connection between conditions under which a causal mechanism exists and is operative/at work and this very mechanism. With respect to the reconstruction of the law of value in *Capital*, Volume I given in chapter 4, and with respect to the concepts of HS explicated above, the descriptions of those conditions should stand for the *action-characteristics of actors*, while the description of mechanism should stand for a conceptual grasping of the *social structural mechanism at work*. This description and conceptual grasp should at the same time enable one to express the reproduction of the actions-characteristics of the actors and of their specific, historical location as social being and the reproduction/recurrence of a specific social structural mechanism. Here I bring in also the claim of Kiser and Hechter that "Marx's work provides important causal mechanisms" (Kiser and Hechter 1991, 16), so that in my reconstruction of the structure of social science laws I draw also on Marx, however, not on his general theory,[236] but on his economic works.[237]

Among the action-characteristics of actors, Marx lists, as seen in chapter 4, the subjective (conscious) aims and the objective social determinants of these actors framing their actions. The structure of a social scientific law as a law of reproduction can be expressed as shown in Figure 7.3.

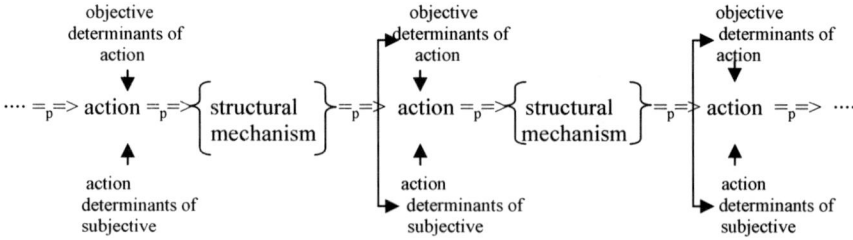

Figure 7.3 Social scientific law as expressing the process of (re)production

---

236 This general theory is sometimes labeled "historical materialism."
237 Here I mean the manuscripts *Grundrisse*, the *Manuscript 1861–63*, the manuscript of Volume 3 of *Capital*, and *Capital*, Volume 1.

In Figure 7.3 "$=_p=>$" stands for production, and where the production leads, through the (re)production of structural mechanism, to the (re)production of both the objective and subjective (conscious) determinants of action.

Let me now characterize the specific features of this type of social scientific law. *First*, contrary to Paige's view, stated earlier, *a scientific law of that type is applicable only and only where the respective actions-characteristics are given, that is, it is universal and at the same time historically conditional.*

Second, it holds that its truth is not dependent on the number and space-time distribution of instances. Even if there would be just one instance of it given only in one space-location and only in a limited time-interval, its description would still have the status of a scientific law.

Third, the set of general concepts of HS stated above can be used for a more detailed characterization of the social science law as represented in Figure 7.3. The process of reproduction is at the same time a process of *self-reinforcement*, where each component involved in reproduction depends as an outcome on the previous component; and any changes in the components of reproduction can lead to a *bifurcation* in the *path* of the system. Such changes can be brought about by factors either *endogenous*[238] or *exogenous* with respect to the reproduction process. Exogenous factor stand for a conjuncture and brings in an irreducible *contingency* into the process of reproduction.

The process of reproduction displays also the following temporal features: *duration* as the time elapsed from the onset of action until the generation of the causal mechanism where, of course, this time can change; *pace/tempo* as the number of actions performed and mechanisms generated in a certain time span; and finally, *cycle* as the time interval elapsed between the reoccurrence of action and/or mechanism of the same kind.

Figure 7.3 expresses the unity of the description of process, action, and structure in a scientific law. But where and how can *history* come in? In that figure it is presupposed that action (and its determinants) and structural mechanism reinforce each other, but then one faces the question where does the respective type of actor(s) come from in the sense that such a type *is as yet, not the outcome of the operation of a social structural mechanism, but only its point of departure.* The answer that suggests itself is that it is the product of *operation* and at the same time of *disaggregation of a structurally different social mechanism*. So, one faces here a change of the *type of both action and of social structural mechanism*. Once Figure 7.3 is widened in such a way, then the concept of *mesostructure* can be applied to it.

---

238 For example Marx considers in *Capital*, VolumeI the process of nonsimple reproduction, so that the quantitative determination of the components changes.

As has been shown, the truth of social scientific law represented by Figure 7.3 does not depend on the number and space-time distribution of its instances, so the method of its presentation stands for a *narrative* and at the same time for a conceptual grasping of the structure of a social scientific law. Here also the *method of comparison* comes in. Initially, when dealing with the process of reproduction as unity of a specific type of action and of a specific type of mechanism, it is a type of *thought-representation of that process as not distorted by an intersection with other types of action and of structural mechanisms.* Marx even assigns to the object of such a type of thought-representation a special term, namely, that it is a thought-object "in its ideal average, as it were" (Marx 1981, 970).

A different type of thought-representation is asked for once there are several instances of the same reproduction process given in different space- and/or time-locations that, however, differ mutually due to the incursion of qualitatively different types of action and of social structural mechanisms. Marx assigns to the object of such mutually different thought-representations in the *Zur Kritik der politischen Ökonomie* of 1859 the term "original types." They stand for the thought-objects involved in what McMichael labels the "diachronic comparative method." What type of thought-object corresponds to what he labels "dyachronic comparative method" has to be found out.

Finally, by drawing on my reconstruction in chapter 4 of Marx's thought-movement based on the concept of value, it becomes evident how *narrative* and *explanation based on a social scientific law* with the structure expressed in Figure 7.3 *stand for a sequence of thought-operations.* Marx initially applies the concept of value to situations where the following three essential conditions are given:

1. the existence of producers privately owning the means of production and purchasing both instruments and objects of labor;
2. the subjective (conscious) aim of the producers to exchange commodities for commodities by means of money (commodity → money → commodity'); and
3. the subjective (conscious) aim of the producers to obtain use-values *via* exchange.

Where these three essential conditions are given, the mechanism of production of value from abstract labor is at work. At the same time Marx states the relation between those three conditions and this mechanism as a relation of mutual reproduction, thus in fact broadening the concept of value to that of law of value. Based on that law he then performs the following thought-operation. He brings in new, additional essential conditions:

1) the existence of producers privately owning the means of production and purchasing in addition to instruments and objects of labor also labor-power
2) the existence of a group of free people selling their labor-power
3) the subjective (conscious) aim of the producers to make money (money → commodity → money').

Under these new three essential conditions both the types of social actions involved and the social causal mechanism which is their outcome are transformed,[239] and so the law of value is transformed as well. Now value is understood as being the outcome of a mechanism involving both (pre)paid labor and unpaid (surplus) labor. And again, Marx views the relation between those new essential conditions and this mechanism as that of mutual reproduction. This means that in Marx's explanation as given in Volume I of *Capital*, the law of value stands both for the explanans and the explanandum, but where the respective types of action and the respective causal mechanisms at work conceptually expressed in those scientific laws are different.

---

[239] The social actions, understood as happenings, which significantly transform structures are labeled by W. Sewell as "events", on this see (Sewell 1996a; 1996b, 262–264).

Chapter 8: Reflections about Metareflections: Roy Bhaskar and Wesley C. Salmon

8.1 Can ontology (of the social world) be a substitute for epistemology and methodology of (social) science?

The aim of subchapter 8.1 is to analyze the contribution of Critical Realism (hereafter, CR) to the philosophy of science from the point of view of ontology, epistemology, and methodology. I will attempt to demonstrate that even if CR claims to provide a developed ontology, it cannot succeed because it does not succeed as an epistemology and methodology. I start by analyzing M. S. Archer's views on linking explanation and understanding, as presented in (Archer 2007). Next, I provide an overview of R. Bhaskar's realist theory of science, together with T. Lawson's contribution to it. Next, I highlight certain deficiencies of the ontological dimension of CR and try to show that in order to overcome them CR is in need of a new epistemology and methodology. Finally, I provide the gist of such an epistemology and epistemology, together with a reconstruction of Marx's way of linking explanation and understanding in *Capital*, Volume I.

8.1.1 M. S. Archer on Linking Explanation and Understanding

What M. S. Archer aims at in her paper is (Archer 2007, 63)

> to examine how agents reproduce ... or transform ... social structures and are themselves reproduced or transformed in the self-same process. Stripped down to its chassis, this approach is based upon two fundamental propositions: that structure necessarily pre-dates the action(s) which transform it; and that structural elaboration necessarily postdates those actions.

In her position one can identify two dimensions of CR's approach: (a) the *epistemological* – to examine, find out, etc., and (b) the grounding of the epistemology in an *ontological* point of view (here, by introducing an ontological claim pertaining to the relation of social structure and action).

The ontological part of the attempt at linking explanation and understanding is based on the view that because causal powers are at work both in social structures and human agency, one can start considering an "interplay" between them. Accordingly, since these causal powers are "*necessarily* intertwined because actions, their conditions and consequences, span the two realms and thus cannot be divorced – which thus implies one story" (Archer 2007, 67), explanation and understanding can be linked.

Such a story has to provide a detailed conceptual grasp, first, "of *how* structural and causal powers impinge upon agents," and second, "of *how* agents use their own personal powers to act 'so rather than otherwise' in such situations" (Archer 2007, 68). And with respect to the latter, this in fact means that it is necessary to investigate "the role played by human subjectivity in general ... [and] in particular ... the part of reflexivity in enabling agents to design and determine their responses to the structured circumstances in which they find themselves – in the light of what they care about most" (Archer 2007, 68–69).

Archer, however, complains that critical realists have put too much emphasis on how structural powers impinge upon agents and neglected the other side of the story, namely, the investigation into how agents, based on their reflexivity, design their acts and then act as a response to the structured circumstances. As a remedy to this she then provides an overview of how she conceptualizes this reflexivity – namely as *Internal Conversation*,[240] where it is understood as a process "which mediates the impact of social forms upon us and determines our response to them" (Archer 2007, 72). Accordingly, the following two aspects of the paper (Archer 2007) are worth mentioning here. First, even if the title of the paper suggests a link between explanation and understanding, in the paper itself is not spell out explicitly what are the meanings of the terms "explanation" and "understanding," understood to be *methods* of social sciences, which are *specific* to CR. The basis for a reconstruction of the method of understanding is however, at least implicitly, given in the conceptual system Archer labels as "Internal Conversation" in the sense that this system could be used for the explication of the main characteristics of the method of understanding.

Second, a more problematic aspect of Archer's paper is that it imputes the view that CR has provided *at least* a reconstruction of the methods of explanation corresponding to CR's ontology of powers, mechanisms, tendencies, and so on. However, as I will show later, this does not hold; furthermore, a reconstruction of the methods of explanation can serve at the same time as a basis for linking the methods of explanation and understanding. In order to do this I have to turn to the views of R. Bhaskar as presented in (Bhaskar 1978) as well as to the views of T. Lawson presented in (Lawson 1997) and (Lawson 2003).

---

240 On this see (Archer 2003).

8.1.2 Bhaskar, Lawson and the Realist Philosophy of Science

R. Bhaskar's aim in his *A Realist Theory of Science* (Bhaskar 1978) was to furnish in a transcendental manner an ontology – and a logic of scientific discovery corresponding to it – that would make it understandable why science (both experimental and theoretical) is possible at all.

Experimental science is viewed by Bhaskar as a practical activity through which scientists, by intervening into the external world, trigger certain mechanisms while at the same time screening off certain other mechanisms so that they could not act as possible countervailing causes. This in turn should enable the triggered mechanism and its activity to become the subject matter of investigation without introducing any distorting influences (Bhaskar 1978, 46). For experimental activity it should thus hold that the (Bhaskar 1978, 53)

experimental scientist must perform two essential functions in an experiment. First, he must trigger the mechanism under study to ensure that it is active; and secondly, he must prevent any inference with the operation of the mechanism. These activities could be designated "experimental production" and "experimental control." The former is necessary to ensure the satisfaction of the antecedent (or stimulus) conditions, the latter to ensure the realization of the consequent, i.e., that closure has been obtained. ... Only if the mechanism is active and the system in which it operates is closed can scientists in general record a unique relationship between the antecedent and consequent of a law-like statement. The aim of an experiment is to get a single mechanism going in isolation and record its effects. Outside a closed system these will normally be affected by the operations of other mechanisms, either of the same or of different kinds, too so that no unique relationship between the variables or precise description of the mode of operation of the mechanism will be possible.

The term "closure" given in the last quote is introduced by Bhaskar in order to explain the fact that "leaving aside astronomy, it is only under conditions that are experimentally produced and controlled, that a closure, and hence a constant conjunction of events, is possible" (Bhaskar 1978, 65). So closure is here understood as an operation by which a system is isolated or closed up in such a way that a regular conjunction of events (a regular event-pattern) occurs.

Based on the concept of closure, Bhaskar can then spell out his ontology of causality and causal laws which he views as a counterpart to the Humean (regularist) approach to causal laws. While a sequence of events appears regularly, according to Bhaskar, with the exception of astronomy, only under the conditions of closure brought about by experimenting scientists in laboratories, causal laws are laws that related to the generative mechanisms one can isolate and trigger in experimental activities taking place in laboratories. These mechanisms are at work both under the conditions of closure and in *open systems* – the latter being characteristic of the world outside the confines of laboratories. The difference between closed and open systems should, according to Bhaskar, enable us to understand that (Bhaskar 1978, 33)

> an experiment is necessary precisely to the extent that the pattern of events forthcoming under experimental conditions would not be forthcoming without it. Thus in an experiment we are a causal agent of the sequence of events, but not of the causal law which the sequence of events, because it has been produced under experimental conditions, enables us to identify. Two consequences flow from this. First, the real basis of causal laws cannot be the sequence of events; there must be an ontological distinction between them. Secondly, experimental activity can only be given a satisfactory rationale if the causal law it enables us to identify is held to prevail outside the contexts under which the sequence of events is generated. ... it implies that causal laws endure and continue to operate in their normal way under conditions which may be characterized as 'open', where no constant conjunction or regular sequence of events is forthcoming.

T. Lawson provides a very similar argument based on the already explicated concept of causal sequence (Lawson 2003, 23–24):

> most event regularities of the causal sequence sort regarded as of interest to natural scientists are actually restricted to conditions of experimental control, whilst ... the results of these experiments are frequently successfully applied outside the experiments *where event regularities are not in evidence*. ... by viewing experimental practitioners as intervening in a sphere of reality and experimentally manipulating it in order that ... the working of a specific intrinsically stable causal mechanisms are ... insulated from the effects of countervailing factors. ... Notice, then, that to make sense of the experimental process, it is essential to recognise that the event regularity produced corresponds to the empirical identification of an underlying causal mechanism.

Only by means of such an interpretation of the process of experimentation can we (Lawson 2003, 24):

> make sense of the observation ... that experimental knowledge is somehow successfully applied outside the laboratory, even in conditions where event regularities do not occur. For the knowledge or insights obtained relate primarily not to the (contingent and experiment-bound) regularity that is produced but to a (experimentally empirically identified) mechanism that, when triggered, operates independently of scientists and their experimental work.

Bhaskar can then bring the explication of his ontology to an end by examining its most important categories: *generative mechanism*, *power*, and *tendencies*. Generative mechanisms are those mechanisms that are capable of regularly producing sequences of events not only because these mechanisms are triggered, but primarily because they *endure*; and they endure, according to Bhaskar, even when the respective sequences of events are not produced due to, say, the intervention of countervailing mechanisms (Bhaskar 1978, 46). That generative mechanisms can be at work outside the conditions of closure is demonstrated by the fact that they have certain *powers*[241] as potentialities that are efficacious in open systems – potentialities being labeled by Bhaskar as *tendency* of a causal mechanism (Bhaskar 1978, 46–49).[242]

Bhaskar can then provide the following typology of necessary and accidental sequences of events E_a, E_b, E_c, etc. "A sequence $E_a.E_b$ is necessary if there is a generative mechanism M such that whenever E_a, E_b tends to be produced; a sequence is accidental if this is not the case" (Bhaskar 1978, 165). This is represented in Figure 8.1 (Bhaskar 1978, 165).[243]

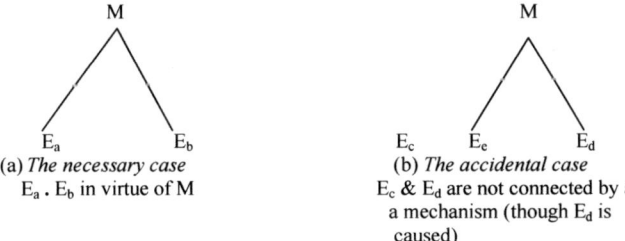

(a) *The necessary case*
$E_a . E_b$ in virtue of M

(b) *The accidental case*
E_c & E_d are not connected by a a mechanism (though E_d is caused)

Figure 8.1 Bhaskar on the necessary and accidental relation between events and its mechanism

Based on such an ontology I can now move to the *epistemology* of CR, which should explicate how knowledge about mechanisms, power, tendencies, and so on, are being produced. According to Bhaskar "there is in science a characteristic kind of dialectic in which a regularity is identified, a plausible explanation for it is invented and the reality of entities and processes postulated in the explanation is then checked" (Bhaskar 1978, 145). This sequence is viewed by Bhaskar as the *logic of scientific discovery* corresponding to his ontology; he illustrates it by means of Figure 8.2 (Bhaskar 1978, 145):[244]

---

[241] For an analysis of Bhaskar's catogory of power see (Kaidesoja 2007)
[242] For an explication of these concepts see (Lawson 1997, 20–23; 2003, 144) and (Brown and Fleetwood and Roberts 2002, 5–6).
[243] For Bhaskar's figures of closed and open systems see (Bhaskar 1978, 165).
[244] In order not to overburden this figure I leave out the names of philosophies dealing with the respective types of scientific activity.

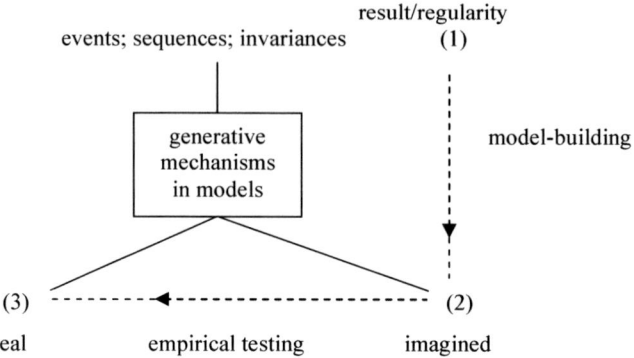

Figure 8.2 Bhaskar's logic of scientific discovery

From the point of view of such a logic of scientific discovery, *statements of causal laws*, that is, scientific causal laws are, according to Bhaskar, *statements about tendencies* and "are neither empirical statements (statements about experiences) nor statements about events. Rather, they are statements about the ways of acting of independently existing and transfactually active things" (Bhaskar 1978, 52) and thus "statements about the forms of activity characteristic of the things of the world" (Bhaskar 1978, 46). Such statements he labels *normic* statements which (Bhaskar 1978, 51)

> do not say what would happen, but what is happening in a perhaps unmanifested way ... a normic statement says that A really is ψ'ing, whether or not its actual (or perceivable) effects are counteracted. They are ... *transfactuals*; they take us to a level at which things are really going on irrespective of the actual outcome. To invoke a causal law is to invoke a normic conditional. ... A normic statement is a transfactual statement, with actual instances in the laboratory that constitute its empirical grounds.

The movement from a registered (perceived) sequence of events, say E_k and E_l stated in what Bhaskar labels as "protolaw" (Bhaskar 1978, 163) to a hypothetized underlying mechanism M is represented by him as follows (Bhaskar 1978, 167):

Figure 8.3 Bhaskar on the case of a protolaw

An important enlargement of the conceptual framework of CR was accomplished by the introduction, by Lawson, of the term "demiregularity" (hereafter, "demi-reg") and "contrastive demiregularity."[245] According to Lawson, in the sphere of nature "the goal of a perfect closure, turning as it does upon system or mechanism insulation, constitutes an ideal scenario that cannot always adequately be engineered; indeed it may very rarely be" (Lawson 1997, 203). But even if this holds, this does not mean that there exist just two possible options: either strict event regularity under conditions of complete closure or a totally random flux. In between these two options he locates *partial regularities* because "over restricted regions of time-space certain mechanisms may come to dominate others and/or shine through: non-spurious, rough and ready, partial regularities may be observed" (Lawson 1997, 204).

These partial event-regularities Lawson labels "demi-regularities" or "demi-laws" and they indicate, according to Lawson, "the occasional, but less than universal actualization of a mechanism or tendency, over a finite region of time-space" (Lawson 1997, 204), and as statements of such demi-regs he gives the following examples: "Autumn leaves do fall to the ground *much* of the time," "Women are *concentrated* in secondary sectors of labour markets," and "Productivity growth in the UK over the last century has *frequently* been slower than that of most other, otherwise comparable, industrial countries" (Lawson 1997, 204). Based on the last example, Lawson introduces also the term "contrastive demi-reg," which stands for a demi-regularity that holds for a certain set of entities with respect to another set, while all these sets are so mutually similar that it is rather surprising that such a contrast comes about at all. This moment of surprise and unexpectedness, once it turns into curiosity then calls for an explanation of *why a strict regularity is violated in the superset of all similar entities but at the same does not collapse completely and is present in one of the sets of these entities.*

8.1.3 A Constructive Critique of Critical Realism

Even though Bhaskar and Lawson claim that "the switch of emphasis in the philosophy of science engendered by Bhaskar ... is away from epistemology towards ontology" (Bhaskar and Lawson 1998, 10), I will now try to prove that *because Bhaskar and Lawson have not provided a detailed epistemology and methodology, this in turn yields an impoverished ontology on their part*. What I have in mind here, and what was stated already in the introduction of this book, is the fact that categories, such as underlying mechanism, power, and tendency, that

---

245 Here I draw on (Lawson 1997; 2003, 105–107).

are *seemingly purely ontological categories*, have at the same time an *epistemic* dimension; they stand for a type of *knowledge* about the real (natural and social) world and not only for knowledge about the *real world*. These categories are a type of appropriation in mind of the external world and at the same type they guide both the creation of thought-projects for the transformation of the external world and the very practical transformation of this world. So, as knowledge given in categories develops, so, too, does the ontological dimension of these categories develop; and it is by means of this that we can then obtain a much broader and deeper understanding of what the world must be like for science to be possible. Categories understood in this way then have, as we will see below, important spin-offs in the realm of methodology. Therefore, the ontology of CR can be regarded neither as a substitute for an epistemology nor for a methodology.

According to Bhaskar, one of his aims – in the framework of his ontology – "is establishing the necessity for an ontological distinction between causal laws and patterns of events" (Bhaskar 1978, 18). And at the level of his epistemology he labels statements about empirical regularities "protolaws" (Bhaskar 1978, 163), that is, they are not scientific laws in their own right. These protolaws, according to Lawson, "are neither empirical statements (statements about experiences) ... nor statements about events or their regularities ..., but precisely statements elucidating structures and their characteristic modes of activity" (Lawson 1997, 24). Such claims, in my view, are based on the following three mistakes.

First, Bhaskar and Lawson refuse to assign ontological status to the knowledge about the events and their regular patterns. This then leads to the negative consequence that one has to view statements about regular event-patterns as *ontologically empty*, in the sense that they do not carry (at least) some potential information about the mechanism underlying them. But, this contradicts the fact, noted by Bhaskar, that statements about regular event-patterns are the point of departure for the discovery of the mechanisms underlying them and for the description of these mechanisms in certain statements; these statements thus state some truth about the real world. Stated otherwise: *from statements that*, according to Bhaskar and Lawson, *should be ontologically vacuous* – that is, should not have any objective content pertaining to causality – one *should not be able to derive statements that have ontological content, that is, that state something about underlying causal mechanisms*.

Second, Bhaskar – framing a set of ontological categories that should enable us to understand why science as a human activity is possible – ontologizes the set of categories he proposes in the sense that he clams that this is the way *the real world must be*. He starts from a certain state of science, then sets up an ontology, and finally claims that *this ontology absolutely correctly states how the real world really is*, while at the same time he refuses to assign any ontological *status to the knowledge about events and their regular patterns.* Bhaskar thus proceeds, from the point of applied philosophical strategy, in a manner that is symptomatic of Humeans. The Humeans move from the state of natural science, fixing statements about event-regularities (produced in experimental activities) to the ontological claim that the *very world itself is composed only of regularly appearing events,* while at the same time *refusing to assign ontological status to any possible knowledge about structure and mechanisms underlying regular event-patterns.* This strategy, while drawing on *ontological-epistemic* categories, nevertheless neglects their epistemic dimension and maintains that the world has to be exactly as it is claimed by the (seeming) purely ontological categories. Stated otherwise, it eliminates the epistemic aspect of categories in the sense that it neglects the fact that categories stand for a type of *knowledge* about the world and that this *world* can be different as compared to what is claimed by these categories. Such a strategy leads one into an "ontological trap."

One negative consequence of Bhaskar's walking into this "ontological trap" is a *systematic derogation* of a type of scientific knowledge in which there is as yet no explicit knowledge about underlying mechanisms, powers, tendencies, and so on. Statements formulated in the framework of this type of knowledge should have, at most, the status of "protolaws" (Bhaskar), or should even be denied the status of a causal law entirely (Lawson). This, in turn, leads to a complete absence, in the framework of Critical Realism, of a reconstruction of the ontologico-epistemic structure of statements of laws expressing regular event-patterns and of the method of explanation based on them.

Yet another negative consequence of Bhaskar's walking into this "ontological trap" is that CR cannot discern the ontologico-epistemic categorial "outfit" of various types of knowledge of underlying mechanisms, powers, and tendencies. Instead of reconstructing the structure of law-statements and methods of explanation based on them which would correspond to those types of knowledge, one is offered only the following consolation (Lawson 2003, 84):

It will be appreciated that ... my overall goal remains highly limited. This ... is necessarily the case. The most I can hope to achieve is a demonstration that successful explanatory work ... is indeed possible. ... The actual relevance of any specific method or approach elaborated will always depend on context ... the aim is not, and in my assessment could not be, to provide a list of highly specific or concrete methodological rules.

This then means that *CR does not propose a differentiated typology of models of scientific explanation which could be regarded as an alternative to Hempel's Humean-inspired D-N model of scientific explanation.* Subsequently, I will, based on chapters 3 and 4 of this book, try to provide such a typology based on categories like universe of discourse, effect, cause/essence/ground, and so forth,. which, even if they are expressed in terms of *categories*, still are *context-specific* and *context-sensitive*.

The *third* mistake of Bhaskar and Lawson can be located in the following claims of theirs. According to the Bhaskar (1978, 33):

> in an experiment we are a causal agent of the sequence of events, but not of the causal law which the sequence of events, because it has been produced under experimental conditions, enables us to identify ... experimental activity can only be given a satisfactory rationale if the causal law it enables us to identify is held to prevail outside the contexts under which the sequence of events is generated.

According to the Lawson (2003, 23–24):

> most event regularities of the causal sequence sort regarded as of interest to natural scientists are actually restricted to conditions of experimental control, whilst ... the results of these experiments are frequently successfully applied outside the experiments *where event regularities are not in evidence*. ... Notice, then, that to make sense of the experimental process, it is essential to recognise that the event regularity produced corresponds to the empirical identification of an underlying causal mechanism. In other words, even in experimental work, i.e. even in the branch of scientific work which is most bound up with the production of event regularities of the causal sequence sort, the primary concern is not with the production of an event regularity *per se*, but with the empirical identification of an underlying mechanism (co-responsible for any regularity produced). ... For the knowledge or insights obtained relate primarily not to the (contingent and experiment-bound) regularity that is produced but to a (experimentally empirically identified) mechanism.

What is presupposed here, both by Bhaskar and Lawson, is, *first*, that one has already discovered the respective causal mechanism and causal law underlying the experimentally discovered regular event-pattern. This thus means that in their epistemology the logic of discovery is not reconstructed as a sequence of self-carrying, relatively closed category-structures (or clusters of categories) standing for different levels/types of knowledge (say, empirical, theoretical, etc.). Rather, it is based on an inconsistent mix-up of categories belonging to different levels/types of cognition, so that the meaning of each of these levels is not given by its own respective categorical "outfit," but by the meaning of consecutive levels/types of cognition. That this is a wrong approach to the reconstruction of the logic of scientific discovery can readily be seen if we take, as an example, Galileo's experiments with the fall of bodies. These experiments were strictly limited to the production of the phenomenon of free fall which should then enable one to quantify the space covered in a certain time and, accordingly, the acceleration of the falling body. Galileo, however, at the same time completely put aside any reflections about the cause underlying acceleration. Thus, the purpose of experimental activity *need not be*, as claimed by Bhaskar and Lawson, to identify some underlying mechanism and its law but simply to remain at the level of regular event-patterns.

Second, while Bhaskar claims that the sequence of events enables one to identify the causal law (Bhaskar 1978, 33), Lawson claims something much stronger, namely, that the experimentally produced event-regularity enables one to identify *empirically* the causal mechanism at work in that regularity (Lawson 2003, 24). But, as we will see later, that mechanism cannot be identified empirically in an experiment because in the movement from (Bhaskar's) "protolaws" to the scientific law fixing the respective causal mechanism the very universe of discourse (i.e., the respective natural-kind-terms) do change; thus *an irreducible heuristic moment is present in the formulation of the law pertaining to a causal mechanism*.

8.1.4 Types of Law-Statements and Methods of Explanations

In order to propose alternatives to Hempel's D-N model of scientific explanation two additional deficits of Bhaskar's philosophy of science as given in (Bhaskar 1978) have to be brought to the fore.

First, even if, in the explication of his logic of scientific discovery, Bhaskar speaks about "a characteristic kind of dialectic in which a regularity is identified, [and] a plausible explanation for it is invented" (Bhaskar 1978, 144), this dialectic is not represented at all in the figure 8.2. Instead of a reconstruction of the movement *from* the identified regularity *to* the identified causal mechanism and *from* the latter, serving as an *explanans*, "*back*" to the regularity derived now as an *explanandum* (i.e., instead of a reconstruction of a *cycle* of cognition) Figure 8.2 represents only a *unidirectional sequence of unidirectional movements* of cognition. The same deficiency is found in the categories employed by Bhaskar; *he does not differentiate at the level of categories between events identified before the discovery of their underlying mechanism and events derived by explanation from that mechanism*. This also brings to the surface a deficit given in the Figures 8.1 and 8.3, namely, that they do not contain arrows indicating the *bi*directionality of movement of cognition: *from* the pair of events E_a, E_b and E_k, E_l *to* their respective mechanism *M* and *from* the mechanism *to* the explanatory derivation of the pair of events E_a, E_b and E_k, E_l. As I will show below the same deficit can be found in the works of Wesley C. Salmon.

Second, CR's philosophy of science lacks a reconstruction of the method by means of which one moves from the knowledge about regular event-patterns to knowledge about the underlying mechanism.

With these points in mind, let me now deal with the ontologico-epistemic structure of a law statement expressing knowledge about a regular event-pattern. As shown already in chapter 3 above, I view it as an *empirical type of scientific law* and regard it as a causal type of scientific law in its own right which contains (potentially) some information that can be used for the thought-movement to the mechanism underlying that pattern. Its structure is, as shown in chapter 3, as follows:

$$(x)(U_1 x \ \& \ Id_{1\text{-}k} x \ \& \ C_1 x \rightarrow E_1 x). \tag{1}$$

Here "U_1" stands for a universe of discourse, that is, for the set of entities to which the law can be applied; "$Id_{1\text{-}k}$" stands for the conjunction of 1 through k idealizations, "$\rightarrow$" stands for the sentential connective "if …, then …", while "C_1" stands for the *cause* and "E_1" for its *effect*. It states in plain English that if entities of kind U_1 undergo, under conditions of closure $Id_{1\text{-}k}$, a certain change C_1, then they will undergo another change E_1. Laws with the structure given in /1/ can serve as the basis of explanation in such a way that based on such laws, together with the introduction of the singular condition $C_1 a$, one derives – as an *explanandum* – a description of a particular event $E_1 a$.

As shown in chapter 3, as a first type of scientific law that already expresses at least some knowledge of the very underlying mechanism, one can view laws with the following structure:

$$(x)[U_2 x \ \& \ Cmod_{1\text{-}k} x = d_{1\text{-}k} \rightarrow f_1(Cx) = E^{(k)} x]. \tag{2}$$

What are the specific features of /2/ worth mentioning with respect to the philosophy of science of Critical Realism? *First*, /2/ fixes the process of thought-movement from an idealized effect to the essence as cause that is the ground of this effect. *Second*, /2/ overcomes CR's reductionist philosophy of science, namely, the absolutistic reduction of all types of scientific laws to just two types: *either* protolaws fixing "only" regular events *or* full-blown causal laws containing all the knowledge about the powers, tendencies, capacities, and so on. How a law with the structure of /2/ breaks up this absolutism becomes readily seen when one compares this structure with Bhaskar's claims about essence, powers, tendencies, and so on. Even if, as shown in chapter 3 in the critique of the views of Ellis, Bigelow, and Lierse, /2/ involves terms like "universe of discourse," "kind of entities," and "essence/ground/cause," and even if the second could be understood as "natural kind" and the last as "essence of the respective kind," scientific laws with the structure of /2/ still do not fit into the ontological framework delineated by Bhaskar. Expressed in ontologico-epistemic categories, theoretical sciences centered around laws with the structure of /2/, as shown in chapter 3, do not relate the respective natural kind to its underlying cause/ground in the sense that individuals sharing a certain essential property belong to one and the same natural kind.

Third, the idealizations in /2/ in the form of $Cmod_{1-k}=d_{1-k}$ stand here for a type of closure that differs in an important aspect from the type of closure given at the level of laws with the structure of /1/. In the latter case, they should enable one to identify a regular event-pattern, in the former, they should enable one to grasp *not a regular event-pattern but an idealized effect whose quantity enables one to determine the quantity of its cause/ground/essence.*

Fourth, a law with the structure of /2/ can serve as the basis of explanation so that on the basis of $f_i(C)$ (identified by means of $E^{(k)}$), one derives various different phenomena-effects. Central here is the procedure of explanation, reconstructed in chapter 3, labeled "gradual concretization," where *via* a gradual abolishment of idealizations $Cmod_{1-k}=d_{1-k}$, (so that $Cmod_i \neq d_i$ holds), one gradually adjusts the law with the structure of /2/ in the *explanans* to the context of the *explanandum* one wants to derive, so that one obtains laws with (k-1), (k-2), ..., (k-j) idealizations with the respective phenomena-effects $E^{(k-1)}, E^{(k-2)}, ..., E^{(k-j)}$. It is worth emphasizing that one can differentiate here, contrary to the philosophy of science of CR, between two *epistemically* different types of phenomena: those that are the point of departure for the thought-movement *to* essence/cause/ground, that is, *appearances*, and those that are derived by gradual concretization *from* the law with the structure of /2/, that is, *manifestations*.

Fifth, the reconstruction of the cyclical movement appearance → ... → manifestations enables one also to understand the role of demi-regs and contrastive demi-regs in this movement. Before the movement from appearances to the essence/cause/ground is accomplished, one has to isolate by abstraction a full-regularity (hereafter, full-reg) from a set of demi-regs. The full-reg then serves as the point of departure to determine the quantity of the idealized phenomenon-effect $E^{(k)}$ and based on it also the quantity $f_1(C)$. Once this is accomplished, the process of a thought-movement by gradual concretization from $f_1(C)$ to $E^{(k-1)}, E^{(k-2)}, ..., E^{(k-j)}$ (which is a thought- movement from a full-reg to the mutually contrastive demi-regs) can set in. Stated otherwise, initially, when there is as yet no knowledge about the underlying essence/cause/ground of the phenomenon, a full-reg is required in order to quantify the essence/cause/ground by means of the phenomenon. Once we know the quantity of the essence/cause/ground, we can, based on this knowledge, derive the quantity of various phenomena as well as their mutual quantitative differences by taking into account their respective contexts, thus we move to various demi-regs.

A second type of scientific law which expresses knowledge about the essence/cause/ground is that with the following ontologico-epistemic structure:

$$(x)[U_2 x \rightarrow Cx = f_2(Sx)]. \qquad /3/$$

Here "U_2" stands here for the universe of discourse fixed in a law with the structure /2/ and "C" stands for the essence/cause/ground, the same that was initially quantified in a law with the structure /2/, but here that quantity is already derived from the knowledge about the quantity of the *substance* (hereafter, *S*) in which C *originates*. This structure I view as exemplified by Ricardo's formulation of the law of value as given in chapter 4. It remains a task of future work to reconstruct the methods of explanations based on laws with the structure of /3/. A scientific law with the structure of /3/ stands for a type of knowledge that abstracts completely (but only temporarily, as we will see later) from all phenomena and, thus, stands for theoretical knowledge that is completely submerged in the "pure" essence/cause/ground.

A scientific law with the structure of /3/ can be viewed as the preparatory phase for the formulation of a law with the following ontologico-epistemic structure:

$$(x)[Ces^{1-r}x \xrightarrow{n} Cx = f_3(Sx)]. \qquad /4/$$

I view /4/ as a reconstruction of a type of scientific law exemplified by the law of value given in Chapter 1 of Capital, Volume I. "Ces^{1-r}" stands here for a conjunction of 1 through *r essential conditions*; "$C = f_3(S)$" stands for the knowledge about the quality of the substance and thus also of the essence/ground/cause and at the same time about their quantity, this knowledge about the quantity being transferred to the structure /4/ from the knowledge given in a structure as expressed in /3/; and "$\xrightarrow{n}$" expresses the necessity of the coming into being of the essence out of its own substance. A law with the structure of /4/ stands for the knowledge of both the conditions under which the essence/cause/ground comes into being out of its substance and how the quantity of the substance determines the quantity of the essence/cause/ground. Next I will show that scientific laws with the structure corresponding to that of /4/ can serve as a basis for linking the methods of explanation and understanding.

Based on the knowledge of a law with the structure of /4/, one can, as the next step, move to a law with the following structure ("$\xrightarrow{n}$" expresses here the necessity of the coming into being of the manifestation out its own essence/cause/ground):

$$(x)[Ces^{1-r}x \& Cmod_{1-k}x = d_{1-k} \xrightarrow{n} E^{(k)}x = f_4(Cx)]. \qquad /5/$$

It stands for the knowledge of why under the conditions of closure $Cmod_{1-k}x = d_{1-k}$, the essence/cause/ground/ C, identified in /3/ and /4/, necessarily manifests itself as $E^{(k)}$. Based on a law with the structure of /5/, one can, then, using the method of gradual concretization described earlier, derive the respective demi-regs and contrastive demi-regs $E^{(k-1)}, E^{(k-2)},\ldots, E^{(k-j)}$. The ontologico-epistemic structure given in /4/ explicates the claim of CR, namely, that the relevant causal mechanism is operative in both closed and open systems. It can be at work in both because the respective essential conditions are at work in both and, thus, the appropriate essence/cause/ground is *given* and *endures*.[246] At the same time, the ontologico-epistemic structure given in /5/ indicates that the manifestation $E^{(k)}$ can exist only if respective essential conditions are at work and the closure $Cmod_{1-k} = d_{1-k}$ holds, while the manifestations $E^{(k-1)}, E^{(k-2)},\ldots, E^{(k-j)}$ can be at work only where the essential and the relevant modification conditions are at work. The views presented here, especially the introduction of the terms "essential conditions" and "modification conditions," can thus be viewed as a refutation of Lawson's claim that "the context in which a mechanism is operative is irrelevant to the law's specification" (Lawson 1997, 29).

I view reconstructions /1/ through /5/ as an epistemico-developmental sequence of the types of scientific laws to which I assign the following catchphrases: *external and external*; (quantity of) *internal of the* (quantity of the) *external*; (quantity of) *internal of the* (quantity of the) *internal*; (quantity and quality of) *internal of the* (quantity and quality of the) *internal*; and (quantity and quality of) *external of the* (quantity and quality of the) *internal*.

How could CR benefit from my explication of those types of scientific laws? My typology, which one need not regard as complete (i.e., as covering all possible types of scientific laws), enlarges the ontological categories of CR in the following ways: the causal mechanism underlying these events has a *quality* and a *quantity*; this mechanism *exists* only under certain essential conditions and it *endures* because these conditions endure as well; and the *manifestations* of this mechanism exist under certain conditions of closure and/or *modification conditions*. At the same time the reconstruction of the law with the structure of /4/ enables one to reconstruct a possible way by means of which one can link the methods of explanation and understanding.

8.1.5 Marx's Linking of Explanation and Understanding

So as in my view the structure /4/ is exemplified by the law of value in Volume I of *Capital*, I will now try to show, by reconstructing the structure of this law and

---

246 In (Hanzel 1999) I explicate this endurance by means of the concept of reproduction.

the method of explanation based on this law, how Marx in Volume I of *Capital*,[247] was able to link the methods of understanding and explanation.

According to Marx, as we saw in chapter 4 of this book, three essential conditions under which the law of value, symbolized in chapter 4 as L_v, is operative are:

1. the existence of producers privately owning the means of production and purchasing both instruments and objects of labor;
2. the subjective (conscious) aim of the producers to exchange commodities for commodities by means of money (commodity → money → commodity'); and
3. the givenness of commodity fetishism as a characteristic of the consciousness of producers.

Marx then, in a next step, brings in, in Chapter 4 of *Capital*, Volume I, a set of new conditions, which he *explicitly* labels as "essential conditions" (Marx 1872/1987b, 184; 1976a, 272):

1. the existence of producers privately owning the means of production purchasing in addition to instruments and objects of labor also labor-power;
2. the existence of a group of free people selling their labor-power; and
3. the subjective (conscious) aim of the producers to make money (money → commodity → money').

Marx's movement from the law of value given in Chapters 1 through 3 of Capital, Volume I, to the law of value given in Chapters 5 through 24 of this volume, I represented in chapter 4 above as follows:

$$L_v \ \& \ Ces^B \ =_{tr}= | \ L_{v(s)}.$$

Now, what are the specific features of the movement from the *explanans*-law L_v to *explanandum*-law $L_{v(s)}$ from the point of view of linking explanation and understanding? *First*, the law of value as the *explanans*-law describes a type of *purposeful* human action. Marx therefore characterizes one of its essential conditions – the aim of obtaining commodities (C → M → C') – as "ultimate purpose" (*Endzweck*) (Marx 1872/1987b, 168; 1976a, 250). Understanding is thus performed by Marx even before the thought-movement from the *explanans* to

---

[247] By this I overcome the deficit of CR philosophy of science, noted in (Brown and Slater and Spencer 2002), namely that it does not provide the means for a reconstruction of the methods of explanation employed by Marx in his economic works. For a detailed methodological reconstruction of the methods of explanation employed by Marx in the manuscript of Vol. III of *Capital* see (Hanzel 1999).

the *explanandum*, namely, in the reconstruction of the agent's understanding of his/her own purposeful activity, that is, of his or her self-understanding. The extra-understanding component in the formulation of the law of value is given, of course, by reconstructing its two other essential conditions and by stating the relation between value and expended labor.

Second, in the next step, understanding is performed by Marx when he brings in, as one of the *explanans*-conditions, the reconstruction of a new type of self-understanding of the agent's own purposeful activity – the aim of making money. Here Marx applies the terms "determining purpose" (*bestimmender Zweck*) and "subjective purpose" (*subjektiver Zweck*) (Marx 1872/1987b, 168, 171; 1976a, 250, 254). The extra-understanding component comes in here when Marx brings two essential conditions – the existence of private producers purchasing instruments of labor, objects of labor, and labor-power, as well the existence of free human beings selling their labor-power.

Third, these two consecutive applications of understanding and extra-understanding components enable Marx to arrive at an *explanandum*-law that involves again, as did the *explanans*-law, a reconstruction of the agent's understanding of his or her own purposeful action plus a reconstruction of the extra-conscious essential conditions together with the statement of the relation between value, labor, and surplus-labor.

Understanding as applied by Marx, thus, enables him, at least in my view, to enter conceptually into the actions of the respective actors from the point of view of the *meanings* these actions have for these actors. But then understanding could be a "method" different from that of explanation only if those meanings could not be grasped conceptually by procedures that are part of methods of construction of scientific theories. In the conclusion of this book I will show that those meanings can be conceptualized by cyclically constructed scientific theories and that social science is in no need of a special "method" of understanding complementing the method of explanation.

8.2 Salmon versus Hegel on causation, principle of common cause, and theoretical explanation

The aim of this subchapter is to analyze the main contributions of Wesley C. Salmon to the philosophy of science in the 1970s and 1980s – that is, his concepts of *causation, common cause,* and *theoretical explanation* – and to provide a critique of them. This critique will be based on a comparison of those Salmonian concepts with categories developed by Hegel in his *Science of Logic* (Hegel 1923;

1969),[248] and which can be applied to the issues treated by Salmon by means of concepts of *causation, common cause,* and *theoretical explanation*. It is my contention that by means of categories it becomes possible to provide a critique of Salmon's philosophy of science.

I start, drawing on the Introduction – in which I explicate what categories are – with an analysis of Salmon's approach to the category of causation, and I show that, instead of reconstructing *categories* that are subsidiary to the category of causation (that is, instead of conceiving the category of causation as a *cluster of categories*), he substitutes for those subsidiary categories *concepts of physics*. In addition, he reduces and in fact impoverishes the category of causation by taking into account only *unidirectional causal action*, while leaving out *causal loops/ closed causal chains*.

Next I show that the absence, on Salmon's part, of treatment of categories as the proper medium and tool for philosophical analyzes of science leads him to incomplete reconstruction of Jean Perrin's computation of Avogadro's constant N. Finally, I show that the absence of treatment of categories leads him to a wrong understanding of what he labels as "principle of common cause" and to a distorted understanding of what he labels as "theoretical explanation."

8.2.1 Salmon vs. Hegel on causation

As one of the basic subsidiary concepts of the category of causality Salmon takes those of *process* (vs. pseudo-process), *mark*, and *mark-transmission* (Salmon 1984, 139–155). Processes are, according to him, entities that, when compared to events, "have much greater temporal duration, and in many cases, much greater spatial extent" (Salmon 1984, 139). The meaning of the term "process" is then specified by him by drawing on the *special theory of relativity from physics*. Based on this theory Salmon states the following (Salmon 1984, 141):

Causal relations must be accorded a fundamental place in the special theory of relativity any given event E_0, occurring at a particular space-time point P_0, has an associated double-sheeted light cone. All events that could have a causal influence upon E_0 are located in the interior or on the surface of the past light cone, and all events upon which E_0 could have any causal influence are located in the interior or on the surface of the future light cone. All such events are *causally connectable* with E_0. Those events that lie on the surface of either sheet of the light cone are said to have a *lightlike separation* from E_0, those that lie within either part of the cone are said to have a *timelike separation* from E_0, and those that are outside of the cone are said to have a *spacelike separation* from E_0.

---

248 For a detailed analysis of Hegel's approach to causation covering all his works see (Vetö 2000).

The Minkowski light cone can, with complete propriety, be called "the cone of causal relevance." ... Special relativity demands that we make a distinction between *causal processes* and *pseudo-processes*. It is a fundamental principle of that theory that light is a *first signal* – that is, no signal can be transmitted at a velocity greater than the velocity of light in a vacuum. There are, however, certain processes that can transpire at arbitrary high velocities – at velocities vastly exceeding that of light. This fact does not violate the basic relativistic principles, however, for these "processes" are incapable of serving as signals or of transmitting signals: Causal processes are those that are capable of transmitting signals; pseudo-processes are incapable of doing so.

As an example, he mentions the case when in a very large circular building – a kind of a super-Astrodome – a rotating spotlight is placed at its center. The light pulses traveling from that spotlight to the walls of the building stand, according to Salmon, for a causal process, while the spotlight traveling on the wall of the building stands for a pseudo-process: the former, but not the latter, can transmit a mark. So, for example, when one places a red filter in front of the rotating spotlight, the change of color is moved from the source of the light up to the wall. But if one places a red filter right in front of the wall, the light is turned red in that location but not in other locations where the spot moves. Based on this analysis, Salmon draws the conclusion that a "causal process is one that transmits energy, as well as information and causal influence" (Salmon 1984, 146).

Salmon's approach to causality is, on the one hand, based on a *specific physical theory*, and the *concepts falling under the category of causality are either taken directly from that theory or, at most, modeled on those concepts*. But, on the other hand, Salmon's endeavor to deal with the category of causality and its subsidiary concepts aims at something more general and at the same time more rich, namely, to provide *a set of categories falling under the category of causality that would cover also cases different from those described in the special theory of relativity*. So, for example, he claims that (Salmon 1984, 146)

Our main concern with causal processes is their role in the propagation of causal influence; radio broadcasting presents a clear example. The transmitting station sends a carrier wave that has a certain structure – characterized by amplitude and frequency, among other things – and modifications of this wave, in the form of modulations of amplitude (AM) or frequency (FM), are imposed for the purpose of broadcasting. ... Such processes are the means by which causal influence is propagated in our world. Causal influence, transmitted by radio, may set your foot to tapping, or induce someone to purchase a different brand of soap.

But this means that the (categorial) structure of the category of causality based on the concepts of special theory of relativity is too narrow to fit causal processes given, for example, in human action. In order to induce a person to tap his/her foot, it is causally irrelevant if the sound waves from the radio move with a high or low velocity or that the speed of that tapping cannot exceed the speed of light. And in the same manner, the speed of the radio waves by which advertisements are broadcasted is causally irrelevant for a person's switch from, say, Karcher soap to Lever 2000 soap. Thus Salmon's understanding of the category of causality does not fulfill one of the central requirements put on categories by Kant, namely, that they stand for an *object in general* (§ 14, B128; Kant 1965, 128) and that this generality should unify in itself the *richness of particular cases/objects*.

Let me now compare Salmon's approach to causality with that of Hegel in (Hegel 1923; 1969).[249] Hegel, in order to access the category of causality, uses as the point of departure the categories of *substance* understood by him initially as a unity of *substantiality* and *accidentality* (Hegel 1923, T. 2, 188; 1969, 557).[250] At the level of appearances there is as yet no difference between them and only at the level of the structure underlying these appearances – Hegel labels it as their *ground* – do they become differentiated in cognition so that, initially, substance is present in the accidents as their inside, while the accidents are just *on the substance* (Hegel 1923, T. 2, 188; 1969, 557).

Based on this, Hegel distinguishes three category clusters pertaining to the category of causality. At the level of *formal causality*, the power of the substance, that is, what it is in itself (*das An-sich*) is manifested by means of its accidents as an effect, while at the same time the very substance stands outside the generation of the effect; it is still the original (*das Ursprüngliche*) and thus as yet not posited (*gesetzt*), that is, *not explained* (Hegel 1923, T. 2, 189–190; 1969, 558–559). From this Hegel draws the conclusion that "the actuality which the substance has as a cause, it has *only in its effect*" (Hegel 1923, T. 2, 190; 1969, 559). It holds, according to him, also that the *effect* is "*necessary* because it is just the manifestation of the cause or this necessity which is the cause ... *independent source of production out of itself*; it must *act*" (Hegel 1923, T. 2, 190; 1969, 559).

Hegel, therefore, also states that once we take the relation of a certain cause to its specific effect, thus, leaving out determinations that are irrelevant for this cause with respect to this specific effect, then "*the effect contains ... nothing whatever that cause does not contain. Conversely, the cause contains nothing that is not in its effect*" (Hegel 1923, T. 2, 190–191; 1969, 559).

At the level of the category cluster of *the determinate causal relation*, Hegel reflects on the relation of cause and effect with respect to the category of substance in such a way that the effect is produced in a substance that is different

---

249 For a more concise treatment of the category of causality see §§ 150–159 in (Hegel 1975).
250 I use in this paper my own translations from the German edition of Hegel's *Science of Logic*.

from that to which its producing cause is related; to the former substance he, therefore, assigns the category of *substrate* (Hegel 1923, T. 2, 197; 1969, 565). In fact, according to Hegel, one has here an *active* substance related to the cause and a *passive* substance related to its effect. What holds here also for such a type of causality is that the produced effect turns into a cause producing yet another effect; one obtains here a causal chain going into infinity, which Hegel qualifies as the "bad infinite" (*das Schlecht-Unendliche*) (Hegel 1923, T. 2, 197; 1969, 565), because here the *cause* (even if it is an effect) produces an effect and this its *effect* (even if it a cause) *differs from its cause.*

The overcoming of the bad infinite determination of causal chains is accomplished according to Hegel in the categories of *action* and *counteraction* (*Wirkung und Gegenwirkung*). Here causality is understood already as a *presupposing acting* (*voraussetzendes Tun*) (Hegel 1923, T. 2, 198; 1969, 566), so that the cause itself is *conditioned* by the effect and, according to Hegel, one has here, not a passive substance and an active substance in which causality should reside, but only one substance, while the reaction *redirects the cause against the first/initial acting cause*. From this it follows that at the level of the category of *conditioned causality* "the cause *relates* in the effect to *itself*" (Hegel 1923, T. 1, 202; 1969, 569). By means of this, Hegel passes, as a conclusion of his categorial reflections on causality, to the category cluster of *mutual action* (*Wechselwirkung*), where "that first cause, which first acts and receives the effect into itself as counteraction, thus reappears as cause, whereby ... action going into badly infinite progress is *bent around* and becomes an into itself returning ... *mutual action*" (Hegel 1923, T. 2, 202; 1969, 569). Simultaneously, substance, initially understood as passive is now understood as *active* and *self-related*; this shift of cognition at the level of categories is expressed by Hegel by means of a shift from the category of substance to that of *substance-subject*.

Worth noting is, first, that Hegel in his reflections on causality *bypasses the use of any concepts different from categories*; *he thus moves exclusively within the framework of category clusters*. Second, by differentiating between the clusters of formal causality and of mutual action, it is possible to approach Salmon's characterization of the category of causality from yet another point of view.

Salmon claims that, in his approach to the concept of causation, he draws on the views of Hans Reichenbach and states that (Salmon 1984, 158, 163)

In his posthumous book, *The Direction of Time* (1956), [Reichenabach] ... enunciated the principle of the common cause, and he attempted to explicate the principle in terms of a statistical structure that he called a *conjunctive fork*. The principle of the common cause states, roughly, that when apparent coincidences occur that are too improbable to be attributed to chance, they can be explained by reference to a common causal antecedent. Reichenbach claimed ... that conjunctive forks possess an important asymmetry. Just as we can have two effects that arise out of a given common cause, so also may we find a common effect resulting from two distinct causes. ... Reichenbach distinguished three situations: (1) a common cause C giving rise to two separate effects, A and B, without any common effect arising from A and B conjointly; (2) two events A and B that, in the absence of a common cause C, jointly produce a common effect E; (3) a combination of (1) and (2) in which the events A and B have both a common cause C and a common effect E.

These three possible situations can be represented as shown in Figure 8.4.

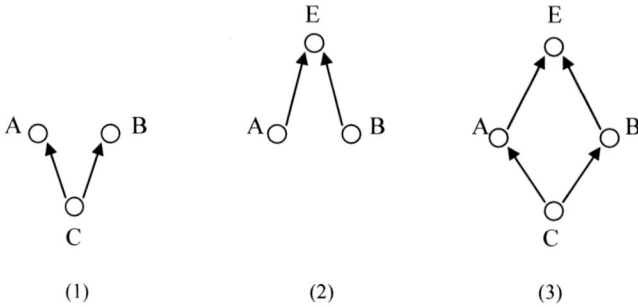

Figure 8.4 Salmon on conjunctive forks

A more detailed look at the works of Reichenbach reveals, however, that Salmon has provided a *highly selective* and in fact a *reductive* reading of Reichenbach's approach to the category of causality, that is, that he did not take into account that Reichenbach *reflected on the possibility of causal loops/closed causal chains and that he endorsed the possibility of the latter.*

Reichenbach in his *Direction of Time* states that "traveling along causal chains we ... make the discovery that we never return to the starting point; or; to put it another way, that there are no closed causal chains" (Reichenbach 1956, 36). He represents a closed causal chain by the scheme shown in Figure 8.5 and accompanies it with the comment that "arrangements of this kind never occur" (Reichenbach 1956, 39).

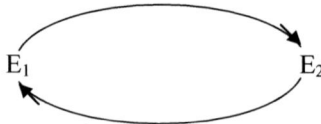

Figure 8.5 Reichenbach on closed causal chains/causal loops

In his *Philosophy of Space and Time* he then clarifies what he means by "never occur." He suggests, as an example, the mechanism of an electric bell, which seemingly functions on the basis of a closed causal chain/causal loop represented in Figure 8.6.

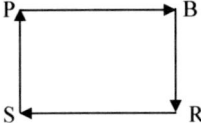

Figure 8.6 Reichenbach on the functioning of an electric bell

Reichenbach states (1958, 139–140):

The pulling (P) of the lever causes a break (B) in the current, which in turn causes a return (R) of the lever; this switches on (S) the current which finally pulls the lever. It appears as though this chain could be diagrammed [by] … a close curve. The mistake in this argument, however, is easily seen. The individual pulls of the lever are different events, i.e., although they are of the same kind, they are not identical events. Consequently, the chain should be diagrammed as … an open chain.

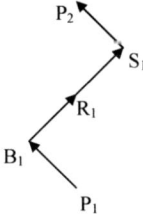

Our principle of the mark forces us to this conception. If we make some change in B_1 by short-circuiting the current and thus preventing the return of the lever, then P_2 will no longer occur. However such inference does not change P_1.

Based on this argument, Reichenbach affirms the possibility of causal loops/ closed causal chains; they return to the same kind of event, not the same singular events. "Now it is clear what we mean by a closed chain, namely a chain that returns *to identically the same event*, not to one of the same kind" (Reichenbach 1958, 140).

I thus arrive at the conclusion that Salmon has not only not reconstructed *categories* falling under the category of causality, but at the same time *by excluding reflections on cyclical causal processes* has from the very beginning severely restricted the possibility of constructing a rich *cluster of categories for the cat-*

369

egory of causality. Hegel, unlike Salmon, by distinguishing between the categorial clusters of formal causality, action and counteraction, and mutual action, has provided an *inherently rich and differentiated network of categories* falling under the category of causality where one can already make reflections on closed causal chains/causal loops.

The importance of such a rich and differentiated network of categories for the very philosophy of science is given by the fact that it provides the framework for a philosophical analysis of scientific theories which explicitly employ the category of closed causal chain/causal loop. An interesting example of such a theory is Manfred Eigen's theory aiming at the explanation of the origin of life on Earth. Already in the introduction of his first paper dealing with this issue, Eigen under the caption "cause and effect" starts with the following claim (Eigen 1971, 465–467):

The question of the origin of life very often appears as a question about "cause and effect." ... It is mainly due to the nature of this question that many scientists believe that our present physics does not offer any obvious explanation for the existence of life. ... As a consequence of the exciting discoveries of "molecular biology," a common version of the above question is: *What came first, the protein or the nucleic acid?* The term "first" is usually meant to define a causal rather than a temporal relationship, and the words "protein" and "nucleic acid" may be substituted by "function" and "information." The question in this form, when applied to the interplay of nucleic acids and proteins as presently encountered in the living cell, leads ad absurdum because "function" cannot occur in any organized manner unless "information" is present and this "information" only acquires its meaning *via* the "function" for which it is coding. Such system may be compared to a closed causal loop. ... The present interplay of nucleic acids and proteins corresponds to a complex hierarchy of "closed loops."

Eigen models such hierarchy by means of various types of hypercycle; its realistic model is represented as shown in Figure 8.7 (Eigen and Schuster 1979, 5):

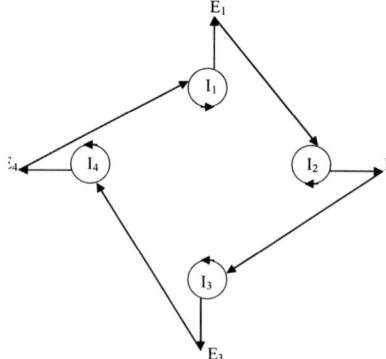

Figure 8.7 M. Eigen's realistic model of hypercycle

Here I_i stands for an information carrier exhibiting two kinds of instruction: one for its own reproduction and the other for the translation into an intermediate (enzyme) E_i, which in turn provides catalytic help for the reproduction of the subsequent information carrier I_{i+1}. The realistic model of hypercycle conforms to the first principle of evolution, which Eigen formulates as follows: *The individuals among which selection is to take place must be self-reproducing entities; they are replicators* (Eigen 1976; Eigen and Schuster 1979; Eigen and Schuster 1982; Eigen and Winkler-Oswatitsch 1992).

Worth mentioning, with respect to this paper, is Eigen's second principle of evolution. *Self-reproduction is subject to errors. This means that some replicators come into existence, not as a result of true copying of an identical parent, but as a consequence of inaccurate copying of one that is closely related* (Eigen 1976; Eigen and Schuster 1979; Eigen and Schuster 1982; Eigen and Winkler-Oswatitsch 1992). The importance of this principle, with respect to the framework of the category cluster of mutual action and with respect to the category of substance-subject, is that it enables one to go beyond Salmon's as well as Reichenbach's understanding of how a causal process can be marked. While Reichenbach in his bell-example and Salmon in his super-Astrodome example suppose that a causal process is marked from the *outside*, by an intersection with another, different causal process, Eigen's second principle of evolution suggests that in a cyclical type causal processses their kinds of constituents can change by themselves; thus causal processes of this type *mark themselves*. Understood in this way, causal processes then correspond to what the category of substance-subject in fact enables: the provision of a thought-representation of a causal process as an *active*, *self-related*, and *self-developing process*.

8.2.2 Salmon on Molecular Reality: A Critique from the Point of View of Hegel's "Retreat into the Ground and the Emergence from it"

Let us now return to Salmon's common-cause principle. It is worth mentioning how Salmon initially introduces this principle. He gives the following example (1984, 158):

> Suppose ... that several members of a traveling theatrical company who have spent a pleasant day in the country together become violently ill that evening. We infer that it was probably due to a common meal of which they all partook. When we find that their lunch included some poisonous mushrooms that they had gathered and cooked, we have the explanation.

This means that he uses common-sense terms and not categories to describe the process of, first, how a common cause (the presence of spoiled food in a hotelrestaurant) is *discovered* based on the prior discovery of the effects of that cause (discovery that the actors were poisoned by spoiled food) and, second, how these effects are, then, *explanatory-derived* from their common cause. This means that Salmon initially introduces the common-cause principle in the framework of *epistemology*, while at the same time he *bypasses here any categorial analysis*. Then, however, he suddenly leaves the framework of epistemology and, by considering the three types of conjunctive forks – Figure 8.4 above – shifts to reflections about the common-cause principle that are located inside the framework of *ontology*. The fact that Salmon shifts from epistemology to ontology and in the former does not embark on the reconstruction of categories leads, at least in my view, on his part also to a distorted reconstruction of the process of how science relates phenomenological quantities to quantities underlying them; the former are labeled by Salmon as "macro-quantities" and the latte, "micro-quantities." As an example for his reconstruction, he has chosen Jean Perrin's experiments enabling the computation of Avogadro's constant N. Perrin described these experiments and performed these computations in his articles published in 1908 (Perrin 1908a; 1908b; 1908c). In order to locate the shortcomings of Salmon's approach let me give, first, an overview of Perrin's experiments and computations.[251]

Jean Perrin's experiments and computations are initially framed by him as follows (Perrin 1908a, 967–968):

1. Any particle placed into a liquid in equilibrium moves in a continuous and perfectly irregular manner, so that the smaller it is, the more vivid it is (Brownian motion).
2. This eternal motion is an essential property of liquids.
3. This motion is an already visible consequence of molecular collisions, produced in an irregular manner, with the particle.

Based on these statements, he prepared an emulsion of natural latex, and after subjecting it to a sequence of treatments, he took one droplet of it as a microscopic preparation the thickness of which was approximately 0.12 mm. This enabled him to investigate into the distribution of the granules according to the height of the preparation. By taking the average of distributions from several thousands of tests, he found that if he represented the concentration of the granules by the numeral 100 for a certain level, then for the levels 25, 50, 75, and 100 microns below that certain level, the concentration is given by the numerals 116, 146, 170, and 200 or 119, 142, 169, and 201, which differ from the previous only due to observation-errors. So as the second series of numerals stands for a geometric progression, Perrin draws the conclusion that (Perrin 1908a, 968–969)

---

251 Here I hold to the system of units used by Perrin.

[t]he distribution of the equilibrium in the preparation (and probably in any colloidal solution) is thus exponential like for a gas in equilibrium under the impact of weight. Only the decrease by half of the concentration produced in the atmosphere at the height of 6 km, is produced here for a height of 1/10 of a millimeter.

In a next step, Perrin proposes the following mathematical treatment of the distribution of the granules in the colloid. He considers identical granules with density ρ and mass m, the concentration of which is n in a unit of volume, and that by their impacts on the wall which stops them produce an osmotic pressure proportional to their concentration, this pressure being kn. So, ndh granules given in a horizontal layer of a thickness dh and a unit cross-section are maintained in suspension by the sum of the Archimedean pressure and the difference of the osmotic pressures on the two surfaces of the layer. This leads him to the differential equation:

$$\frac{dn}{n} = \frac{1}{k} g\, dh \left(1 - \frac{1}{\rho}\right) m$$

Its integration in the boundaries from 0 to h yields:

$$2.3 \log_{10} \frac{n_0}{n} = \frac{1}{k} m g h \left(1 - \frac{1}{\rho}\right)$$

Based on this equation, Perrin computes, first, the value of the constant k based on the knowledge of m and ρ. In order to compute the value of the mass m of the granule he applied Stokes's law,[252] to an experiment, in which a vertical column of several centimeters of height of an emulsion with the granules is placed into a capillary tube. The granules from the upper layers fall like droplets from a cloud, according to Perrin experiments 0.97 mm per day. By the application of Stokes's law, relating the velocity of the spherical droplet with its radius and the viscosity of the medium, he computed that m equals 9.86×10^{-15} which in turn yields a k equal to 360×10^{-16} and, thus, the osmotic pressure p equals to $n \times 360 \times 10^{-16}$. Finally, Perrin draws on the kinetic theory of gases, according to which "this pressure is equal to that which would be produced in the same volume by the same number of molecules of any ideal gas, that is, equal to $n\frac{RT}{N}$," (Perrin 1908c, 531), where R stands for the universal gas constant, T for absolute temperature, and N for the number of molecules in one gram-mole. Based on the relation between p, on the one hand, and N, R, and T, on the other, Perrin computes N as equal to 6.7×10^{23}.

Salmon, when dealing with Perrin's experiments and computations makes the following three claims that (i) the number N "is *the* link between the microcosm and the macrocosm" (Salmon 1984, 214); (ii) "the mass of the Brownian par-

---

252 On this see (Perrin 1908b).

ticle and the average velocities of the molecules are quite directly measurable" (Salmon 1990, 125); and (iii) with respect to the application of the kinetic theory, Perrin's computation of N was based on the fact that "during the nineteenth century, the ideal gas law $PV = nRT$ was derived from the kinetic theory of gases" (Salmon 1984, 218).

However, if one takes a closer look at Perrin's computations, one finds out that at their basis and as their framework are the following crucial presuppositions:

1. The Brownian movement of the particles of the colloids is a visible consequence of the collisions of the particles of the liquid, by themselves unobservable, with the particles of the colloid.
2. There exists an analogy between the behavior of gas molecules and the behavior of the particles of the colloid, therefore the kinetic theory of gases can be applied to the Brownian movement.
3. One can apply Stokes's law to the movement of the particles of the colloid.

This thus means, *first*, that there is no way to directly compute the microquantity of the mass of a Brownian particle. In order to compute it, one has to apply Stokes's law; it is thus a microquantity which can be *measured only indirectly*. *Second*, the relations of phenomenological thermodynamics are not derivable in the full sense of "derivable" from the kinetic theory of gases. In order to derive from kinetic theory relations, initially given in the framework of phenomenological thermodynamics, one needs *in advance* the latter and then to unify them with the kinetic theory. As an example let us take the equation $p = n\frac{R}{N}T$ used by Perrin in the computation of N.

Let m_i and v_i stand for the mass and velocity, respectively, of a molecule in an ideal gas with n as the total number of molecules. For the total energy of such a gas it holds that $\sum_{i=1}^{n} \frac{1}{2} m_i v_i^2 = $ const. Let us now introduce the mean square speed, which is the speed all molecules of an ideal gas should have in order for its total energy to be equal to its real energy. Let us suppose that all molecules have the same mass m, and that n_1 molecules move with the speed of v_1, n_2 molecules move with the speed of v_2, and so on, so that $\sum_{i=1}^{n} n_i = n$. The total energy of all molecules then is $\frac{1}{2} n_1 m v_1^2 + \frac{1}{2} n_2 m v_2^2 + \ldots = \frac{1}{2} m \sum_{i=1}^{n} n_i v_i^2$. If all n molecules would have the same speed C, then the energy of the gas would be $\frac{1}{2} nmC^2$, and C would be determined by the following condition $\frac{1}{2} m \sum_{i=1}^{n} n_i v_i^2 = \frac{1}{2} nmC^2$. Thus we have obtained for C^2 as the mean square speed so that it holds that $C^2 = \frac{1}{n} \sum_{i=1}^{n} n_i v_i^2$. The values for C^2

can be found out by considering an ideal gas with N_0 molecules closed in a spherical container with a radius of r. It can be proved that a molecule with mass m acts on the wall of the container with a mean force equal to mC^2/r. From the total number N_0 of molecules a unit of area is hit by $N_0/4\pi r^2$ molecules whose resultant force acting on the unit of area is equal to the effect of the pressure acting on the wall; for this pressure it holds that $p = \dfrac{N_0 m C^2}{4\pi r^3}$ as well as $p = \dfrac{M_0 C^2}{4\pi r^3}$, where $M_0 = N_0 m$ stands for the total mass of the gas. If we multiply these equations with the volume v_m of one kilo-mole of the ideal gas, that is, $v_m = \tfrac{4}{3}\pi r^3$ kmol⁻¹, we obtain the relations

$$pv_m = \frac{1}{3} NmC^2, \tag{8}$$

$$pv_m = \frac{1}{3} MC^2. \tag{9}$$

where N is Avogadro's constant. If we now compare /8/ with the equation of state $pv_m = RT$ from phenomenological thermodynamics, we obtain

$$\frac{1}{3} NmC^2 = RT. \tag{10}$$

By multiplying both sides of the equation /10/ by , we have

$$u = \frac{1}{2} mC^2 = \frac{3}{2} \frac{R}{N} T. \tag{11}$$

The last equation states that u, the *mean energy of a molecule of a gas (composed of identical atoms) is proportional to the absolute temperature.*

By dividing both sides of /8/ by v_m, we have

$$p = \frac{1}{3} nmC^2 \tag{12},$$

where n stands for the number of molecules in a unit of volume. Finally, by substituting /11/ into /12/, we obtain the equations

$$p = \frac{2}{3} nu, \tag{13}$$

$$p = n \frac{R}{N} T. \tag{14}$$

From this, I draw the conclusion that the derivation of the equation /14/ is based on (a) the employment of mechanics to gases and fluids (via the concept of pressure understood as force F exerted per unit of area S); (b) the employment of the concept of force from mechanics inside kinetic theory; and (c) the employment of the equation of state from phenomenological thermodynamics inside the kinetic theory. In the derivation of the equation /14/, we thus have the sequence of employed theories represented in Figure 8.8.

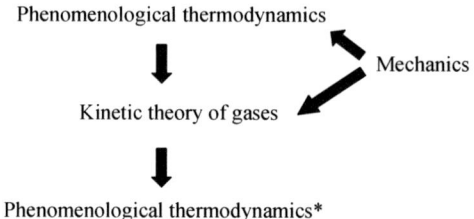

Figure 8.8 Sequence of theories employed in the derivation of $p = n\frac{R}{N}T$

The star appearing at the term "Phenomenological thermodynamics" stands for the difference between the concepts belonging initially to phenomenological thermodynamics and concepts now obtained from kinetic theory; for example, the concept of pressure, initially introduced in phenomenological thermodynamics as $p = F/S$, now becomes $p = n\frac{R}{N}T$.

It is worth noting that Hegel provides a cluster of categories that fits the sequence represented in Figure 8.8. He introduces the category of *ground*, which he characterizes as the essential structure underlying the respective phenomena under investigation and at the same time distinguishes at the level of categories, as shown in chapter 3, between the *phenomena that are at the beginning of the sequence* represented in Figure 8.8 and the *phenomena that are at the end of this sequence*. To the former he assigns the category of *appearance*, while to the latter the category *manifestation*.

At the same time, Hegel in a *strictly categorial language* expresses the epistemic/cognitive movement from appearances via ground to manifestations as the "retreat into the ground and the emergence from it" (1923, T. 2, 83; 1969, 462). Worth noting is also that Hegel not only provides a categorial reconstruction of the movement appearance → ground → manifestation, but differentiates, as shown in chapter 3, between two categories of ground. Based on this differentiation one can evaluate Salmon's views on the principle of common cause and on theoretical explanation.

8.2.3 Salmon on the principle of common-cause and theoretical explanation versus Hegel on formal and real ground

Let me return to the equations /11/ and /13/. They display a highly specific epistemic/cognitive feature, namely, that they unify macroquantities and microquantities in a specific way. The equation $u = \frac{3}{2}\frac{R}{N}T$ relates the macroscopic feature of the absolute temperature of a gas with the kinetic energy of the chaotic movement of its microparticles in such a way that *by measuring the former one can compute the latter*. The same holds also for $p = \frac{3}{2}nu$: by measuring p, and by simultaneously employing the results of computation of u, based on $u = \frac{3}{2}\frac{R}{N}T$, one can compute n. Thus the measurement of macro-quantities T and p enables one to compute the microquantities u and n, which *cannot be measured directly* in the framework of kinetic theory. This fact was expressed by Hegel by introducing the category of *formal ground*, which stands for such a *level of knowledge about the ground of phenomena as appearances, when we are as yet not able to quantify/measure the ground independently of the quantification/measurement of the phenomena as appearances*.

And again, as before, Hegel assigns a whole cluster of categories falling under a category; with respect to formal ground they are *ground* and *grounded* (*Begründetes*), relation of the ground (*Grundbeziehung*), posited (*Gesetzte*) and basis (*Grundlage*). According to Hegel, the description of the phenomena as appearances "is the identical basis of the ground and grounded" (Hegel 1923, T. 2, 79; 1969, 459); thus for the formal ground it holds that it "has no other content than the phenomenon itself" (Hegel 1923, T. 2, 79; 1969, 459). The reasonable requirement that "the ground should have *another content* than that what should be explained" (Hegel 1923, T. 2, 80; 1969, 459) is not fulfilled at the level of knowledge characterized by the category of formal ground, because (Hegel 1923, T. 2, 80; 1969, 459)

in this way of explanation the two *opposite directions* of the relation of the ground are given without being recognized in their determinate relation. The ground is, on the one hand, ground ... of the phenomenon which it grounds, on the other hand, it is the posited. It is that from which the phenomenon is to be understood; but *conversely*, it is from the latter from whom the ground is inferred and understood ... the ground, instead of being ... independent, is therefore rather the posited and derived.

At the level of knowledge characterized by the category of formal ground "one finds oneself in a kind of a vicious circle in which ... ground and grounded ... are mixed up in an indiscriminate company and enjoy equal rank with one another" (Hegel 1923, T. 2, 82; 1969, 461), and in order to escape this vicious circle in knowledge one should move from the category of formal to that of *real ground*;

here the ground is posited by knowledge that is already independent of the knowledge of phenomena as appearances (Hegel 1923, T. 2, 83; 1969, 462).

If I now compare Hegel's categorial distinction between formal and real ground with what Salmon labels as "theoretical explanation," it becomes readily seen that Salmon provides a rather incomplete understanding of theoretical explanation. While he views Perrin's computation of N as a case of *theoretical explanation* (Salmon 1984, 214), *via* the categories of formal and real ground it becomes evident that it is a case of indirect measurement that in a *noneliminable manner depends on empirical (phenomenological) knowledge*. Therefore, the movement of knowledge *from* the phenomenological characteristics of gases as well as their measurement *to* the computation of Avogadro's constant N, exemplifies the movement from the phenomena as appearances to the formal ground. Therefore, *it does not stand for theoretical knowledge in the proper sense of this term but for, as yet, only empirico-theoretical knowledge. Only knowledge in the framework of the category of real ground can be viewed as genuine theoretical knowledge.*

For what would the knowledge at the level of real ground stand for in the case of the computation of Avogadro's constant N? Salmon, drawing on Perrin, shows that this constant can be computed by the following five ways: by the phenomenon of Brownian motion, as shown earlier; as well as by the phenomenon of alpha decay, X-ray diffraction, black-body radiation, and electrolysis (Salmon 1984, 217–219). Once N is computed on the basis of these phenomena, the next step should be to find out *why* N has the value it actually has. Thus, the aim in the long run should be to compute N by using only microquantities while completely bypassing the usage of macroquantities in this computation.

With this in mind it becomes readily apparent that Salmon has misunderstood that which he labels as *principle of common cause*. He states, in respect to Perrin, that "[t]he claim I should like to make about the argument ..., stated ... by Perrin, is that it relies upon the principle of common cause – indeed that it appeals to a conjunctive fork" (Salmon 1984, 220), while at a more general level he states (Salmon 1984, 158):

There is a familiar pattern of causal reasoning that we all use every day. ... Confronted with what appears to be an improbable coincidence, we seek a common cause. If the common cause can be found, it is invoked to explain the coincidence, it is invoked to explain the coincidence.

From my analysis, based on the categories of ground, formal ground, and real ground, of Perrin's computation of N, I draw the conclusion that Salmon *has not differentiated between the phase of cognition where one moves from phenomena as appearances to their ground and the phase where one moves, once that phase has been accomplished, from the ground to its (phenomena as) manifestations.* Perrin's computation of N corresponds to the former movement. However, only the *whole* cyclical movement from phenomena as appearances to phenomena as manifestations corresponds to what Salmon understands under the principle he labels *common cause principle* and as *conjunctive fork*. Nor has Salmon considered the possible situation, given in the case of Perrin's computation of N, that the *movement, after the movement from appearances to the ground has already been accomplished, has to stop at the level of ground without going back to the phenomena as manifestations, because the ground has been grasped as yet only as a formal ground.* The latter cannot serve as a sufficient explanatory basis for the movement from the ground to its manifestations. I, therefore, restate Salmon's claim, quoted earlier, that "if the common cause can be found, it is invoked to explain the coincidence" as follows: *If the common cause is found, first, as a formal ground and, then, as a real ground, it can be invoked to explain the coincidence.*

The reason why Salmon has misunderstood the principle of common cause and what the appeal to a conjunctive fork stands for is given by the fact that he has not reflected the categories ground, formal, and real ground, as well as appearance and manifestation. That this is so becomes evident when one compares my critique of Salmon with my critique of Harper in subchapter 3.1. Neither Salmon nor Harper have realized that the appeal to a common cause and to a conjunctive fork can stand at the level of categories either, in the case of the *formal* ground, for a thought-movement from appearance to the ground/cause, or, in the case of the *real* ground, for a full cycle characterized by Hegel as a *retreat from certain phenomena as appearances into their ground and coming out from it to the positing of these same phenomena as manifestations* (Hegel 1923, T. 2, 83; 1969, 462).

Conclusion

In chapters 3 through 8, I tried to show, both at the level of categories and at the level of analyses of particular scientific theories that a cyclically built (natural or social science) theory fulfills a critical function. By this I have, I hope, answered two of the three questions stated in (Fay and Moon 1977), quoted in the introduction of this book.

Let me now deal with the third question of Fay and Moon: What is the *relation between interpretation and explanation in social science?* In order to answer this question let me state first another question, which I regard as more fundamental: *Why does this question arise at all in the framework of philosophy and methodology of social science?* Fay and Moon give the following reason for this (Fay and Moon 1977, 210):

actions differ from mere movements in that they are intentional and rule-governed: they are performed in order to achieve a particular purpose, and in conformity to some rules. These purposes and rules constitute what we shall call the "semantic dimension" of human behavior.

Thus, because human actions are structured by meanings, sciences about human actions require a method (or a set of methods) that differs from the methods employed by sciences that investigate into object (e.g., electrons, genes, molecules etc.), which are devoid of meaning. But this then means that both the question about the relation between the method of explanation and the method of interpretation as well as the claim that social science is in need of a special method of interpretation is based on the supposition that the *meaning context of science and the meaning context of action, are, with respect to each other, somehow or even completely different* in the sense that while the former is *objective* the latter is *not objective*.

Once one accepts this supposition then one has to view *social* science as displaying a special feature, namely, that it is an *objective meaning-context the subject matter of which is the subjective meaning-context*. But if this holds, then, social science faces a fundamental problem already stated by Alfred Schutz in 1932: "*How are sciences of subjective meaning-context possible at all?*" (Schutz 1967, §43, 223).

My view, however, is that what Schutz labels as "subjective meaning context" is *not devoid of objective content*. In order to show this, I will draw at the level of *categories* on the views of Hegel and at the level of *particular scientific theories* on Marx's views expressed in *Capital*, Volume I.

As shown already in chapter 4, Hegel targets the central problem of epistemology by the following question: "How do we, as subjects, get over to the objects" (Hegel 1842, §246, Zusatz, 13; 1970, 8), and then he gives the following solution to this problem. To appropriate a thing means to think it; "Intelligence familiarizes itself with things ... by thinking them, it sets their content into itself" (Hegel 1842, §246, Zusatz, 16; 1970, 9). By thinking *real singular objects*, we "add to them, so to say, the form of universality ... This universal of things is not a subjective belonging to us, but rather a noumenon opposed to the transitory phenomena; the true, the objectivity, the real of the things themselves" (Hegel 1842, §246, Zusatz, 16–17; 1970, 9).

Stated otherwise: "Thinking [is] the activity of the universal ... which with respect to its content exists solely in the object and its determinations. ... Thinking is with respect to its content only true insofar as it is immersed into the object" (Hegel 1840, §23, 44; 1991, 55); and via phenomena thinking is capable of penetrating the *inside* of things (Hegel 1928, 111; 1977, 86–87). Hegel, therefore, speaks about "*objective* thoughts" (Hegel 1840, §24, 45; 1991, 56) and about objective thinking (Hegel 1923, T.1, 31; 1969, 49) which, because it is *objective*, can serve for a "more exact determination and discovery of *objective* relations" (Hegel 1923, T. 1, 13; 1969, 35).

According to Hegel there is even a hierarchy of forms given in human consciousness that can be represented as shown in Figure 9.1 (Hegel 1840, §20, 35–36; 1991, 49–50).

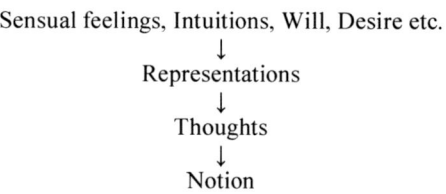

Figure 9.1 Hegel on the forms of human consciousness

With respect to the content of these forms he states: "The content which fills our consciousness, independently of what type it is, makes up the *determinations* of our feelings, intuitions, images, representations, of our purposes, duties, etc., and of our thoughts and notions" (Hegel 1840, §3, 6; 1991, 26). At the same time he puts emphasis on the *activity* of thinking: "Let it be feeling, belief, or representations – *thinking in general* has not been inactive. Its own activity and production are *present and contained in it*" (Hegel 1840, §2, 5; 1991, 25). Finally, this activity of thinking he relates to its *objective content* as follows: "By means of thinking something is altered in the way in which the content initially is feeling, intuition or representation. Thus, via *change* does the *true* nature of the object come to our consciousness" (Hegel 1840, §22, 42; Hegel 1991, 54).

So, the moral that can be drawn from Hegel's reflection on thinking, its activity and its content is that phenomena with the epistemic status of *appearances* still refer to certain characteristics of the reality; they have an *objective content*. If they would be devoid of the latter then they could not serve a point of departure of a thought-movement yielding knowledge about their underlying ground. Once the knowledge of the latter, as the result of thought-movement, has an objective content, then also the knowledge about appearances as point of departure of that movement has to have objective content. *From untrue types of knowledge one cannot derive true types of knowledge.*

Marx held rigorously to this Hegelian view in *Capital*, Volume I. Worth noting are here his views on the two types of "false" consciousness/knowledge he deals with in *Capital*, Volume I. First, when dealing with false consciousness at the level of commodity production described in Chapter 1 and labeled as "commodity fetishism," he characterizes the concepts involved in it as "crazy forms" (Marx 1872/1987b, 106; 1976a, 169). But at the same time he states about such *crazy* forms the following that "the categories of bourgeois economics consist precisely of forms of this kind. They are forms of thought which are socially valid, and therefore objective" (Marx 1872/1987b, 106 – 107; 1976a, 169). Thus even *false* forms of thinking in economics/political economy are *objective*, because they have as *referents* the state of affairs in the social world.

Second, when passing from chapter 1 through 3, via chapter 4, to chapters 5 through 23 in the second edition of *Capital*, Volume I, Marx brings in the formula $M \to C \to M'$, which should express the subjective aim of *making money* ($\Delta M = M' - M$), and which he, in the manuscript of Volume III of *Capital*, then derives as *manifestations* labeled "(industrial) profit," "rent," and "interest." That aim then has, with respect to the overall architecture[253] of *Capital* as a scientific theory, the status of an *appearance*. But at the same time, Marx claims that this appearance has an *objective content* (Marx 1872b/1987b, 171; 1976a, 254).

How then does the cyclical method of theory construction fulfill, from the point of view of method, the function that is wrongly, I claim now, assigned to a "method of interpretation"? It fulfills it by grasping conceptually the social cognitive "mechanisms" that lead to the production of the meaning contexts as given in the forms of knowledge that have the status of appearances. Thus *as a*

---

253 This overall architecture was only partially realized by Marx himself.

method that method of theory construction answers the central question of any science: *How and why are those meanings produced in the process of thought-appropriation of the external (natural and social) world by actors operating in that world?* So, at least in my view, it holds that if we require that a social theory should be built by a cyclical method then interpretation is not method specific to social science. The problem of a social-science "method" of interpretation surfaces only if one does not approach the structure of scientific theories as a cyclical one and instead holds to the more superficial view that the term "theory," refers "to *systematic, unified explanations of a diverse range of social phenomena*" (Fay and Moon 1977, 216) and that "theories provide a systematic account of a diverse set of phenomena by showing that the events in question all result from the operation of a few basic principles. ... Theories ... explain the underlying causal interconnections among phenomena of a widely divergent sort" (Fay and Moon 1977, 219, 227).

There exists, however, at least in my view, interpretation that is not a method peculiar to social science but a *skill* potentially given to any socializable human and actualized by any socialized human using meanings of language expressions in the discourse with other humans about the world. It spans a space, so to say, that is delineated by two axes: the vertical *intersubjective*, and the horizontal *expressive/determinational*. Two (or several) subjects S_i and S_j by using *language expressions* that *express meanings* determining in turn the *denotations* of the expressions and where these denotations are the intersubjectively shared counterparts of the *references* of the expressions in the *world* about which they communicate. That space can be represented as shown in Figure 9.2.

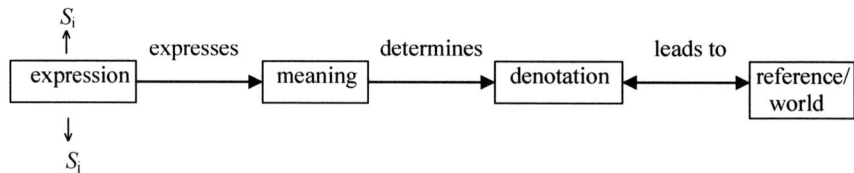

Figure 9.2 The space of the skill of interpretation

The conceptual reconstruction of this space I view as the task of a theory about the *use* of language; it should thus have the status of a *pragmatics* as theory. So as it would enable to reconstruct how any social actor can grasp the meanings of any language expressions, it would also enable one to reconstruct *how scientists can access the meanings of those terms that serve social actors to express the meanings of terms that for the scientists fulfill the function of names of appear-*

ances. In this sense, and only in this sense, such a pragmatics could be viewed as a contribution to the methodology of science.

References

(References in the text are to reprints and translations where these are given with the exception of Hegel's and Marx's works, where both the German original and the translation are referred to. In the case of Marx, the quotes are from the *Marx-Engels Gesamtausagbe*, hereafter, MEGA)

Abbott, Andrew 1988. Transcending General Linear Reality. *Sociological Theory* 6: 169–86.
Actes du congrès international de philosophie scientifique.1936. Paris: Hermann.
Ackermann, Robert. 1965. Discussion. *Philosophy of Science* 32: 155–67.
Ackermann, Robert and Alfred Stenner.1966. A Correct Model of Explanation. *Philosophy of Science* 33: 168–71.
Aiton, Eric J. 1955. The Contribution of Newton, Bernoulli and Euler to the Theory of the Tides. *Annals of Science* 11: 206–23.
Aiton, Eric J. 1964. The Inverse Problems of Central Forces. *Annals of Science* 20: 81–99.
Aiton, Eric J. 1969. Kepler's Law of Planetary Motion. *Isis* 60: 75–90.
Aiton, Eric J. 1973. "Infinitesimals and the Area Law." In *International Kepler-Symposium*, ed. Fritz Krafft, 285–305. Hildesheim: H. A. Gerstenberg.
Aiton, Eric J. 1975a. The Elliptical Orbit and the Area Law. *Vistas in Astronomy* 18: 573–83.
Aiton, Eric J. 1975b. Johannes Kepler and the Astronomy without Hypotheses. *Japanese Studies in the History of Science* 14: 49–71.
Aiton, Eric J. 1988. The Solution of the Inverse Problem of Central Forces in Newton's *Principia*. *Archives internationales d'histoire des sciences* 31: 271–76.
Aiton, Eric J. 1989. "The Contributions of Isaac Newton, Johann Bernoulli and Jakob Hermann to the Inverse Problem of Central Forces." In *Der Ausbau des Calculus durch Leibniz und die Brüder Bernoulli*, eds. Heinz J. Hess and Fritz Nagel, 48–58. Stuttgart: F. Steiner.
Alexander, Jeffrey C. 1982. *Theoretical Logic in Sociology*, Vol. 1. Berkeley: University of California Press.
Aminzade, Ron. 1992. Historical Sociology and Time. *Sociological Methods and Research* 20: 456–80.
Anderson, Douglas. R. 1986. The Evolution of Peirce's Concept of Abduction. *Transactions of the Charles S. Peirce Society* 22: 145–64.

Aoki, Shinko. 1992. The Moon-Test in Newton's *Principia*. *Archive for History of Exact Sciences* 44: 147–90.

Archer, Margaret S. 2003. *Structure, Agency and the Internal Conversation*. Cambridge: Cambridge University Press.

Archer, Margaret S. 2007. "The Long March with a Key Problem." In *Immigration and Race Relations*, eds. Tahir Abbas and Frank Reeves, 61–76. *Immigration and Race Relations*. New York: I. B. Tauris.

Armstrong, David M. 1978. *Universals and Scientific Realism*. Cambridge: Cambridge University Press.

Armstrong, David M. 1983. *What is a Law of Nature?* Cambridge: Cambridge University Press.

Arthur, Christopher J. 1993. "Hegel's Logic and Marx Capital." In: Moseley, ed. 1993, 63–89.

Arthur, Christopher J. 2004. "Money and the Value Form." In: Bellofiore and Taylor eds. 2004, 35–62.

Arthur, Christopher J. 2005. "Value and Money." In: Moseley, ed. 2005, 111–123.

Bailey, Samuel. 1825. *A Critical Dissertation on the Nature, Measurement and Causes of Value*. London: R. Hunter (reprinted New York 1967: A. M. Kelley).

Barr, William. F. 1971. A Syntactical and Semantical Analysis of Idealizations in Physics. *Philosophy of Science* 38: 258–72.

Barr, William F. 1974. A Pragmatic Analysis of Idealizations in Physics. *Philosophy of Science* 41: 48–61.

Bellofiore, Riccardo ed.1998. *Marxian Economics*, Vol. 2. New York: Palgrave.

Bellofiore, Riccardo. 2004. "Marx and the Macro-Monetary Foundation of Microeconomics." In: Bellofiore and Taylor eds. 2004, 170–216.

Bellofiore, Riccardo and Nicola Taylor eds. 2004. *The Constitution of Capital*. New York: Palgrave.

Berger, Joseph and David G. Wagner and Morris Zelditch Jr. 1992. "A Working Strategy for Constructing Theories." In *Methatheorizing*, ed. George Ritzer, 107–23. London: SAGE.

Berka, Karel. 1982. *Measurement*. Dordrecht: Reidel.

Bhaskar, Roy. 1978. *A Realist Theory of Science*, 2nd Edition. Brighton: Harvester.

Bhaskar, Roy and Lawson, Tony. 1998. Introduction. In *Critical Realism*, ed. Margaret Archer and Roy Bhaskar and Andrew Collier and Tony Lawson, 3–15. London: Routledge.

Bigelow, John and Ellis, Brian and Lierse, Caroline. 1992. The World as One of a Kind. *British Journal for the Philosophy of Science* 43: 371–88.

Blumer, Herbert. 1940. The Problem of the Concept in Social Psychology. *American Journal of Sociology* 45: 707–19.

Blumer, Herbert. 1954. What is wrong with Social Theory? *American Sociological Review* 19: 3–10.

Blumer, Herbert. 1956. Sociological Analysis and the "Variable." *American Sociological Review* 21: 683–90.
Blumer, Herbert. 1962. "Society as Symbolic Interaction." In: *Human Behavior and Social Processes*, ed. Arnold M. Rose, 179–92. Boston: Houghton Mifflin.
Blumer, Herbert. 1966. Sociological Implications of the Thought of George Herbert Mead. *American Journal of Sociology* 71: 707–19.
Blumer, Herbert. 1969. The Methodological Position of Symbolic Interactionism. In Blumer, Herbert. *Symbolic Interactionism*, 1–60. Englewood Cliffs (NJ): Prentice Hall.
Bos, Henk J. M. and Rudwick, Martin J. S. and Snelders. H. A. M. and Visser, Robert P. W., ed. 1980. *Studies on Christian Huygens*. Lisse: Swets and Zeitlinger.
Brackenridge. J. Bruce. 1995. *The Key to Newtonian Dynamics*. Berkeley: University of California Press.
Brackenridge. J. Bruce. 2003. Newton's Easy Quadratures 'Omitted for the Sake of Brevity'. *Archives for History of Exact Sciences* 57: 313–36.
Brannen, Julia ed. 1992. *Mixing Methods*. Brookfield: Avebury.
Brezinski, Jerzy ed. 1990. *Idealization II*. Amsterdam: Rodopi.
Brinton, Daniel G. 1885. The Lineal Measures of the Semi-Civilized Nations of Mexico and Central America. *Proceedings of the APA* 22: 194–207.
Bromberger, Sylvain. 1966. Why-Questions. In *Mind and Cosmos*, ed. Robert G. Colodny, 86–111. Pittsburgh: University of Pittsburgh Press.
Brown, Andrew and Fleetwood, Steven and Roberts John M. 2002. The Marriage of Critical Realism and Marxism. In *Critical Realism and Marxism.*, ed. Andrew Brown and Steven Fleetwood and John M. Roberts, 1–21. London: Routledge.
Brown, Andrew and Slater, Gary and Spencer, David A. 2002. Driven to Abstraction? *Cambridge Journal of Economics* 26: 773–88.
Bryant, Antony and Charmaz, Kathy ed. 2007. *The Sage Handbook of Grounded Theory*. London: SAGE.
Bryman, Alan. 1988. *Quantity and Quality in Social Research*. London: Routledge.
Bryman, Alan. 1992. Quantitative and Qualitative Research. In: Brannen, ed. 1992, 57–79.
Bulmer, Martin. 1979. Concepts in the Analysis of Data. *Sociological Review* 27: 651–77.
Bunge, Mario. 1964. Phenomenological Theories. In *The Critical Approach to Science and Philosophy*, ed. Mario Bunge, 234–54. New York: The Free Press.
Burawoy, Martin. 1989. Two Methods in Search of Science. *Theory and Society* 18: 759–805.

Calhoun, Craig. 1998, Explanation in Historical Sociology. *American Journal of Sociology* 104: 846–71.
Carlson, Mathieu J. 1994. The Epistemological Status of Ricardo's Theory of Value. *History of Political Economy* 26: 629–47.
Carnap, Rudolf. 1929. *Abriss der Logistik*. Wien: Julius Springer.
Carnap, Rudolf. 1932a. Überwindung der Metaphysik durch logische Analyse der Sprache. *Erkenntnis* 2: 219–41. Translated as Carnap 1959.
Carnap, Rudolf. 1932b. Die physikalische Sprache als Universalsprache der Wissenschaft. *Erkenntnis* 2: 432–65. Translated as Carnap 1963c.
Carnap, Rudolf. 1934a. Formalwissenschaft und Realwissenschaft. *Erkenntnis* 5, 30–7.
Carnap, Rudolf. 1934b. *Die Aufgabe der Wissenschaftslogik*. Wien: Gerold & Co.
Carnap, Rudolf. 1934c. *Die logische Syntax der Sprache*. Wien: Julius Springer.
Carnap, Rudolf. 1936a. Von der Erkenntnistheorie zur Wissenschaftslogik. In: *Actes* 1936, Fasc. I: 36–41.
Carnap, Rudolf. 1936b. Wahrheit und Bewährung. In: *Actes* 1936, Fasc. IV: 18–23.
Carnap, Rudolf. 1936/37. Testability and Meaning. *Philosophy of Science* 3: 420–71; 4: 2–40.
Carnap, Rudolf. 1937a. *The Logical Syntax of Language*. London: Routledge and Kegan Paul.
Carnap, Rudolf. 1937b. *Notes on Symbolic Logic*. Chicago (IL): The University of Chicago Bookstore.
Carnap, Rudolf. 1938. Logical Foundations of the Unity of Science. *International Encyclopaedia of Unified Science*, Vol. I/1, 42–62. Reprinted in Feigl and Sellars, ed. 1949, 408–23.
Carnap, Rudolf. 1939a. Science and Analysis of Language. *Erkenntnis* 9: 221–26.
Carnap, Rudolf. 1939b. *Foundations of Logic and Mathematics*. Chicago: The University of Chicago Press.
Carnap, Rudolf. 1940. Science and Analysis of Language. *Erkenntnis* 9: 221–26.
Carnap, Rudolf. 1942a. Semiotic; Theory of Signs. In: Runes, ed. 1942, 288–89.
Carnap, Rudolf. 1942b. *Introduction to Semantics*. Cambridge (MA): Harvard University Press.
Carnap, Rudolf. 1947. *Meaning and Necessity*. Chicago: The University of Chicago Press.
Carnap, Rudolf. 1952. Meaning Postulates. *Philosophical Studies* 3: 65–73.
Carnap, Rudolf. 1956. The Methodological Character of Theoretical Concepts. In *Minnesota Studies in the Philosophy of Science*, Vol. 1, ed.: Herbert Feigl and Michael Scriven, 38–76. Minneapolis: University of Minnesota Press.
Carnap, Rudolf. 1958a. *Introduction to Symbolic Logic*. New York: Dover.

Carnap, Rudolf. 1958b. Beobachtungssprache und theoretische Sprache. *Dialectica* 12: 236–48.
Carnap, Rudolf. 1959. The Elimination of Metaphysics through Logical Analysis of Language. In *Logical Positivism*, Alfred J. Ayer, 60–81. Glencoe (IL): The Free Press.
Carnap, Rudolf. 1959/2000. Theoretical Concepts in Science. *History and Philosophy of Science* 31: 158–72.
Carnap, Rudolf. 1961. On the Use of Hilbert's ε–Operator in Scientific Theories. In *Essays on the Foundations of Mathematics*, ed. Yehoshua Bar-Hillel, 156–64. Amsterdam: North-Holland.
Carnap, Rudolf. 1963a Intellectual Autobiography. In: Schilpp, ed. 1963, 3–84.
Carnap, Rudolf. 1963b. Replies and Systematic Expositions. In: Schilpp, ed. 1963, 859–1013.
Carnap, Rudolf. 1963c. The Physical Language as the Universal Language of Science. In *Readings in Twentieth-Century Philosophy*, ed. William P. Alston, 393–424. New York: The Free Press.
Carnap, Rudolf. 1966. *Philosophical Foundations of Physics*. New York: Basic Books.
Carnap, Rudolf. 1975. Observational Language and Theoretical Language. In *Rudolf Carnap*, ed. Jaakko Hintikka, 75–85. Dordrecht: Reidel.
Chang, Hasok. 1995. Circularity and Reliability in Measurement. *Perspectives in Science* 3: 153–72.
Chang, Hasok. 2004. *Inventing Temperature*. Oxford: Oxford University Press.
Charmaz, Kathy. 1983. Grounded Theory Method. In *Contemporary Field Research*, ed. Robert M. Emerson, 109–26. Boston (MA): Little Brown.
Charmaz, Kathy. 1995. Between Positivism and Postpositivism. *Studies in Symbolic Interaction* 17: 43–72.
Charmaz, Kathy. 2000. Grounded Theory. In *Handbook of Qualitative Research*, 2nd edition, ed. Norman Denzin and Yvonna S. Lincoln, 509–35. Thousand Oaks (CA): SAGE.
Charmaz, Kathy. 2006. *Constructing Grounded Theory*. Thousand Oaks (CA): SAGE.
Church, Alonzo. 1936. *Mathematical Logic* (mimeographed). Princeton (NJ): Princeton University Press.
Church, Alonzo. 1940. A Formulation of the Simple Theory of Types. *Journal of Symbolic Logic* 5: 56–68.
Church, Alonzo. 1941. *The Calculus of Lambda Conversion*. Princeton (NJ): Princeton University Press.
Church, Alonzo. 1942a. On Sense and Denotation. *Journal of Symbolic Logic* 7: 47.

Church, Alonzo. 1942b. Abstraction. In: Runes, ed. 1942: 3.
Church, Alonzo. 1942c. Function. In: Runes, ed. 1942: 113–14.
Church, Alonzo. 1942d. Proposition. In: Runes ed. 1942: 256.
Church, Alonzo. 1942e. Propositional Function. In: Runes, ed. 1942: 257.
Church, Alonzo. 1942f. Signification. In: Runes, ed. 1942: 292.
Church, Alonzo. 1942g. Truth, semantical. In: Runes, ed. 1942: 322.
Church, Alonzo. 1942h. *Elementary Topics in Mathematical Logic*. Brooklyn (NY): Gallois Institute of Mathematics.
Church, Alonzo. 1943a. Review of Willard W. Quine. Notes on Existence and Necessity. *Journal of Symbolic Logic* 8: 45–7.
Church, Alonzo. 1943b. Carnap's 'Introduction to Semantics', *Philosophical Review* 52: 298–304.
Church, Alonzo. 1944a. *Introduction to Mathematical Logic*, Part I. Princeton (NJ): Princeton University Press.
Church, Alonzo. 1944b. Formalization of Logic. By Rudolf Carnap. *Philosophical Review* 53: 493–98.
Church, Alonzo. 1946. A formulation of the logic of sense and denotation. *Journal of Symbolic Logic* 11: 31.
Church, Alonzo. 1951. A Formulation of the Logic of Sense and Denotation. In *Structure, Method and Meaning*, ed. Paul Henle and Horace M. Kallen and Susan K. Langer, 3–24. New York: The Liberal Arts Press.
Church, Alonzo. 1956. *Introduction to Mathematical Logic*, Vol. I. Princeton (NJ): Princeton University Press.
Church, Alonzo. 1973. Outline of a Revised Formulation of the Logic of Sense and Denotation (Part I). *Noûs* 7: 24–33.
Church, Alonzo. 1974. Outline of a Revised Formulation of the Logic of Sense and Denotation (Part II). *Noûs* 8: 135–56.
Church, Alonzo. 1993. A Revised Formulation of the Logic of Sense and Denotation. *Noûs* 27: 141–57.
Cockshott, Paul and Cottrell, Allin and Michelson, Greg. 1995. Testing Marx: Some New Results from UK Data. *Capital and Class* 55: 103–29.
Cohen, Isaac B. 1980. *The Newtonian Revolution*. Cambridge: Cambridge University Press.
Cohen, Isaac B. 1998. Newton's Determinations of the Masses and Densities of the Sun, Jupiter, Saturn and Earth. *Archive for the History of Exact Sciences* 53: 83–95.
Cohen, Isaac B. 1999. An Introduction to Newton's *Principia*. In: Newton 1999, 1–370.
Coleman, James. S. 1990. *Foundations of Social Theory*. Cambridge (MA): The Belknap Press of Harvard University Press.

Corbin, Juliet M. 1998. Alternative Interpretations. *Theory and Psychology* 8: 121–28.
Corbin, Juliet and Cisneros-Puebla, Cesar A. 2007. To Learn Think Conceptually. *Historical Social Research*, Supplement 19: 80-92.
Danto, Arthur and Morgenbesser, Sidney, ed. 1960. *Philosophy of Science*. New York: Meridian Books.
Danziger, Kurt. 1990. *Constructing the Subject*. Cambridge: Cambridge University Press.
Davis, Fred. 1961. Deviance Disavowal. *Social Problems* 9: 120-32.
De Gandt, François. 1995. *Force and Geometry in Newton's Principia*. Princeton (NJ): Princeton University Press.
Denzin, Norman and Lincoln, Yvonna. S. ed.1994. *Handbook of Qualitative Research*. SAGE, London.
Dey, Ian. 1999. *Grounding Grounded Theory*. San Diego: Academic Press.
Drake, Stillman. 1973. Galileo's Discovery of the Law of Free Fall. *Scientific American* 228: 84–92.
Drewery, Alice. E. 2004. A Note on Science and Essentialism. *Theoria* 19: 311–20.
Drewery, Alice. E. 2005. Essentialism and the Necessity of the Laws of Nature. *Synthese* 144: 381–96.
Ducasse, Curt J. 1941. *Philosophy as Science*. New York: O. Priest.
Earman, John. 1993. In Defense of Laws. *Philosophy and Phenomenological Research* 53: 413–19.
Earman, John and Roberts, John T. 1999. Ceteris Paribus. *Synthese* 118: 439–78.
Earman, John and Roberts, John T. and Smith, Sheldon. 2002. Ceteris Paribus Lost. *Erkenntnis* 57: 281–301.
Eigen, Manfred. 1971. Selforganization of Matter and the Evolution of Biological Macromolecules. *Naturwissenschaften* 58: 465–523.
Eigen, Manfred. 1976. Wie ensteht Information? *Berichte der Bunsen Gesellschaft* 80, 1059–81.
Eigen, Manfred and Schuster, Peter. 1979. *The Hypercycle*. Berlin: Springer.
Eigen, Manfred and Schuster, Peter. 1982. Stages of Emerging Life. *Journal of Molecular Evolution* 19: 47–61.
Eigen, M Manfred and Winkler-Oswatitsch, Ruth. 1992. *Steps Towards Life*. Oxford: Oxford University Press.
Einstein, Albert. 1905. Zur Elektrodynamik bewegter Körper. *Annalen der Physik* 17: 891–921.
Ellis, Brian. 1960. Some Fundamental Problems of Direct Measurement. *Australasian Journal of Philosophy* 38: 37–47.
Ellis, Brian. 1963. Universal and Differential Forces. *The British Journal for the Philosophy of Science* 14: 177–94.

Ellis, Brian. 1965. The Origin and Nature of Newton's Laws of Motion. In *Beyond the Edge of Certainty*, ed. Robert G. Colodny, R. G, 29–68. Englewood Cliffs (NJ): Prentice Hall.

Ellis, Brian. 1976. The Existence of Forces. *Studies in the History and Philosophy of Science* 7: 171–85.

Ellis, Brian. 1992. Idealization in Science. In: *Idealization IV*, ed. Craig Dilworth, 265–82. Amsterdam: Rodopi.

Engels, Fridrich. 1988. *Herrn Eugen Dührings Umwälzung der Wissenschaft.* Leipzig 1878: Genossenschaftsdruckerei. MEGA I/27, 221–483. Berlin: Dietz.

Fann, Kerry T. 1970. *Peirce's Theory of Abduction.* The Hague: M. Nijhoff.

Fay, Brian and Moon, J. Donald. 1977. What Would an Adequate Philosophy of Science Look Like? *Philosophy of Social Sciences* 7: 209–27.

Feigl, Herbert. 1970. The 'Orthodox' View of Theories. In *Minnesota Studies in the Philosophy of Science*, Vol. IV, ed. Michael Radner and Stephan Winokur, 3–16. Minneapolis: University of Minnesota Press.

Feigl, Herbert and Sellars, Wilfrid, ed. 1949. *Readings in Philosophical Analysis.* New York: Appleton – Century – Crofts.

Ferejohn, John and Satz, Debra 1996. Unification, Universalism and Rational Choice Theory. In: Friedman, ed. 1996, 71–84.

Fine, Arthur and Forbes, Michael and Wessels, Linda, eds. 1991. *PSA 1990*, Vol. 2. East Lansing (MI): Philosophy of Science Association.

Fisk, Milton. 1970. Are there Necessary Connections in Nature? *Philosophy of Science* 37: 385–404.

Frege, G.ottlob. 1967a. Funktion und Begriff. In: Frege 1967b, 125–42.

Frege, G.ottlob. 1967b. *Kleine Schriften.* Hildesheim: G. Olms.

Frege, G.ottlob. 1984a. Function and Concept. In: Frege 1984b, 137–56.

Frege, G.ottlob. 1984b. *Collected Papers on Mathematics, Logic and Philosophy.* Oxford: Blackwell.

Friedman, Jeffrey, ed.1996. *The Rational Choice Controversy.* New Haven: Yale University.

Frisch, Mathias 2004. Laws and Initial Conditions. *Philosophy of Science* 71: 696–706.

Galileo, Galilei. 1974. *Two New Sciences.* Madison: University of Madison Press.

Gerson, Elihu M. 1991. Supplementing GT. In *Social Organization and Social Process*, ed. David R. Maines, 285–301. New York: Aldine de Gruyter.

Gerstein, Ira. 1986. Production, Circulation and Value. In *The Value Dimension*, ed. Ben Fine, 45–93. London: Routledge and Kegan Paul.

Glaser, Barney G. 1965. The Constant Comparative Method of Qualitative Analysis. *Social Problems* 12: 436–45.

Glaser, Barney G. 1978. *Theoretical Sensitivity*. Mill Valley (CA): The Sociology Press.

Glaser, Barney G. 1992. *Emergence vs. Forcing*. Mill Valley (CA): The Sociology Press.

Glaser, Barney G. 1998. *Doing Grounded Theory*. Mill Valley (CA): The Sociology Press.

Glaser, Barney G. 2005. *The Grounded Theory Perspective III*. Mill Valley (CA): The Sociology Press.

Glaser, Barney G. 2007. Doing Formal Theory. In: Bryant and Charmaz, ed. 2007, 96–113.

Glaser, Barney G. and Strauss, Anselm L. 1964. The Social Loss of a Dying Patient. *American Journal of Nursing* 64: 119–21.

Glaser, Barney G. and Strauss, Anselm L. 1967. *The Discovery of Grounded Theory*. Chicago: Aldine.

Goldstone, Jack. A. 1997. Methodological Issues in Comparative Sociology. *Comparative Social Research* 16: 121–32.

Goldstone, Jack. A. 1998. Initial Conditions, General Laws, Path Dependence and Explanation in Historical Sociology. *American Journal of Sociology* 104: 829–45.

Gooding, David C. 1990. *Experiment and the Making of Meaning*. Dordrecht: Kluwer.

Green, Donald P. and Shapiro, Ian. 1994. *Pathologies of Rational Choice Theory*. New Haven: Yale University Press.

Griffin, Larry J. 1992. Temporality, Events and Explanation in Historical Sociology. *Sociological Methods and Research* 20: 403–27.

Griffin, Larry J. 1993. Narrative, Event-Structure and Interpretation in Historical Sociology. *American Journal of Sociology* 98: 1094–1133.

Guba, Egon G. 1987. What We Have Learned About Naturalistic Evaluation? *Evaluation Practice* 8: 23–43.

Guba, Egon. G. and Lincoln, Yvonna S. 1982. Epistemological and Methodological Bases of Naturalistic Inquiry. *Educational Communication and Technology Journal* 30, 233–52.

Haase, Michaela. 1995. *Galileische Idealisierung*. Berlin: de Gruyter.

Hall, Peter M. 1987. Interaction and the Study of Social Organization. *Sociological Quarterly* 28: 1–22.

Hall, Peter M. 1991. In Search of the Meso Domain. *Symbolic Interaction* 14: 129–34.

Hammersley, Martyn. 1989. *The Dilemma of Qualitative Method*. New York: Routledge.

Hammersley, Martyn. 1992. Deconstructing the Qualitative – Quantitative Divide. In: Brannen, ed. 1992, 39–55.

Hammersley, Martyn. 1996. The Relationship between Qualitative and Quantitative Research. In: Richardson, ed. 1996, 159–74.
Hammersley, Martyn. 2007. What is wrong with Quantitative Research? (manuscript)
Hammersley, Martyn. 2008. *Questioning Qualitative Inquiry*. Los Angeles: SAGE.
Hanzel, Igor. 1999. *The Concept of Scientific Law in the Philosophy of Science and Epistemology*. Dordrecht: Kluwer.
Hanzel, Igor. 2006. Frege, the Identity of *Sinn* and Carnap's Intension. *History and Philosophy of Logic* 27: 229–48.
Hanzel, Igor. 2010. Mistranslations of "Schein" and Erscheinung". *Science and Society* 74: 509–537.
Hanzel, Igor and Černík, Václav and Viceník, Jozef. 1994. What is a Category? *Metaphilosophy* 25: 181–93.
Harper, William L. 1993. Reasoning from Phenomena. In *Action and Reaction*, ed. Paul Theerman and Adele F. Seef, 144–82. Wilmington: University of Delaware Press.
Harper, William. L. 1999. The First Six Propositions in Newton's Argument for Universal Gravitation. *The St. John's Review* XLV: 74–93.
Harper, William. L. and Smith, George. 1995. Newton's Way of Inquiry. In: *The Creation of New Ideas in Physics*, ed. Jarrett Leplin, 113–66. Dordrecht: Kluwer.
Harré, Rom and Secord, Paul F. 1972. *The Explanation of Social Behavior*. Oxford: Blackwell.
Harvey, Herbert R. and Williams, Barbara J. 1980. Aztec Arithmetic: Positional Notation and Area Calculation. *Science* 210: 499–505.
Hechter, Michael and Kanazawa, Satoshi. 1997. Sociological Rational Choice Theory. *Annual Review of Sociology* 23: 191–214.
Hegel, Georg W. F. 1840. *Encyclopädie der philosophischen Wissenschaften im Grundrisse. Erster Theil. Die Logik*. Berlin: Duncker und Humblot.
Hegel, Georg W. F. 1842. *Vorlesungen über die Naturphilosophie als der Encyclopädie der philosophischen Wissenschaften im Grundrisse. Zweiter Theil. Die Logik*. Berlin: Duncker und Humblot.
Hegel, Georg W. F. 1923. *Wissenschaft der Logik*, T. 1, 2. Leipzig: Felix Meiner.
Hegel, Georg W. F. 1928. *Phänomenologie des Geistes*. Leipzig: Felix Meiner.
Hegel, Georg W. F. 1930. *Encyclopädie der philosophischen Wissenschaften*. Leipzig: Felix Meiner.
Hegel, Georg W. F. 1969. *The Science of Logic*. Atlantic Highlands: Humanities Press.
Hegel, Georg W. F. 1970. *Philosophy of Nature*. Oxford: Clarendon.

Hegel, Georg W. F. 1977. *The Phenomenology of Spirit*. Oxford: Clarendon.
Hegel, Georg W. F. 1975. *Logic*. Oxford: Clarendon.
Hegel, Georg W. F. 1991. *Encyclopaedia Logic*. Indianapolis: Hackett.
Hegel, Georg W. F. 2001. *Vorlesungen über Logik – 1831*. Hamburg: Meiner.
Hempel, Carl G. 1935. Some Remarks on "Facts" and "Propositions." *Analysis* 2: 93–5.
Hempel, Carl G. 1937. Le problème de la vérité. *Theoria* 3: 206–46.
Hempel, Carl G. 1942. The Function of General Laws in History. *Journal of Philosophy* 39: 35–48. Reprinted in Hempel 1965b, 231–43.
Hempel, Carl G. 1952. *Fundamentals of Concept Formation in Empirical Science*. Chicago: The University of Chicago Press.
Hempel, Carl G. 1958. The Theoretician's Dilemma. In *Midwest Studies in the Philosophy of Science*, Vol. 2, ed. Herbert Feigl and Michael Scriven, 37–98. Minneapolis: University of Minnesota Press. Reprinted in Hempel 1965b, 173–226.
Hempel, Carl G. 1959. The Logic of Functional Analysis. In *Symposium on Sociological Theory*, ed. Llewellyn Gross, 271–307. Evanston: Row.. Reprinted in Hempel 1965b, 297–330.
Hempel, Carl G. 1963. Implications of Carnap's Work for the Philosophy of Science. In: Schilpp, ed. 1963, 685–709.
Hempel, Carl G. 1964. *Postscript (1964) to* Studies in the Logic of Explanation. In: Hempel 1965b, 291–95.
Hempel, Carl G. 1965a. Aspects of Scientific Explanation. In: Hempel 1965b, 331–496.
Hempel, Carl G. 1965b. *Aspects of Scientific Explanation and Other Essays in the Philosophy of Science*. New York: The Free Press.
Hempel, Carl G. 1966. *The Philosophy of Natural Sciences*. Englewood Cliffs (NJ): Prentice Hall,.
Hempel, Carl G. 1970. On the Standard Conception of Scientific Theories. In *Minnesota Studies in the Philosophy of Science*, Vol. IV, ed. Michael Radner and Stephan Winokur, 142–63. Minneapolis: University of Minnesota Press.
Hempel, Carl G. 1973. The Meaning of Theoretical Terms. In *Logic, Methodology and Philosophy of Science IV*, ed. Patrick Suppes, 367–78. Amsterdam: North Holland.
Hempel, Carl G. and Oppenheim, Paul. 1948. Studies in the Logic of Explanation. *Philosophy of Science* 15: 135–75. Reprinted in Hempel 1965b, 245–90.
Henwood, Karen L. 1996. Qualitative Inquiry. In: Richardson, ed. 1996, 25–41.
Herivel, John. 1965. *The Background to Newton's Principia*. Oxford: Clarendon.
Hitchcock, Christopher and Woodward, James. 2003. Explanatory Generalizations, Part I, II. *Noûs* 37: 1–24, 181–99.

Holton, Judith. 2007. The Coding Process and its Challenges. In Bryant and Charmaz, ed. 2007, 237–60.

Horstmann, Rolf–Peter. 1995. What's Wrong with Kant's Categories, Professor Hegel? In *Proceedings of the Eighth International Kant Conference*, Vol. I.2., ed. Hoke Robinson, 1005–15. Milwaukee (WI): Marquette University Press.

Howe Kenneth R. 1988. Against the Quantitative-Qualitative Incompatibility or Dogmas Die Hard. *Educational Research* 7: 10–16.

Huygens, Christian. 1986. The Pendulum Clock. Ames: Iowa State University Press.

Isaac, Larry W. 1997. Transforming Localities. *Historical Methods* 30: 4–12.

Jenks, Chris. ed. 1998. *Core Sociological Dichotomies*. Thousand Oaks (CA): SAGE.

Joas, Hans 1978. G. H. Mead. In *Klassiker des soziologischen Denkens*, ed. Dirk Käsler, 7–39, 417–24, 509–14. München: Beck.

Joas, Hans. 1992. *Pragmatismus und gesellschaftliche Theorie*. Frankfurt am Main: Suhrkamp,.

Kaidesoja, Tuukka. 2007. Exploring the Concept of Causal Power in a Critical Realist Tradition. *Journal for the Theory of Social Behavior* 37: 63–87.

Kant, Immanuel. 1965. *The Critique of Pure Reason*. St Martin's Press, New York.

Kearney, Margaret H. 2001. New Directions in Formal Grounded Theory. In: Schreiber and Stern, ed. 2001, 227–46.

Kearney, Margaret H. 2007. From the Sublime to the Meticulous: The Continuing Evolution of Formal Grounded Theory. In: Bryant and Charmaz, ed. 2007, 99–122.

Kelle, Udo. 1994. *Empirisch begründete Theoriebildung*. Weinheim: Deutscher Studien Verlag.

Kelle, Udo. 1996. Die Bedeutung theoretischen Vorwissens in der Methodologie der Grounded Theory. In *Wahre Geschichten?*, ed. Strobl, Rainer and Böttger, Andreas, 23–45. Baden-Baden: Nomos.

Kim, Jaegwon. 1963. Discussion. *Philosophy of Science* 30: 286–91.

Kiser, Edgar and Hechter, Michael. 1991. The Role of General Theory in Comparative-historical Sociology. *American Journal of Sociology* 97: 1–30.

Kiser, Edgar and Hechter, Michael. 1998. The Debate on Historical Sociology. *American Journal of Sociology* 104: 785–816.

Kistler, Max. 2006. Lois, exceptions et dispositions. In *Les dispositions en philosophie et en science*, ed. Bruno Gnassounou and Max Kistler, 175–94. Paris: CNRS.

Klement, Kevin C. 2002. *Frege and the Logic of Sense and Reference*. New York: Routledge.

Kokoszyńska, Maria.1936a. Syntax, Semantik und Wissenschaftslogik. *Actes* 1936, Fasc. III: 9–14.
Kokoszyńska, Maria. 1936b. Über den absoluten Wahrheitsbegriff und einige andere semantische Begriffe. *Erkenntnis* 6: 143–65.
Kokoszyńska, Maria. 1936c. Logiczna składnia języka, semantyka i logika wiedzy. *Przeglad filozoficzny* 39: 38–49.
Kokoszyńska, Maria. 1937. Filozofia nauki w Kole Wiedeńskim. *Kwartalnik filozoficzny* 13, 151–65, 181–94.
Kokoszyńska, Maria. 1937/38. Bemerkungen über die Einheitswissenschaft. *Erkenntnis* 7: 325–30.
Kollerstrom, Nicholas. 2000. *Newton's Forgotten Lunar Theory*. Santa Fe: Greenlion Press.
Koslow, Arnold. 1982. Quantity and Quality: Some Aspects of Measurement. *PSA 1982*, Vol. 1. Philosophy of Science Association, East Lansing (MI), 183–98.
Koslow, Arnold. 1992. Quantitative Nonnumerical Relations in Science. In *Philosophical and Foundational Issues in Measurement Theory*, ed. C. Wade Savage and Philip Ehrlich, 139–65. Hillsdale (NJ): Lawrence Erlbaum.
Kosso, Peter. 1988. Dimensions of Observability. *British Journal for the Philosophy of Science* 39: 449–67.
Kosso, Peter. 1989. *Observability and Observation in Physical Science*. Dordrecht: Kluwer.
Kreiser, Lothar. 1986. *Deutung und Bedeutung*. Berlin: Akademie-Verlag.
Kusá, Zuzana. 2005. Kvantitatívne a kvalitatívne stratégie v spoločenskovednom výskume. In *Zákon, explanácia a interpretácia v spoločenských vedách*, ed. Václav Černík and Jozef Viceník, 225–38. Bratislava: IRIS.
Kyburg, Henry E. Jr. 1965. Comments. *Philosophy of Science* 32: 147–51.
Kyburg, Henry E. Jr. 1997. Quantities, Magnitudes, and Numbers. *Philosophy of Science* 64: 377–410.
Landgrebe, Ludwig. 1991. Reflections on the Schutz-Gurwitsch Correspondence. *Human Studies* 14: 107–27.
Lange, Marc. 1993. Natural Laws and the Problem of Provisos. *Erkenntnis* 38: 233–48.
Lange, Marc. 2002. Who's Afraid of Ceteris-Paribus Laws? *Erkenntnis* 57: 407–23.
Lawson, Tony. 1997. *Economics and Reality*. London: Routledge.
Lawson, Tony. 2003. *Reorienting Economics*. London: Routledge.
Laymon, Ronald. 1985. Idealizations and the Testing of Theories by Experimentation. In *Observation, Experiment, and Hypothesis in Modern Physical Science*, ed. Peter Achinstein and Owen Hannaway, 147–73. Cambridge (MA): MIT Press.

Laymon, Ronald. 1989. Cartwright and the Lying Law of Physics. *Journal of Philosophy* 86: 353–71.

Lazarsfeld, Paul F. 1958. Evidence and Social Research. *Daedalos* 87: 99–130.

Lazarsfeld, Paul F. 1966. Concept Formation and Measurement. In *Concepts, Theory and Explanation in the Behavioral Sciences*, ed. Gordon J. DiRenzo, 144–202. New York: Random House.

Lieberson, Stanley. 1991. Small N's and Big Conclusions. *Social Forces* 70, 307–20.

Lieberson, Stanley. 1994. More on the Uneasy Case of Using Mill-Type Methods in Small N Comparative Studies. *Social Forces* 72: 1225–37.

Lincoln, Yvonne. S. and Guba, Egon. G. 1985. *Naturalistic Inquiry*. Beverly Hills (CA): SAGE.

Lind, Hans. 1993. A Note on Fundamental Theory and Idealizations in Economics and Physics. *British Journal for the Philosophy of Science* 44: 493–503.

Little, Daniel. 1991. *Varieties of Social Explanation*. Boulder (CO): Westview Press.

Liu, Chang. 1999. Approximation, Idealization, and Laws of Nature. *Synthese* 118: 229–56.

Liu, Chang. 2004a. Laws and Models in a Theory of Idealization. *Synthese* 138: 363–85.

Liu, Chang. 2004b. Idealization and Models in Statistical Mechanics. *Erkenntnis* 60: 235–63.

Little, Daniel. 1991. *Varieties of Social Explanation*. Boulder: Westview Press.

Lomborg, Kirsten, and Kirkevold, Marit. 2003. Truth and Validity in Grounded Research. *Nursing Philosophy* 4: 189–200.

MacDonald, Marjorie. 2001. Finding a Critical Perspective in Grounded Theory. In: Schreiber and Stern, ed. 2001, 113–57.

MacDonald, Paul K. 2003. Useful Fiction or Miracle Maker. *American Political Science Review* 97: 551–65.

Mach, Ernst. 1960. *The Science of Mechanics*. La Salle (IL): Open Court.

Mahoney, James. 1999. Nominal, Ordinal and Narrative Appraisal in Macrosocial Analysis. *American Journal of Sociology* 104: 1154–96.

Mahoney, James. 2000. Path Dependence in Historical Sociology. *Theory and Society* 29: 507–48.

Mahoney, James. 2003. Strategies of Causal Assessment in Comparative Historical Analysis. In: Mahoney and Rueschemeyer, ed. 2003: 337–72.

Mahoney, James. 2004. Revisiting General Theory in Historical Sociology. *Social Forces* 83: 459–89.

Mahoney, James and Rueschemeyer, Dietrich. 2003. Comparative Historical Analysis. In: Mahoney and Rueschemeyer, ed. 2003, 3–40.

Mahoney, James and Rueschemeyer, Dietrich, ed. 2003. *Comparative Historical Analysis in the Social Sciences.* Cambridge: Cambridge University Press.

Mahoney, James and Snyder, Richard. 1999. Rethinking Agency and Structure in the Study of Regime Change. *Studies in Comparative International Development* 34: 3–32.

Mahoney, Michael S. 1980. Christian Huygens. In Bos and Rudwick and Snelders and Visser, ed. 1980, 234–70.

Maines, David R. 1979. Mesostructure and Social Process. *Contemporary Sociology* 8: 524–27.

Maines, David R. 1982. In Search of Mesostructure. *Urban Life* 11: 267–79.

Mandelbaum, Maurice. 1964. Newton, Boyle and the Problem of Transdiction. In: Mandelbaum, Maurice: *Philosophy, Science and Sense Perception,* 61–117. Baltimore (MD): Johns Hopkins Press.

Martin, Edward Jr. 1982. Referentiality in Frege's Grundgesetze. *History and Philosophy of Logic* 3: 151–64.

Marx, Karl. 1973. *Grundrisse.* Penguin: Harmondsworth.

Marx, Karl. 1976a. *Capital,* Vol. I. Penguin: Harmondsworth.

Marx, Karl. 1976b. *Ökonomische Manuskripte 1857/58* (Grundrisse), Teil 1. MEGA II/1.1. Berlin: Dietz Verlag.

Marx, Karl. 1976c. Results of the Immediate Process of Production. In: Marx 1976a, 948–1084.

Marx, Karl. 1978. *Zur Kritik der politischen Ökonomie* (Manuskript 1861-63). MEGA II/3.3. Berlin: Dietz Verlag.

Marx, Karl. 1979. *Zur Kritik der politischen Ökonomie* (Manuskript 1861-63). MEGA II/3.4. Berlin: Dietz Verlag.

Marx, Karl. 1980a. *Zur Kritik der politischen Ökonomie,* Urtext. In Marx 1980c, 19–94.

Marx, Karl. 1980b. *Zur Kritik der politischen Ökonomie,* Erstes Heft. Franz Duncker: Berlin 1858. In Marx 1980c, 99–255.

Marx, Karl. 1980c. *Ökonomische Manuskripte und Schriften 1858-1861.* MEGA II/2. Berlin: Dietz Verlag.

Marx, Karl. 1980d. *Zur Kritik der politischen Ökonomie* (Manuskript 1861-63). MEGA II/3.5. Berlin: Dietz Verlag.

Marx, Karl. 1981a. *Ökonomische Manuskripte 1857/58* (*Grundrisse*), Teil 2. MEGA II/1.2. Berlin: Dietz Verlag.

Marx, Karl. 1981b. *Capital,* Vol. III. Pinguin, Harmondsworth.

Marx, Karl. 1982. *Zur Kritik der politischen Ökonomie* (Manuskript 1861-63). MEGA II/3.6. Berlin: Dietz Verlag.

Marx, Karl. 1867/1983. *Das Kapital,* Erster Band. Otto Meissner, Hamburg. MEGA II/5. Berlin: Dietz Verlag.

Marx, Karl. 1984. The Future Results of British Rule in India. MEGA I/12. Dietz, Berlin, 248–53.

Marx, Karl. 1987a. [Ergänzungen und Veränderungen zum ersten Band des „Kapitals"]. MEGA II/6. Berlin: Dietz Verlag, 1–54.

Marx, Karl. 1872/1987b. *Das Kapital*, Erstes Band. Otto Meissner, Hamburg. MEGA II/6. Berlin: Dietz Verlag.

Marx, Karl. 1988a. Resultate des unmittelbaren Produktionsprozesses. In: Marx 1988b, 24–135.

Marx, Karl. 1988b *Ökonomische Manuskripte 1863-1867*. MEGA II/4.1. Berlin: Dietz Verlag.

Marx, Karl. 1992. *Das Kapital (Ökonomische Manuskripte 1863-1865). Drittes Buch*. MEGA II/4.2. Berlin: Dietz – Internationales Institut für Sozialgeschichte.

Materna, Pavel. 1995. *Svět pojmů a logika*. Praha: Filosofia.

Materna, Pavel. 1998. *Concepts and Objects*. Helsinki: Societa Philosophica Fennica.

Materna, Pavel. 2005. *Conceptual Systems*. Berlin: Logos Verlag.

Maynard, Douglas W. and Wilson, Thomas P. 1980. On the Reification of Social Structure. *Current Perspectives in Social Theory* 1: 287–322.

McDonald, Terrence J., ed. 1996. *The Historic Turn in the Social Sciences*. Ann Arbor: University of Michigan Press.

McGuire, John E. 1968. The Origin of Newton's Doctrine of Essential Qualities. *Centaurus* 12: 233–80.

McKinsey, John J. C. and Suppes, Patrick 1955. On the Notion of Invariance in Classical Mechanics. *British Journal for the Philosophy of Science* 5: 290–302.

McMichael, Phillip. 1992. Rethinking Comparative Analysis in Post-Developmental Context. *International Social Science Journal* 44: 351–65.

Mey, Günter and Mruck, Katja. 2007. Grounded Theory Methodologie. *Historical Social Research*; Supplement 19:11–39.

Mills, Jane and Bonner, Anne and Francis, Karen. 2006a. The Development of Constructivist Grounded Theory. *International Journal of Qualitative Methods* 5: 1–10.

Mills, Jane and Bonner, Anne and Francis, Karen. 2006b. Adopting a Constructivist Approach to Grounded Theory. *International Journal of Nursing Practice* 12, 8–13.

Mills, Jane and Chapman, Ysanne and Bonner, Anne and Francis, Karen. 2006: Grounded Theory. *Journal of Advanced Nursing* 58: 72–9.

Mitchell, Sandra. D. 1997. Pragmatic Laws. *Philosophy of Science* 64: S466–S479.

Mitchell, Sandra. D. 2000. Dimensions of Scientific Law. *Philosophy of Science* 67: 242–65.

Mongin, Philippe. 1989. Les deux théories marxiennes de la valeur-travail et le problème de la mesure immanente. *Archives de philosophie* 52: 247–66.

Morse, Jane M. 2004. Constructing Qualitatively Derived Theory. *Qualitative Health Research* 14: 1387–95.

Moseley, Fred. 1991. *The Declining Rate of Profit in the US Economy.* Basingstoke: Macmillan.

Moseley, Fred, ed. 1993. *Marx's Method in Capital.* Atlantic Highlands (NJ): Humanities Press.

Moseley, Fred. 1999. Marx's Theory of Value and Money (manuscript), http://www.mtholyoke.edu/~fmoseley/Working_Papers_PDF/measure.pdf).

Moseley, Fred. 2003. A Critique of Benneti's Critique of Marx's Derivation of the Necessity of Money (manuscript), http://www.mtholyoke.edu/~fmoseley/working%20papers/BENETTI.pdf)

Moseley, Fred, ed. 2005. *Marx's Theory of Money.* New York: Palgrave.

Murray, Patrick. 1990 *Marx's Theory of Scientific Knowledge.* Atlantic Highlands (NJ): Humanities Press.

Murray, Patrick. 1993. The Necessity of Money. In: Moseley, ed. 1993, 37–62.

Nagel, Ernst 1942. Introduction to Semantics. Rudolf Carnap. *Journal of Philosophy* 39: 468–73.

Naylor, Ronald H. 1974a. Galileo's Simple Pendulum. *Physis* 16: 23–46.

Naylor, Ronald H. 1974b, Galileo and the Problem of Free Fall. *British Journal for the History of Science* 7: 105–34.

Neurath, Otto. 1935. Erster international Kongress für Einheit der Wissenschaft. *Erkenntnis* 5: 377–406.

Newton, Isaac. 1687/1960. *Philosophia Naturalis Principia Mathematica.* Londini: Josephi Streater. (Reprint London: William Dawson & Sons).

Newton, Isaac. 1946. *Mathematical Principles of Natural Philosophy and the System of the World.* Berkeley: University of California Press.

Newton, Isaac. 1965a The Motion of Revolving Bodies. In: Herivel, ed. 1965, 277–92.

Newton, Isaac. 1965b. The Motion of Spherical Bodies in Fluids. In: Herivel, ed. 1965, 299–303.

Newton, Isaac. 1965c. On the Motion of Bodies in Uniformly Yielding Media. In: Herivel, ed. 1965, 309–15.

Newton, Isaac. 1965d. On the Motion of Bodies. In: Herivel, ed. 1965, 317–20.

Newton, Isaac. 1999. *The Mathematical Principles of Natural Philosop* Berkeley: University of California Press.

O'Brien, Patricia J. and Christiansen, Hanne D. 1986. An Ancient Maya Measurement System. *American Antiquity* 51: 136–51.

Nowak, Leszek. 1972. Laws of Science, Theories, Measurement. *Philosophy of Science* 39: 533–48.
Nowak, Leszek. 1980. *The Structure of Idealization*. Dordrecht: Kluwer.
Nowak, Leszek. 1992. The Idealizational Approach to Science. In *Idealization III*, ed. Jerzy Brzeziñski and Leszek Nowak, 9–63. Amsterdam: Rodopi.
Nowakowa, Isabella. 1994. *Idealization V.* Amsterdam: Rodopi.
Nowakowa, Isabella and Nowak, Leszek. 2000. *Idealization X.* Amsterdam: Rodopi.
Nye, Mary J. 1972. *Molecular Reality.* London: Macmillan.
Paige, Jeffery M. 1999. Conjuncture, Comparison and Conditional Theory in Macrosocial Inquiry. *American Journal of Sociology* 105: 781–800.
Pargetter, Robert 1984. Laws and Modal Realism. *Philosophical Studies* 46: 335–47.
Parsons, T. 2001. The Logic of Sense and Denotation. In *Logic, Meaning and Computation*, ed. Anthony Anderson, C. and Michael Zelĕny, 507–44. Dordrecht: Kluwer.
Pawson, Ray. 1989. *A Measure for Measures.* London: Routledge.
Peach, Terry. 1993. *Interpreting Ricardo.* Cambridge: Cambridge University Press.
Peirce, Charles. S. 1960–1966. *Collected Papers*, Vol. 1.–7. Cambridge (MA): Belknap Press of Harvard University Press.
Perrin, Jean. 1908a. L'agitation moléculaire et le movement brownien. *Comptes rendus* 146 : 967–70.
Perrin, Jean. 1908b. La loi de Stokes et le movement brownien. *Comptes rendus* 147: 475–76.
Perrin, J. 1908c. Le origine de mouvement brownien. *Comptes rendus* 147: 530–32.
Pestello, Frances G. and Voydanoff, Patricia. 1991. Search of Mesostructure in the Family. *Symbolic Interaction* 14: 105–128.
Pietroski, Paul and Ray, Georges. 1995. When Other Things Aren't Equal. *British Journal for the Philosophy of Science* 46, 81–110.
Plichtová, Jana 2002. *Metódy sociálnej psychológie zblízka.* Bratislava: Média.
Popper, Karl R. 1948. What Can Do Logic for Philosophy? *Proceedings of the Aristotelian Society*; Supplementary Vol. XXII: 141–54.
Popper, Karl R. 1964. *The Poverty of Historicism.* New York: Harper and Row.
Popper, Karl R. 1968. *The Logic of Scientific Discovery.* New York: Harper and Row.
Popper, Karl R. 1983. *Realism and the Aim of Science.* Totowa (NJ): Rowman and Littlefield.
Potts, T. C. 1979. "The Grossest Confusion Possible?" – Frege and the Lambda-Calculus. *Revue internationale de philosophie* 33: 761–85.

Pourciau, Bruce. 2007. Force, Deflection and Time. *Historia Mathematica* 34, 140–72.
Putnam, Hillary 1974. 'Scientific Explanation' (An Editorial Summary-Abstract). In *The Structure of Scientific Theories*, ed. Frederick Suppe, 424–33. Urbana: University of Illinois Press.
Quaas, Fridrun and Quaas, Georg, ed. 1997. *Elemente der Kritik der Wertheorie.* Frankfurt am Main: Peter Lang.
Quaas, Georg. 1984. Zum Verhältnis von Wert und Preis aus mathematischer Sicht. *Wirtschaftswissenschaft* 32, 1649–58.
Quaas, Georg. 1997. Werttheoretische Rekonstruktion der Konkurrenz als Ursache der Unterentwicklung. In: Quaas and Quaas, ed. 1997, 243–61.
Quaas, Georg. 2001. *Arbeitsquantentheorie.* Frankfurt am Main: Peter Lang.
Quadagno, Jill and Knapp, Stan. 1992. Have Historical Sociologists Forsaken Theory? *Sociological Method and Research* 20: 481–507.
Quine, Willard W. O. 1964. Ontological Reduction and the World of Numbers. *Journal of Philosophy* 64: 209–16.
Quine, Willard W. O. and Carnap, Rudolf. 1990. *Dear Carnap, Dear Van.* Berkeley: University of California Press.
Reichardt, Charles and Cook, Thomas D. 1979. Beyond qualitative *versus* quantitative methods. In *Qualitative and Quantitative Methods in Evaluation Research*, ed. Thomas D. Cook and Charles Reichardt, 7–32. Beverly Hills (CA): SAGE.
Reichenbach, Hans. 1956. *The Direction of Time.* Berkeley: University of California Press.
Reichenbach, Hans. 1958. *The Philosophy of Space and Time.* New York: Dover.
Reuten, Geert. 1993. The Difficult Labor of a Theory of Social Value. In: Moseley, ed. 1993, 89–114.
Reuten, Geert. 2005. Money as Constituent of Value. In: Moseley, ed. 2005, 78–92.
Ricardo, David. 1951–1973. *The Works and Correspondence of David Ricardo.* Cambridge: Cambridge University Press.
Richardson, John T. E., ed. 1996. *Handbook of Qualitative Research Methods for Psychology and Social Sciences.* Leicester: The British Psychological Society.
Rosenberg, Alan. 1995. Laws, Damn Laws, and Ceteris Paribus Clauses. *Southern Journal of Philosophy* 36: 183–224.
Ruben, Peter. 1997. Vom Problem der ökonomischen Messung und seiner möglichen Lösung. In: Quaas and Quaas, ed. 1997, 53–75.
Runes, Dagobert D. ed.1942. *The Dictionary of Philosophy.* New York: Philosophical Library.
Rupert, Robert D. 2007. *Ceteris Paribus* Laws, Component Forces, and the Nature of Special-Science Properties. *Noûs* 42: 349–80.

Russell, James L. 1964. Kepler's Laws of Planetary Motion. *The British Journal for the History of Science* 2: 1–22.

Salmon, Wesley C. 1984. *Scientific Explanation and the Causal Structure of the World*. Princeton (NJ): Princeton University Press.

Salmon, Wesley C. 1990. *Four Decades of Scientific Explanation*. Minneapolis: University of Minnesota Press,.

Savolainen, Jukka. 1994. In Defense of Mill's Methods. *Social Forces* 72: 1217–24.

Schilpp, Paul A. ed. 1963. *The Philosophy of Rudolf Carnap*. La Salle (IL): Open Court.

Schlesinger, George 1964. The Terms and Sentences of Empirical Science. *Mind* 73: 394–406.

Schmitt, G. 2005. Zum ägyptischen Hohlmaßsystem. *Zeitschrift für ägyptische Sprache und Altertumskunde* 132: 55–72.

Schreiber, Rita S. and Stern, Phyllis N., eds. 2001. *Using Grounded Theory in Nursing*. New York: Springer.

Schurz, Gerhard. 2001. Pietroski and Ray on Ceteris Paribus Laws. *British Journal of the Philosophy of Science* 52: 359–70.

Schutz, Alfred. 1967. *The Phenomenology of the Social World*. Evanston (IL): Northwestern University Press.

Sewell, William H. Jr. 1992. A Theory of Structure. *American Journal of Sociology* 98: 1–29.

Sewell, William H. Jr. 1996a. Historical Events as Transformation of Structure. *Theory and Society* 25: 841–81.

Sewell, William H. Jr. 1996b. Three Temporalities. In: McDonald, ed. 1996, 24–80.

Shaikh, Anwar M. 1984. The Transformation from Marx to Sraffa. In *Ricardo, Marx, and Sraffa*, ed. Ernst Mandel, 43–84. London: Verso.

Shaikh, Anwar M. and Tonak, Ertugrul A. 1994. *Measuring the Wealth of Nations*. Cambridge: Cambridge University Press.

Shalin, Dmitri N.1986. Pragmatism and Social Interactionism. *American Sociological Review* 51: 9–29.

Silverman, David. 1998. Qualitative/Quantitative. In: Jenks, ed. 1998, 78–95.

Sintonen, Matti. 2005. Scientific Explanation. *Synthese* 143: 179–205.

Sklar, Lawrence 1991. How Free Are Initial Conditions? In: Fine and Forbes and Wessels, ed. 1991, 551–64.

Skocpol, Theda. 1979. *States and Social Revolutions*. Cambridge: Cambridge University Press.

Skocpol, Theda. 1984a. Sociology's Historical Imagination. In: Skocpol, ed. 1984b, 1–21.

Skocpol, Theda, ed. 1984b. *Vision and Method in Historical Sociology.* Cambridge: Cambridge University Press.

Skocpol, Theda. 1994a. Reflections on Recent Scholarship about Social Revolutions and How to Study Them. In: Skocpol, 1994b, 301–43.

Skocpol, Theda. 1994b. *Social Revolutions in the Social World.* Cambridge: Cambridge University Press.

Skocpol, Theda and Somers, Margaret. 1980. The Uses of Comparative History in Macrosocial Inquiry. *Comparative Studies in History and Society* 22: 174–97.

Smith, George S. 1999a. Newton and the Problem of the Moon's Motion. In: Newton, 1999, 252–57.

Smith, George S. 1999b. The Motion of Lunar Apsis. In: Newton, 1999, 257–64.

Smith, Sheldon. 2002. Violated Laws, Ceteris Paribus Clauses, and Capacities. *Synthese* 130: 235–64.

Somers, Margaret. 1992. Narrativity, Narrative Identity and Social Action. *Social Science History* 16: 591–630.

Somers, Margaret. 1996. Where is Sociology after the Historical Turn? In: McDonald, ed. 1996, 53–90.

Somers, Margaret. 1998. We're No Angels. *American Journal of Sociology* 104: 722–84.

Spohn, Willfried. 1996. Zur Programmatik und Entwicklung der neuen historischen Soziologie. *Berliner Journal für Soziologie* 6: 361–73.

Stegmüller, Wolfgang. 1969. *Wissenschaftliche Erklärung und Begründung.* Berlin: Julius Springer.

Steinmetz, George. 2007. The Historical Sociology of Historical Sociology. *Sociologica*: No. 3: 1–28.

Stephenson, Bruce. 1987. *Kepler's Planetary Astronomy.* New York: Julius Springer.

Stinchcombe, Arthur L. 1973. Theoretical Domains and Measurement: I, II. *Acta Sociologica* 16: 3–12, 79–94.

Strauss, Anselm L. 1970. Discovering New Theory from Previous Theory. In *Human Nature and Collective Behavior*, ed. Tamotsu Shibutani, 46–53. Englewood Cliffs (NJ): Prentice-Hall.

Strauss, Anselm L. 1987. *Qualitative Analysis for Social Scientists.* Cambridge: Cambridge University Press.

Strauss, Anselm L. 1995. Notes on the Nature and Development of General Theories. *Qualitative Inquiry* 1: 7–18.

Strauss, Anselm L. and Corbin, Juliet. 1990. *Basics of Qualitative Research.* Newbury Park (CA): SAGE.

Strauss, Anselm L. and Corbin, Juliet. 1994. Grounded Theory Methodology. In: Denzin and Lincoln, ed. 1994, 273–85.

Strauss, Anselm L. and Corbin, Juliet. 2008. *Basics of Qualitative Research*. 3d. edition. Los Angeles (CA): SAGE.
Strübing, Jörg. 2004. *Grounded Theory*. Wiesbaden: Verlag für Sozialwissenschaften.
Strübing, Jörg. 2007a. Glaser vs. Strauss? *Historical Social Research*; Supplement 19: 157–74.
Strübing, Jörg. 2007b. Research as Pragmatic Problem Solving. In: Bryant and Charmaz, ed. 2007, 581–601.
Such, Jan 1978. Idealization and Concretization in the Natural Sciences. In *Aspects of the Growth of Science*, ed. Wladyslaw Krajewski, 49–73. Amsterdam: Grüner.
Swaan De, Abram. 1996. Rational Choice and Process. *Netherlands Journal of Social Sciences* 32: 3–15.
Sydenham, Peter H. 1979. *Measuring Instruments*. Stevenage: Peregrinus.
Tarski, Alfred. 1936. Grundlegung der wissenschaftlichen Semantik. In: *Actes* 1936, Fasc. III: 1–8.
Tarski, Alfred. 1956. *Logic, Semantics, Metamathematics*. Oxford: Clarendon.
Teller, Paul. 2004. The Law-Idealization. *Philosophy of Science* 71: 730–41.
Ter Haar, Dirk. 1971. *Elements of Hamiltonian Mechanics*. New York: Pergamon.
Thalos, Mariam. 1999. In Favor of Being only Humean. *Philosophical Studies* 93: 265–98.
Tichý, Pavel. 1988. *The Foundations of Frege's Logic*. Belrin: de Gruyter,.
Tilly, Charles 1984. *Big Structures, Large Processes, Big Comparisons*. New York: Russell Sage Foundation.
Tooley, Michael 1977. The Nature of Laws. *Canadian Journal of Philosophy* 7: 667–98.
Tugan-Baranowski, Michael I. 1905. *Theoretische Grundlagen des Marxismus*. Leipzig: Duncker und Humblot.
Tuomela, Raimond. 1972. Deductive Explanation of Scientific Laws. *Journal of Philosophical Logic* 1: 369–82.
Van Fraassen, Bas C. 1980. *Scientific Image*. New York: Oxford University Press.
Van Fraassen, Bas C. 1989. *Laws and Symmetry*. Oxford: Clarendon.
Vetö, Michael 2000. Les métamorphoses de la causalité dans la logique de Hegel. *Revue philosophique de Louvain* 98 : 519–48.
Waff, Craig B. 1976. Isaac Newton, the Motion of the Lunar Apogee, and the Establishment of the Inverse Square Law. *Vistas in Astronomy* 20: 99–103.
Walsh, William H. 1953/1954. Categories. *Kant-Studien* 45: 274–85.
Whewell, William. 1834 On the Nature and Truth of the Laws of Motion. *Transactions of the Cambridge Philosophical Society* 5: 149–72. Reprint in *William Whewell's Theory of Scientific Method*, ed. Robert E. Butts, 79–100. Pittsburgh: University of Pittsburgh Press.

Whiteside, Derek T. 1974. Astronomical Eggs – Laid and Unlaid – in Kepler's Planetary Theories. *Journal for the History of Astronomy* 5: 1–21.

Whiteside, Derek T. 1975. Newton's Lunar Theory. *Vistas in Astronomy* 19: 317–28.

Wilson, Curtis A. 1968. Kepler's Derivation of the Elliptical Path. *Isis* 59: 5–25.

Wilson, Curtis A. 1970. From Kepler's Laws, So-Called – to Universal Gravitation. *Archive for History of Exact Sciences* 6: 121–36.

Wilson, Curtis A. 1972a. How Did Kepler Discover his First Two Laws? *Scientific American* 226: 92–105.

Wilson, Curtis A. 1972b. The Inner Planets and the Keplerian Revolution. *Centaurus* 17: 205–48.

Wilson, Curtis A. 1989. The Newtonian Achievement in Astronomy. In *Planetary Astronomy from the Renaissance to the Rise of Astrophysics*, Part A: Tycho de Brahe to Newton, ed. Rene Taton and Curtis A. Wilson, 233–74. Cambridge: Cambridge University Press.

Wilson, Curtis A. 1999. Redoing Newton's Experiment for Establishing the Proportionality of Mass and Weight. *The St. John's Review* XLV: 64–73.

Wilson, Curtis A. 2001. Newton on the Moon's Variation and Apsidal Motion. In *Newton's Natural Philosophy*, ed. Jed Z. Buchwald and I. Bernard Cohen, 139–88. Cambridge (MA): MIT Press.

Wilson, Mark 1991 Law along the Frontier. In: Fine and Forbes and Wessels, ed. 1991, 565–575.

Wilson, Thomas P. 1981. Qualitative "or" Quantitative Methods in Social Research. (manuscript)

Wilson, Thomas P. 1982. Qualitative "oder" quantitative Methoden in der Sozialforschung. *Kölner Zeitschrift für Soziologie und Sozialpsychologie* 34: 487–508.

Woodward, James. 1979. Scientific Explanation. *British Journal for the Philosophy of Science* 30: 41–67.

Zachariah, Dave. 2006. Labor Value and Equalisation of Profit Rates. *Indian Developmental Review* 4: 1–20.

Gijsbert van den Brink

Philosophy of Science for Theologians
An Introduction

Frankfurt am Main, Berlin, Bern, Bruxelles, New York, Oxford, Wien, 2009.
299 pp.
Contributions to Philosophical Theology.
Edited by Vincent Brümmer, Gijsbert van den Brink and Marcel Sarot. Vol. 12
ISBN 978-3-631-56951-1 · hardback € 49.80*

This book tells the story of the philosophy of science from its inception in the aftermath of the first World War to its current stage, and relates this story to the status of theology. In doing so, it fills a remarkable gap in the literature. The unexpected resurgence of religious issues in often heated discussions since the beginning of the 21st century gave a new urgency to the question of the academic treatment of religion(s). Is it still adequate to allow for the academic study of religion only in a distanced and matter-of-fact way, without people's own views of life being brought into play and confronted with each other? Or can we also have a viable form of theology that starts from a basic religious commitment, but nevertheless fully satisfies academic standards? There is a wide debate on topics like these – but seldom this debate is conducted in a way that is informed by the state of the art in the philosophy of science.

Contents: History of the philosophy of science · Theology as an academic discipline · Science and religion · The idea of progress · Logical positivism · Karl Popper · Thomas Kuhn · Imre Lakatos · Theory-ladenness of observation · Perspectivity · Theory of modal spheres · Christian faith and spirituality · Wolfhart Pannenberg · Nancey Murphy · Ian Barbour

Frankfurt am Main · Berlin · Bern · Bruxelles · New York · Oxford · Wien
Distribution: Verlag Peter Lang AG
Moosstr. 1, CH-2542 Pieterlen
Telefax 00 41 (0) 32 / 376 17 27

*The €-price includes German tax rate
Prices are subject to change without notice
Homepage http://www.peterlang.de